Monthly Digest
of Statistics

Editor: Dilys Rosen

Office for National Statistics

ISBN 978-0-230-52601-3

ISSN 0308–6666

Contact points

For enquiries about this publication, contact
the Editor
Tel: 01633 812903
E-mail: monthly.digest@ons.gsi.gov.uk

For general enquiries, contact the National Statistics
Customer Contact Centre on: 0845 601 3034
(minicom: 01633 812399)
E-mail: info@statistics.gsi.gov.uk
Fax: 01633 652747
Post: Room 1015, Government Buildings,
Cardiff Road, Newport NP10 8XG

You can also find National Statistics on the Internet at:
www.statistics.gov.uk

Subscriptions

To subscribe to this publication, contact Palgrave Macmillan at:
www.palgrave.com/ons or on 01256 302611.

About the Office for National Statistics

The Office for National Statistics (ONS) is the government agency responsible for compiling, analysing and disseminating many of the United Kingdom's economic, social and demographic statistics, including the retail prices index, trade figures and labour market data, as well as the periodic census of the population and health statistics. It is also the agency that administers the statutory registration of births, marriages and deaths in England and Wales. The Director of ONS is also the National Statistician and the Registrar General for England and Wales.

A National Statistics publication

National Statistics are produced to high professional standards set out in the National Statistics Code of Practice. They undergo regular quality assurance reviews to ensure that they meet customer needs. They are produced free from any political influence.

Contents

Contents

Contents

Units of Measurement

Length

1 millimetre (mm)		= 0.03937 inch
1 centimetre (cm)	= 10 millimetres	= 0.3937 inch
1 metre (m)	= 1,000 millimetres	= 1.094 yards
1 kilometre (km)	= 1,000 metres	= 0.6214 mile
1 inch (in.)		= 25.40 millimetres or 2.540 centimetres
1 foot (ft.)	= 12 inches	= 0.3048 metre
1 yard (yd.)	= 3 feet	= 0.9144 metre
1 mile	= 1,760 yards	= 1.609 kilometres

Area

1 square millimetre (mm²)		= 0.001550 square inch
1 square metre (m²)		= 1.196 square yards
1 hectare (ha)	= 10,000 square metres	= 2.471 acres
1 square kilometre (km²)		= 247.1 acres
1 square inch (sq. in.)		= 645.2 square millimetres or 6.452 square centimetres
1 square foot (sq. ft.)	= 144 square inches	= 0.09290 square metre or 929.0 square centimetres
1 square yard (sq. yd.)	= 9 square feet	= 0.8361 square metre
1 acre	= 4,840 square yards	= 4,046 square metres or 0.4047 hectare
1 square mile (sq. mile)	= 640 acres	= 2.590 square kilometres or 259.0 hectares

Volume

1 cubic centimetre (cm³)		= 0.06102 cubic inch
1 cubic decimetre (dm³)	= 1,000 cubic centimetres	= 0.03531 cubic foot
1 cubic metre (m³)		= 1.308 cubic yards
1 cubic inch (cu.in.)		=16.39 cubic centimetres
1 cubic foot (cu. ft.)	= 1,728 cubic inches	= 0.02832 cubic metre or 28.32 cubic decimetres
1 cubic yard (cu. yd.)	= 27 cubic feet	= 0.7646 cubic metre

Capacity

1 litre (l)	= 1 cubic decimetre	= 0.2200 gallon
1 hectolitre (hl)	= 100 litres	= 22.00 gallons
1 pint		= 0.5682 litre
1 quart	= 2 pints	= 1.137 litres
1 gallon	= 8 pints	= 4.546 litres
1 bulk barrel	= 36 gallons (gal.)	= 1.637 hectolitres

Weight

1 gram (g)		= 0.03527 ounce avoirdupois
1 hectogram (hg)	= 100 grams	= 3.527 ounces or 0.2205 pound
1 kilogram (kg)	= 1,000 grams or 10 hectograms	= 2.205 pounds
1 tonne (t)	= 1,000 kilograms	= 1.102 short tons or 0.9842 long ton
1 ounce avoirdupois (oz.)	= 437.5 grains	= 28.35 grams
1 pound avoirdupois (lb.)	= 16 ounces	= 0.4536 kilogram
1 hundredweight (cwt.)	= 112 pounds	= 50.80 kilograms
1 short ton	= 2,000 pounds	= 907.2 kilograms or 0.9072 tonne
1 long ton (referred to as ton)	= 2,240 pounds	= 1,016 kilograms or 1.016 tonnes
1 ounce troy	= 480 grains	= 31.10 grams

Energy

British thermal unit (Btu)	= 0.2520 kilocalorie	(kcal) = 1.055 kilojoule (kj)
Therm	= 10^5 British thermal	units = 25,200 kcal = 105,506 kj
Megawatt hour (MWh)	= 10^6 watt hours (Wh)	
Gigawatt hour (GWh)	= 10^6 kilowatt hours = 34,121 therms	

Food and drink

Butter	23,310 litres milk	= 1 tonne butter (average)
Cheese	10,070 litres milk	= 1 tonne cheese
Condensed milk	2,550 litres milk	= 1 tonne full cream condensed milk
	2,953 litres skimmed milk	= 1 tonne skimmed condensed milk
Milk	1 million litres	= 1,030 tonnes
Milk powder	8,054 litres milk	= 1 tonne full cream milk powder
	10,740 litres skimmed milk	= 1 tonne skimmed milk powder
Eggs	17,126 eggs	= 1 tonne (approximate)
Sugar	100 tonnes sugar beet	= 92 tonnes refined sugar
	100 tonnes cane sugar	= 96 tonnes refined sugar

Shipping

Gross tonnage	= total volume of all the enclosed spaces of a vessel, the unit of measurement being a 'ton' of 100 cubic feet.
Deadweight tonnage	= Deadweighttonnage is the total weight in tons of 2,240 lb. that a ship can legally carry, that is the total weight of cargo, bunkers, stores and crew.

Introduction

This publication has been prepared by the Office for National Statistics (ONS) in collaboration with a number of government departments and other organisations. The assistance provided by them is gratefully acknowledged.

The name of the department or organisation providing the statistics is shown under each table, additionally, on some tables this is followed by a contact telephone number.

All the data series published in the *Monthly Digest* are contained on an ONS database, and nearly all are stored with a four letter identification code (e.g. ABMZ). These codes appear at the start of columns or rows so that they can be quoted if you contact us requiring any further information.

The latest Annual Supplement to *Monthly Digest* was published in the January 2007 edition. This gives detailed definitions and explanatory notes and includes an index of sources.

Definitions and classifications

The following general definitions should be noted in using the *Digest*:

Area covered. Except where otherwise stated, all statistics relate to the United Kingdom of Great Britain and Northern Ireland.

Seasonality. Except where otherwise stated, all statistics are not adjusted to take account of seasonal factors.

The UK Standard Industrial Classification 1992 is used in a number of tables in this digest to split economic activity. Full details are available from *UK Standard Industrial Classification of Economic Activities 1992*, and *Indexes to the UK Standard Industrial Classification of Economic Activities 1992*, both available from Palgrave Macmillan.

Regional classification is based on the Government Office Regions.

Symbols and conventions used

Change of basis. Where consecutive figures have been compiled on different bases and are not strictly comparable, a footnote is added indicating the nature of the difference. Also, a line may be drawn across a column between two consecutive figures indicating that the figures above and below the line have been compiled on different bases.

Units of measurement. The various units of measurement used in this digest are listed on the opposite page.

Symbols. The following symbols have been used throughout:

.. = not available (also information suppressed to avoid disclosure)

- = nil or less than half the final digit shown

† = indicates that the data have been revised

p = provisional data

since the last edition: the period marked is the earliest in the table to have been revised

Also, some tables have symbols specific to them. These will be explained in the footnotes to those tables.

Rounding of figures. In tables where figures have been rounded to the nearest final digit, there may be a slight discrepancy between the sum of the constituent items and the total as shown.

Provisional data

Some figures are provisional and may be subject to revision in later editions. This applies par ticularly to data for the most recent time periods. Where data has been revised a dagger symbol, as previously mentioned, will appear.

National Statistics Online: www.statistics.gov.uk

Web-based access to time series, cross sectional data and metadata from across the Government Statistical Service (GSS), available using the site search and index functions from the homepage. Download many datasets, in whole or in part, or consult directory information for all GSS statistical resources, including censuses, surveys, periodicals and enquiry services. Information is posted as PDF electronic documents or in XLS and CSV formats, compatible with most spreadsheet packages.

Time Series Data

Access to around 40,000 time series, of primarily macro-economic data, drawn from the main tables in a range of our major economic and labour market publications. Download complete releases, or view and download your own customised selection of individual time series.

Complete copies of this publication are available to download free of charge on the following web page. www.statistics.gov. uk/monthlydigest

Web: www.palgrave.com/ons, email: ons@palgrave.com

Acknowledgements

Contributors

The Editor wishes to thank all her colleagues in ONS, the rest of the Government Statistical Service and all contributors in other organisations for their generous support and helpful comments.

1 National accounts

1.1 Gross domestic product and gross national income

£ million

	At current prices					Chained volume measures			
	Gross national income at market prices	Net income from abroad[1]	Gross domestic product at market prices	*less* Basic price adjustment[2]	Gross value added at basic prices	Gross domestic product at market prices	*less* Basic price adjustment[1]	Gross value added at basic prices	Gross value added at factor cost
	ABMZ	CAES	YBHA	NTAP	ABML	ABMI	NTAO	ABMM	YBHH
1997	811 797	603	811 194	90 570	720 624	936 717	103 014	833 944	820 939
1998	869 706	8 910	860 796	97 116	763 680	968 040	105 165	863 147	849 467
1999	904 737	−1 830	906 567	105 956	800 611	997 295	107 873	889 722	875 372
2000	954 004	777	953 227	112 248	840 979	1 035 295	112 020	923 583	908 587
2001	1 005 313	8 326	996 987	114 234	882 753	1 059 648	116 584	943 186	927 875
2002	1 069 839	21 072	1 048 767	118 470	930 297	1 081 469	121 657	959 811	943 960
2003	1 132 938	22 642	1 110 296	124 738	985 558	1 110 296	124 738	985 558	969 067
2004	1 202 075	25 548	1 176 527	132 362	1 044 165	1 146 523	128 660	1 017 863	1 000 888
2005	1 251 567	26 228	1 225 339	137 471	1 087 868	1 168 713	130 613	1 038 100	1 020 466
2006	..	..	..	..	..	1 200 614	..	1 067 254	..
Seasonally adjusted									
1997 Q1	197 126	−568	197 694	21 414	176 280	231 356	25 286	206 128	202 986
Q2	202 256	1 293	200 963	22 245	178 718	233 166	25 664	207 562	204 322
Q3	205 717	947	204 770	23 165	181 605	235 045	26 028	209 079	205 785
Q4	206 698	−1 069	207 767	23 746	184 021	237 150	26 036	211 175	207 846
1998 Q1	211 244	662	210 582	24 040	186 542	239 311	26 041	213 337	209 945
Q2	214 567	1 207	213 360	23 971	189 389	240 634	26 162	214 540	211 148
Q3	221 773	4 353	217 420	24 341	193 079	243 161	26 396	216 833	213 402
Q4	222 122	2 688	219 434	24 764	194 670	244 934	26 566	218 437	214 972
1999 Q1	220 523	−976	221 499	25 385	196 114	245 774	26 682	219 157	215 658
Q2	224 191	−905	225 096	25 813	199 283	247 398	26 642	220 839	217 255
Q3	228 802	−79	228 881	26 913	201 968	250 676	27 005	223 752	220 145
Q4	231 221	130	231 091	27 845	203 246	253 447	27 544	225 974	222 314
2000 Q1	235 382	735	234 647	27 659	206 988	256 262	27 989	228 340	224 648
Q2	236 889	−330	237 219	28 227	208 992	258 132	27 954	230 250	226 534
Q3	240 987	1 314	239 673	28 160	211 513	259 667	27 877	231 881	228 111
Q4	240 746	−942	241 688	28 202	213 486	261 234	28 200	233 112	229 294
2001 Q1	248 119	1 774	246 345	28 373	217 972	263 631	28 568	235 130	231 322
Q2	250 036	1 978	248 058	28 696	219 362	263 935	28 828	235 149	231 338
Q3	252 418	2 971	249 447	28 492	220 955	265 519	29 323	236 213	232 374
Q4	254 740	1 603	253 137	28 673	224 464	266 563	29 865	236 694	232 841
2002 Q1	261 806	4 438	257 368	29 317	228 051	267 948	29 957	237 993	234 098
Q2	264 279	3 251	261 028	29 402	231 626	269 392	30 346	239 044	235 077
Q3	271 124	7 075	264 049	29 733	234 316	271 368	30 620	240 747	236 737
Q4	272 630	6 308	266 322	30 018	236 304	272 761	30 734	242 027	238 048
2003 Q1	278 360	7 442	270 918	30 341	240 577	274 119	30 740	243 381	239 321
Q2	279 619	4 489	275 130	30 692	244 438	275 712	31 096	244 616	240 522
Q3	284 071	4 047	280 024	31 504	248 520	278 748	31 396	247 351	243 211
Q4	290 888	6 664	284 224	32 201	252 023	281 717	31 506	250 210	246 013
2004 Q1	292 547	5 573	286 975	32 806	254 169	283 725	31 832	251 893	247 727
Q2	299 293	6 173	293 120	32 972	260 148	286 307	32 148	254 159	249 919
Q3	300 666	4 668	295 998	33 209	262 789	287 400	32 311	255 089	250 813
Q4	309 569	9 134	300 434	33 375	267 059	289 091	32 369	256 722	252 429
2005 Q1	308 503	6 760	301 743	33 960	267 783	289 943	32 375	257 568	253 215
Q2	313 734	9 327	304 407	34 121	270 286	291 280	32 531	258 749	254 350
Q3	313 053	6 403	306 650	34 839	271 811	292 753	32 824	259 929	255 506
Q4	316 277	3 738	312 539	34 551	277 988	294 737	32 883	261 854	257 395
2006 Q1	320 431	6 403	314 028	35 325	278 703	296 869	32 968	263 901	259 399
Q2	327 888	8 656	319 232	36 119	283 113	299 084	33 168	265 916	261 351
Q3	331 344	6 310	325 034	36 545	288 489	301 126	33 478	267 648	263 035
Q4	..	..	..	..	..	303 535	..	269 789	..

1.1 Gross domestic product and gross national income

continued

2003 = 100

	Value indices at current prices		Chained volume indices			Implied deflators[3]		
	Gross domestic product at market prices	Gross value added at basic prices	Gross domestic product at market prices	Gross value added at basic prices	Gross national disposable income at market prices	Gross domestic final expenditure	Gross domestic product at market prices	Gross value added at basic prices
	YBEU	YBEX	YBEZ	CGCE	YBFP	YBFV	YBGB	CGBV
1997	73.1	73.1	84.4	84.6	82.0	88.1	86.6	86.4
1998	77.5	77.5	87.2	87.6	85.9	89.9	88.9	88.5
1999	81.7	81.2	89.8	90.3	87.7	91.6	90.9	90.0
2000	85.9	85.3	93.2	93.7	90.8	93.0	92.1	91.1
2001	89.8	89.6	95.4	95.7	93.8	95.2	94.1	93.6
2002	94.5	94.4	97.4	97.4	97.2	97.3	97.0	96.9
2003	100.0	100.0	100.0	100.0	100.0	100.0	100.0	100.0
2004	106.0	105.9	103.3	103.3	103.4	102.4	102.6	102.6
2005	110.4	110.4	105.3	105.3	104.5	105.4	104.8	104.8
2006	..	..	108.1	108.3	..	..	..	..
Seasonally adusted								
1997 Q1	71.2	71.5	83.3	83.7	80.4	87.1	85.5	85.5
Q2	72.4	72.5	84.0	84.2	82.2	87.6	86.2	86.1
Q3	73.8	73.7	84.7	84.9	82.6	88.7	87.1	86.9
Q4	74.9	74.7	85.4	85.7	82.9	89.0	87.6	87.1
1998 Q1	75.9	75.7	86.2	86.6	83.9	89.3	88.0	87.4
Q2	76.9	76.9	86.7	87.1	85.3	89.6	88.7	88.3
Q3	78.3	78.4	87.6	88.0	87.4	90.0	89.4	89.0
Q4	79.1	79.0	88.2	88.7	86.9	90.5	89.6	89.1
1999 Q1	79.8	79.6	88.5	88.9	86.1	90.9	90.1	89.5
Q2	81.1	80.9	89.1	89.6	87.2	91.5	91.0	90.2
Q3	82.5	82.0	90.3	90.8	88.4	91.9	91.3	90.3
Q4	83.3	82.5	91.3	91.7	89.0	92.2	91.2	89.9
2000 Q1	84.5	84.0	92.3	92.7	90.2	92.5	91.6	90.6
Q2	85.5	84.8	93.0	93.4	90.6	92.7	91.9	90.8
Q3	86.3	85.8	93.5	94.1	91.7	93.2	92.3	91.2
Q4	87.1	86.6	94.1	94.6	90.8	93.8	92.5	91.6
2001 Q1	88.7	88.5	95.0	95.4	93.2	94.5	93.4	92.7
Q2	89.4	89.0	95.1	95.4	93.4	94.8	94.0	93.3
Q3	89.9	89.7	95.7	95.9	94.5	95.5	93.9	93.5
Q4	91.2	91.1	96.0	96.1	94.2	95.8	95.0	94.8
2002 Q1	92.7	92.6	96.5	96.6	95.9	96.6	96.1	95.8
Q2	94.0	94.0	97.1	97.0	96.3	97.1	96.9	96.9
Q3	95.1	95.1	97.8	97.7	98.4	97.5	97.3	97.3
Q4	95.9	95.9	98.3	98.2	98.3	98.0	97.6	97.6
2003 Q1	97.6	97.6	98.8	98.8	99.4	98.8	98.8	98.8
Q2	99.1	99.2	99.3	99.3	98.9	99.8	99.8	99.9
Q3	100.9	100.9	100.4	100.4	100.0	100.4	100.5	100.5
Q4	102.4	102.3	101.5	101.6	101.7	101.0	100.9	100.7
2004 Q1	103.4	103.2	102.2	102.2	101.9	101.0	101.1	100.9
Q2	105.6	105.6	103.1	103.2	103.2	102.3	102.4	102.4
Q3	106.6	106.7	103.5	103.5	103.0	102.8	103.0	103.0
Q4	108.2	108.4	104.1	104.2	105.4	103.4	103.9	104.0
2005 Q1	108.7	108.7	104.5	104.5	104.3	104.0	104.1	104.0
Q2	109.7	109.7	104.9	105.0	105.5	104.8	104.5	104.5
Q3	110.5	110.3	105.5	105.5	103.9	106.0	104.7	104.6
Q4	112.6	112.8	106.2	106.3	104.3	106.7	106.0	106.2
2006 Q1	113.1	113.1	107.0	107.1	105.5	106.9	105.8	105.6
Q2	115.0	114.9	107.7	107.9	107.0	107.9	106.7	106.5
Q3	117.1	117.1	108.5	108.6	107.4	108.7	107.9	107.8
Q4	..	..	109.4	109.5	..	..	..	..

1 Includes employment, entrepreneurial and property income.
2 Taxes on products *less* subsidies on products.
3 Derived from expenditure components.

Source: Office for National Statistics: 020 7533 6031

1.2 Gross domestic product: by category of expenditure

£ million[1]

	Domestic expenditure on goods and services at market prices										Statistical discrepancy (expenditure)	Gross domestic product at market prices
	Final consumption expenditure			Gross capital formation								
	House-holds	Non-profit instit-utions [2]	General government	Gross fixed capital formation	Change in inven-tories[3]	Acquisi-tions less disposals of valuables	Total	Total exports	Total final expend-iture	less Total imports		
At current prices												
	ABPB	ABNV	NMRK	NPQX	ABMP	NPJO	YBIJ	KTMW	ABMD	KTMX	GIXM	YBHA
1996	472 711	18 129	148 626	126 593	1 771	−160	767 670	225 158	992 828	227 676	−	765 152
1997	501 290	19 372	150 554	133 620	4 621	−27	809 430	234 019	1 043 449	232 255	−	811 194
1998	534 153	20 837	156 409	151 083	5 026	429	867 937	232 034	1 099 971	239 175	−	860 796
1999	567 994	21 874	169 520	156 344	6 060	229	922 021	239 782	1 161 803	255 236	−	906 567
2000	600 826	23 169	181 851	161 468	5 271	3	972 588	267 602	1 240 190	286 963	−	953 227
2001	632 496	24 720	194 503	165 472	6 189	396	1 023 776	273 140	1 296 916	299 929	−	996 987
2002	664 562	25 968	212 464	173 525	2 909	214	1 079 642	276 511	1 356 153	307 386	−	1 048 767
2003	697 160	27 185	232 699	178 751	3 983	−37	1 139 741	285 397	1 425 138	314 842	−	1 110 296
2004	732 531	28 953	250 708	194 491	4 856	−37	1 211 502	298 694	1 510 196	333 669	−	1 176 527
2005	759 714	31 588	269 017	206 636	4 071	−377	1 270 649	325 871	1 596 520	370 464	−717	1 225 339
Unadjusted												
2002 Q2	164 150	6 455	53 001	40 918	385	67	264 976	71 010	335 986	78 983		
Q3	168 281	6 517	53 530	42 995	2 690	75	274 088	71 145	345 233	79 523		
Q4	174 355	6 596	54 117	46 657	−2 634	13	279 104	67 694	346 798	75 754		
2003 Q1	164 428	6 679	56 739	45 416	2 612	−15	275 859	70 334	346 193	76 165		
Q2	171 826	6 731	58 158	41 917	−1 079	105	277 658	70 387	348 045	77 433		
Q3	177 152	6 837	58 449	43 719	3 700	−75	289 782	71 728	361 510	81 178		
Q4	183 754	6 938	59 353	47 699	−1 250	−52	296 442	72 948	369 390	80 066		
2004 Q1	173 338	7 101	61 166	48 625	3 383	107	293 720	69 837	363 557	77 865		
Q2	180 921	7 193	62 020	46 193	239	−80	296 486	73 410	369 896	82 348		
Q3	185 550	7 279	63 028	48 959	2 084	−104	306 796	76 089	382 885	86 802		
Q4	192 722	7 380	64 494	50 714	−850	40	314 500	79 358	393 858	86 654		
2005 Q1	179 986	7 746	65 457	51 160	2 071	−171	306 249	74 383	380 632	84 170		
Q2	187 144	7 848	66 701	48 479	−334	101	309 939	81 256	391 195	91 393		
Q3	192 103	7 950	68 108	52 529	3 413	−224	323 879	82 563	406 442	97 790		
Q4	200 481	8 044	68 751	54 468	−1 079	−83	330 582	87 669	418 251	97 111		
2006 Q1	187 196	8 228	71 505	55 212	338	−155	322 324	93 938	416 262	107 534		
Q2	194 966	8 323	70 186	53 106	−597	254	326 238	99 560	425 798	114 366		
Q3	200 751	8 400	71 411	55 990	4 326	−46	340 832	87 492	428 324	102 529		
Seasonally adjusted												
	ABJQ	HAYE	NMRP	NPQS	CAEX	NPJQ	YBIL	IKBH	ABMF	IKBI		
2002 Q2	165 458	6 455	52 931	42 932	553	67	268 396	71 108	339 504	78 476	−	261 028
Q3	166 634	6 517	53 589	43 779	692	75	271 286	69 926	341 212	77 163	−	264 049
Q4	168 609	6 596	54 190	45 374	384	13	275 166	67 002	342 168	75 846	−	266 322
2003 Q1	170 937	6 679	56 026	43 973	−550	−15	277 050	72 531	349 581	78 663	−	270 918
Q2	173 555	6 731	57 776	44 100	−555	105	281 712	70 700	352 412	77 282	−	275 130
Q3	175 502	6 837	58 818	44 304	2 238	−75	287 624	70 821	358 445	78 421	−	280 024
Q4	177 166	6 938	60 079	46 374	2 850	−52	293 355	71 345	364 700	80 476	−	284 224
2004 Q1	180 057	7 101	60 824	47 112	−159	107	295 042	71 906	366 948	79 973	−	286 975
Q2	182 611	7 193	62 189	48 619	995	−80	301 527	74 030	375 557	82 437	−	293 120
Q3	184 131	7 279	63 183	49 363	1 182	−104	305 034	75 188	380 222	84 224	−	295 998
Q4	185 732	7 380	64 512	49 397	2 838	40	309 899	77 570	387 469	87 035	−	300 434
2005 Q1	187 533	7 746	65 063	50 001	1 898	−171	312 070	76 651[†]	389 109	86 979[†]	−227	301 743
Q2	188 498	7 848	66 587	50 560	164	101	313 758	80 208	394 402	89 357	−227	304 407
Q3	190 617	7 950	68 285	52 758	1 029	−224	320 415	80 663	401 978	94 610	−179	306 650
Q4	193 066	8 044	69 082	53 317	980	−83	324 406	85 761	411 031	97 743	−84	312 539
2006 Q1	194 134	8 228	69 871	54 158	1 331	−155	327 567	93 721	422 394	108 323	278	314 028
Q2	196 882	8 323	70 966	55 486	1 253	254	333 164	96 934	430 938	111 421	395	319 232
Q3	199 321	8 400	71 784	56 747	1 679	−47	337 885	86 059	424 975	99 571	479	325 034

1.2 Gross domestic product: by category of expenditure

continued

£ million[1]

| | Domestic expenditure on goods and services at market prices | | | | | | | | | | Statistical discrepancy (expenditure) | Gross domestic product at market prices |
| | Final consumption expenditure | | | Gross capital formation | | | | | | | | |
	Households	Non-profit institutions [2]	General government	Gross fixed capital formation	Changes in inventories [3]	Acquisitions less disposals of valuables	Total	Total exports	Gross final expenditure	less Total imports		
Chained volume indices												
	ABPF	ABNU	NMRU	NPQR	ABMQ	NPJP	YBIK	KTMZ	ABME	KTNB	GIXS	ABMI
1997	558 064	23 391	196 353	139 064	3 394	−35	918 661	231 494	1 150 396	218 613	–	936 717
1998	579 342	25 092	198 592	158 525	4 291	30	965 970	238 344	1 203 987	238 834	–	968 040
1999	606 648	25 023	205 853	163 039	5 803	–	1 006 378	247 289	1 253 258	257 809	–	997 295
2000	633 662	27 177	212 265	167 486	4 648	−28	1 045 373	269 830	1 315 374	281 081	–	1 035 295
2001	653 326	27 155	217 359	171 639	5 577	342	1 075 760	277 694	1 353 632	294 449	–	1 059 648
2002	676 833	27 130	224 868	178 066	2 289	183	1 109 596	280 593	1 390 217	308 706	–	1 081 469
2003	697 160	27 185	232 699	178 751	3 982	−37	1 139 741	285 397	1 425 138	314 842	–	1 110 296
2004	721 434	27 327	240 129	189 492	4 597	−42	1 182 937	299 289	1 482 225	335 703	–	1 146 523
2005	730 994	28 132	247 489	195 913	3 611	−354	1 205 785	322 792	1 528 577	359 179	−685	1 168 713
2006	..	..	..	..	..	..	..	..	..	..	..	1 200 614
Unadjusted												
2002 Q2	167 128	6 756	55 411	42 040	73	48	271 426	71 375	342 873	78 708		
Q3	171 224	6 793	56 241	43 845	2 642	62	281 010	72 412	353 470	79 914		
Q4	176 748	6 819	56 901	47 463	−2 601	7	285 438	69 143	354 527	76 886		
2003 Q1	165 903	6 843	58 099	46 880	2 644	−8	280 509	70 460	350 960	76 806		
Q2	172 040	6 779	57 436	42 110	−1 247	−12	277 848	70 218	348 072	77 276		
Q3	176 448	6 790	58 001	43 129	3 793	−14	287 655	71 428	359 081	80 246		
Q4	182 769	6 773	59 163	46 632	−1 208	−3	293 729	73 291	367 025	80 514		
2004 Q1	171 913	6 830	60 335	48 601	3 248	112	291 040	71 101	362 141	79 844		
Q2	178 308	6 805	59 021	45 087	−154	−90	288 976	73 901	362 877	83 080		
Q3	182 480	6 826	59 766	47 333	2 385	−96	298 695	76 134	374 829	86 480		
Q4	188 733	6 866	61 007	48 471	−882	32	304 226	78 153	382 378	86 299		
2005 Q1	175 209	7 040	61 201	49 254	1 934	−158	294 481	73 446	367 927	83 327		
Q2	180 537	7 013	61 602	46 174	−373	86	295 039	80 414	375 453	89 368		
Q3	183 843	7 028	62 176	49 281	3 232	−201	305 359	83 038	388 397	93 406		
Q4	191 405	7 051	62 510	51 204	−1 182	−81	310 906	85 894	396 800	93 078		
2006 Q1	177 692	7 194	62 876	52 016	7	−128	299 658	91 950	391 608	102 046		
Q2	183 679	7 191	61 902	49 170	−448	233	301 727	97 456	399 183	108 195		
Q3	187 851	7 205	63 092	51 805	4 231	−29	314 155	84 487	398 643	95 907		
Seasonally adjusted												
	ABJR	HAYO	NMRY	NPQT	CAFU	NPJR	YBIM	IKBK	ABMG	IKBL		
2002 Q3	169 715	6 793	56 429	44 765	511	62	278 337	71 056	349 422	78 006	–	271 368
Q4	170 727	6 819	56 395	46 393	346	7	280 820	68 564	349 337	76 624	–	272 761
2003 Q1	171 828	6 843	57 099	44 934	−571	−8	280 285	72 662	352 958	78 836	–	274 119
Q2	174 146	6 779	57 684	44 161	−644	94	282 367	70 611	352 971	77 283	–	275 712
Q3	175 140	6 790	58 445	43 924	2 264	−68	286 503	70 334	356 830	78 089	–	278 748
Q4	176 046	6 773	59 471	45 732	2 934	−55	290 586	71 790	362 379	80 634	–	281 717
2004 Q1	178 197	6 830	59 969	47 256	−381	112	291 983	73 389	365 373	81 648	–	283 725
Q2	180 362	6 805	59 530	47 102	1 050	−90	294 759	74 861	369 620	83 313	–	286 307
Q3	181 032	6 826	60 002	47 813	1 025	−96	296 603	75 097	371 700	84 300	–	287 400
Q4	181 843	6 866	60 628	47 321	2 903	32	299 592	75 942	375 532	86 442	–	289 091
2005 Q1	182 294	7 040	60 974	48 171	1 754	−158	300 076	75 931	376 007	85 846	−218	289 943
Q2	182 222	7 013	61 737	48 162	177	86	299 397	80 048	379 445	87 949	−217	291 280
Q3	182 723	7 028	62 232	49 663	835	−201	302 280	82 027	384 307	91 383	−171	292 753
Q4	183 755	7 051	62 546	49 917	845	−81	304 032	84 786	388 818	94 001	−79	294 737
2006 Q1	184 218	7 194	62 657	50 919	1 585	−128	306 446	92 701	399 147	102 540	263	296 869
Q2	185 902	7 191	63 027	51 173	1 332	233	308 858	95 847	404 705	105 991	370	299 084
Q3	186 632	7 205	63 510	52 116	1 512	−29	310 946	84 149	395 096	94 413	444	301 126
Q4	..	..	..	..	..	..	..	..	..	..	..	303 535

1 Estimates given to nearest million but cannot be regarded as accurate to that degree.
2 Non-profit institutions serving households.
3 Quarterly alignment adjustment included in this series.

Source: Office for National Statistics: 020 7533 6031

1.3 Gross domestic product: by category of income

£ million[1]

	Compen-sation of employees	Gross operating surplus of corporations				Other income[3]	Gross value added at factor cost	Taxes on production less subsidies	Statistical discrepancy (income)	Gross domestic product at market prices
		Non-financial corporations		Financial corporations	Total					
		Public	Private[2]							
At current prices										
	HAEA	NRJT	NRJK	NQNV	CGBY	CGBW	CGCA	GCSC	GIXQ	YBHA
1996	403 887	8 815	159 721	19 056	187 592	76 301	667 780	97 372	–	765 152
1997	429 967	7 249	171 338	17 385	195 972	80 449	706 388	104 806	–	811 194
1998	466 080	7 754	174 846	18 430	201 030	81 806	748 916	111 880	–	860 796
1999	495 793	7 678	178 939	15 976	202 593	86 723	785 109	121 458	–	906 567
2000	532 179	7 188	185 198	12 398	204 784	87 842	824 805	128 422	–	953 227
2001	564 194	6 892	185 942	12 052	204 886	97 352	866 432	130 555	–	996 987
2002	587 396	6 657	189 906	32 230	228 793	97 468	913 657	135 110	–	1 048 767
2003	616 893	7 265	202 479	39 936	249 680	102 494	969 067	141 229	–	1 110 296
2004	648 717	6 653	219 738	46 020	272 411	106 183	1 027 311	149 216	–	1 176 527
2005	684 825	7 853	228 030	38 408	274 291	113 374	1 072 490	152 463	386	1 225 339
Unadjusted										
2002 Q2	144 631	1 727	47 333	4 747	53 807	28 111	226 549	33 538		
Q3	144 964	1 704	45 857	11 023	58 584	22 656	226 204	34 225		
Q4	149 508	1 469	52 186	8 815	62 470	24 401	236 379	34 810		
2003 Q1	155 541	1 888	47 530	12 109	71 470 / 61 527	12 633 / 22 576	239 644	33 231		
Q2	151 315	1 788	48 831	8 838	69 376 / 59 457	18 336 / 28 255	239 027	34 747		
Q3	153 142	1 721	50 917	10 974	63 612	23 983	240 737	35 704		
Q4	156 895	1 868	55 201	8 015	65 084	27 680	249 659	37 547		
2004 Q1	165 466	1 992	53 583	10 950	66 525	25 143	257 134	35 868		
Q2	159 003	1 514	53 952	10 554	66 020	28 258	253 281	36 907		
Q3	159 427	1 476	54 708	12 563	68 747	25 953	254 127	37 619		
Q4	164 821	1 671	57 495	11 953	71 119	26 829	262 769	38 822		
2005 Q1	174 597	1 840	54 773	10 319	66 932	26 645	268 174	36 066		
Q2	167 481	1 747	56 824	8 575	67 146	30 407	265 034	37 841		
Q3	168 487	2 162	56 662	8 808	67 632	27 885	264 004	39 029		
Q4	174 260	2 104	59 771	10 706	72 581	28 437	275 278	39 527		
2006 Q1	186 708	1 997	56 863	9 322	68 182	28 173	283 063	37 352		
Q2	176 129	2 154	60 274	8 688	71 116	31 434	278 679	40 068		
Q3	176 474	2 029	59 499	12 203	73 731	28 283	278 488	40 829		
Seasonally adjusted										
	DTWM	CAEQ	CAER	NHCZ	CGBZ	CGBX	CGCB	CMVL		
2002 Q2	146 088	1 716	47 349	6 878	55 943	25 451	227 482	33 546	–	261 028
Q3	147 571	1 759	47 615	9 032	58 406	24 201	230 178	33 871	–	264 049
Q4	149 755	1 468	47 060	9 806	58 334	24 077	232 166	34 156	–	266 322
2003 Q1	150 812	1 855	48 689	10 938	61 482	24 127	236 421	34 497	–	270 918
Q2	152 768	1 807	49 468	10 029	61 304	26 243	240 315	34 815	–	275 130
Q3	155 699	1 802	52 340	9 437	63 579	25 128	244 406	35 618	–	280 024
Q4	157 614	1 801	51 982	9 532	63 315	26 996	247 925	36 299	–	284 224
2004 Q1	159 239	1 813	52 656	9 624	64 093	26 712	250 044	36 931	–	286 975
Q2	160 899	1 524	55 385	11 667	68 576	26 509	255 984	37 136	–	293 120
Q3	162 653	1 593	55 940	11 163	68 696	27 199	258 548	37 450	–	295 998
Q4	165 926	1 723	55 757	13 566	71 046	25 763	262 735	37 699	–	300 434
2005 Q1	167 514	1 887	57 107	9 605	68 599	27 765	263 878	37 702	163	301 743
Q2	169 925	1 836	56 165	10 049	68 050	28 471	266 446	37 814	147	304 407
Q3	172 452	2 011	56 822	7 641	66 474	29 008	267 934	38 627	89	306 650
Q4	174 934	2 119	57 936	11 113	71 168	28 130	274 232	38 320	–13	312 539
2006 Q1	178 266	2 075	56 995	8 524	67 594	29 402	275 262	39 122	–356	314 028
Q2	178 815	2 111	58 980	9 892	70 983	29 828	279 626	40 079	–473	319 232
Q3	180 790	2 093	61 773	11 037	74 903	29 362	285 055	40 537	–558	325 034

1 Estimates given to the nearest million but cannot be regarded as accurate to that degree
2 Quarterly alignment adjustment included in this series.
3 Includes mixed income and the operating surplus of non-corporate sector less the adjustment for financial intermediation services indirectly measured (FISIM)

Source: Office for National Statistics: 020 7533 6031

1.4 Index numbers: gross domestic product, chained volume indices at basic prices: by industry of output

2003 = 100

	Output at basic prices[1]								
				Service industries					
	Agriculture, hunting forestry and fishing	Total production industries[2]	Construc-tion	Distribu-tion, hotels and catering; repairs	Transport, storage and communi-cation	Business services and finance	Government and other services	Total services	Gross domestic product[3]
2003 weights[1,4]	10	186	61	153	78	277	235	744	1000
	GDQA	CKYW	GDQB	GDQE	GDQH	GDQN	GDQU	GDQS	YBEZ
1999	101.1	101.9	89.8	86.3	83.4	85.8	91.0	87.2	89.8
2000	100.3	103.8	90.2	89.1	92.9	90.7	93.1	91.3	93.2
2001	90.9	102.3	92.2	92.1	97.0	94.4	95.3	94.5	95.4
2002	102.1	100.3	95.5	96.4	98.2	96.3	97.7	96.9	97.4
2003	100.0	100.0	100.0	100.0	100.0	100.0	100.0	100.0	100.0
2004	99.0	100.8	104.0	105.2	102.5	105.1	102.0	103.9	103.3
2005	101.3	98.9	105.6	106.3	106.7	109.6	104.1	106.9	105.3
2006	99.5	98.8	106.8	109.8	109.8	115.5	106.3	110.8	108.1
Seasonally adjusted									
1997 Q3	96.6	99.8	88.2	81.1	69.5	76.4	88.9	80.2	84.7
Q4	96.2	99.6	90.0	82.4	71.2	78.0	89.1	81.4	85.4
1998 Q1	98.3	100.5	92.6	82.7	73.4	78.7	89.7	82.2	86.2
Q2	99.3	100.6	88.5	82.9	74.9	80.8	89.5	83.1	86.7
Q3	97.2	100.6	88.6	83.8	77.3	82.2	90.4	84.4	87.6
Q4	96.7	100.4	89.0	84.9	79.2	83.3	90.8	85.4	88.2
1999 Q1	101.0	100.7	89.9	85.1	80.1	84.2	90.3	85.7	88.5
Q2	101.0	101.0	88.6	85.9	82.5	85.2	90.5	86.6	89.1
Q3	100.7	102.8	90.5	86.8	84.8	86.0	91.2	87.6	90.3
Q4	101.8	103.1	90.3	87.6	86.3	87.6	91.9	88.8	91.3
2000 Q1	101.1	103.6	92.1	88.2	89.1	88.6	92.7	89.8	92.3
Q2	100.6	104.1	89.8	88.6	92.2	90.1	92.9	90.9	93.0
Q3	101.6	103.7	88.8	89.5	94.6	91.6	93.3	92.0	93.5
Q4	98.0	104.0	89.9	89.9	95.6	92.4	93.5	92.5	94.1
2001 Q1	91.6	104.0	91.5	91.2	97.2	93.5	94.3	93.7	95.0
Q2	90.2	102.5	91.7	91.3	97.2	94.2	94.9	94.1	95.1
Q3	89.8	102.4	92.3	92.4	96.5	94.9	95.5	94.7	95.7
Q4	92.1	100.5	93.3	93.6	97.1	95.1	96.4	95.4	96.0
2002 Q1	101.0	100.5	94.8	95.3	98.0	94.7	96.9	95.9	96.5
Q2	102.6	100.5	94.4	95.5	96.9	96.1	97.5	96.5	97.1
Q3	102.8	100.2	95.8	96.7	98.4	97.0	97.9	97.4	97.8
Q4	102.0	100.2	97.0	98.0	99.3	97.3	98.3	98.0	98.3
2003 Q1	99.7	99.9	97.0	98.2	99.2	98.5	98.8	98.6	98.8
Q2	99.3	99.4	98.9	99.4	99.8	98.9	99.5	99.3	99.3
Q3	100.1	100.0	101.7	100.6	100.3	100.4	100.3	100.4	100.4
Q4	100.9	100.8	102.4	101.8	100.7	102.2	101.3	101.7	101.5
2004 Q1	99.1	100.9	102.8	103.6	100.7	103.4	101.4	102.5	102.2
Q2	98.3	101.3	103.4	105.2	102.2	104.3	102.2	103.6	103.1
Q3	99.3	100.3	104.4	106.0	103.1	105.6	102.0	104.3	103.5
Q4	99.2	100.6	105.4	105.9	104.1	106.9	102.5	105.0	104.1
2005 Q1	100.8	99.7	106.0	105.7	105.8	107.6	103.1	105.6	104.5
Q2	102.3	99.4	106.4	105.7	105.9	108.8	103.6	106.2	104.9
Q3	101.3	98.6	105.1	106.3	106.7	110.1	104.6	107.2	105.5
Q4	100.9	98.1	104.8	107.6	108.3	111.9	104.9	108.4	106.2
2006 Q1	100.5	98.7	105.8	108.5	108.7	113.1	105.6	109.3	107.0
Q2	99.5	98.8	106.3	109.4	109.4	114.8	106.0	110.4	107.7
Q3	99.4	99.0	107.0	109.6	109.7	116.5	106.5	111.2	108.5
Q4	98.9	98.9	107.9	111.6	111.3	117.6	107.1	112.4	109.4

1 Components of output are valued at basic prices, which excludes taxes *less* subsidies on products, whereas GDP is valued at market prices.

2 The latest data for the index of production (series CKYW) are presented in Table 7.1. The figures given in this table are consistent with the figures for gross value added .

3 Includes an implicit discrepancy compared with the sum of the previous columns because the GDP aggregate takes account of other information based on incomes and expenditures.

4 The weights shown are in proportion to total gross value added (GVA) in 2003 and are used to combine the industry output indices to calculate the totals for 2004 and 2005. For 2003 and earlier, totals are calculated using the equivalent weights for the previous year (eg total for 2002 use 2001 weights).

Source: Office for National Statistics: 020 7533 6031

1.5 Households sector[1]: allocation of primary income account

£ million

	RESOURCES					USES			
	Gross operating surplus including gross mixed income	Wages and salaries	Employers' social contributions	Property Income received	Total resources	Property Income paid	Balance of primary incomes, gross	Total uses	Households' share of gross national income[2]
	RVGJ	QWLW	QWLX	QWME	QWMF	QWMI	QWMJ	QWMF	RVGG
1996	90 215	351 322	52 658	108 635	602 830	38 485	564 345	602 830	74.0
1997	94 187	374 510	55 540	118 339	642 576	42 078	600 498	642 576	74.0
1998	100 465	406 548	59 522	124 863	691 398	51 435	639 963	691 398	73.7
1999	106 929	431 795	64 199	120 966	723 889	47 649	676 240	723 889	74.8
2000	111 765	462 505	69 824	128 800	772 894	53 090	719 804	772 894	75.5
2001	121 204	491 044	73 216	132 918	818 382	52 356	766 026	818 382	76.2
2002	128 315	508 681	78 782	119 736	835 514	51 729	783 785	835 514	73.3
2003	137 057	527 689	89 263	120 914	874 923	53 796	821 127	874 923	72.5
2004	144 667	550 654	98 134	127 180	920 635	62 901	857 734	920 635	71.4
2005	153 737	576 866	108 024	146 636	985 263	71 200	914 063	985 263	73.1
Unadjusted									
2001 Q4	30 920	123 514	19 036	32 829	206 299	12 690	193 609	206 299	75.5
2002 Q1	31 538	127 957	20 237	28 570	208 302	12 650	195 652	208 302	74.9
Q2	31 888	125 928	18 738	32 004	208 558	12 670	195 888	208 558	75.3
Q3	32 251	126 065	19 003	29 176	206 495	13 009	193 486	206 495	71.1
Q4	32 638	128 731	20 804	29 986	212 159	13 400	198 759	212 159	71.9
2003 Q1	33 249	133 142	22 310	27 438	216 139	13 340	202 799	216 139	72.4
Q2	34 137	130 600	20 754	33 397	218 888	13 264	205 624	218 888	74.8
Q3	34 577	131 180	22 053	29 536	217 346	13 417	203 929	217 346	71.5
Q4	35 094	132 767	24 146	30 543	222 550	13 775	208 775	222 550	71.3
2004 Q1	35 556	139 478	25 893	28 277	229 204	14 367	214 837	229 204	72.5
Q2	35 901	135 934	23 117	32 250	227 202	15 019	212 183	227 202	72.4
Q3	36 353	136 101	23 428	32 804	228 686	16 396	212 290	228 686	70.8
Q4	36 857	139 141	25 696	33 849	235 543	17 119	218 424	235 543	69.8
2005 Q1	37 534	146 836	27 647	33 943	245 960	17 550	228 410	245 960	74.5
Q2	38 206	142 553	24 977	38 146	243 882	17 545	226 337	243 882	73.0
Q3	38 742	142 528	26 067	37 928	245 265	18 023	227 242	245 265	72.5
Q4	39 255	144 949	29 333	36 619	250 156	18 082	232 074	250 156	72.3
2006 Q1	39 621	153 439	33 152	36 272	262 484	18 226	244 258	262 484	76.2
Q2	40 019	149 105	27 067	39 512	255 703	18 160	237 543	255 703	73.4
Q3	40 502	149 067	27 508	39 163	256 240	19 166	237 074	256 240	71.8
Seasonally adjusted									
	NRJN	ROYJ	ROYK	ROYL	ROYR	ROYT	ROYS	ROYR	NRJH
2001 Q4	30 920	123 889	18 912	31 902	205 623	12 560	193 063	205 623	75.8
2002 Q1	31 538	124 971	19 019	30 312	205 840	12 671	193 169	205 840	73.8
Q2	31 888	126 664	19 443	30 405	208 400	12 823	195 577	208 400	74.0
Q3	32 251	127 816	19 778	29 795	209 640	12 970	196 670	209 640	72.5
Q4	32 638	129 230	20 542	29 224	211 634	13 265	198 369	211 634	72.8
2003 Q1	33 358	129 933	20 894	29 701	213 886	13 292	200 594	213 886	72.1
Q2	33 938	131 181	21 610	31 556	218 285	13 422	204 863	218 285	73.3
Q3	34 479	132 790	22 920	29 360	219 549	13 328	206 221	219 549	72.6
Q4	35 282	133 785	23 839	30 297	223 203	13 754	209 449	223 203	72.0
2004 Q1	35 556	134 980	24 274	30 521	225 331	14 319	211 012	225 331	72.1
Q2	35 901	136 807	24 124	30 570	227 402	15 200	212 202	227 402	70.9
Q3	36 353	138 323	24 347	32 422	231 445	16 289	215 156	231 445	71.6
Q4	36 857	140 544	25 389	33 667	236 457	17 093	219 364	236 457	70.9
2005 Q1	37 534	142 185	25 335	35 878	240 932	17 527	223 405	240 932	72.4
Q2	38 206	143 431	26 524	36 588	244 749	17 789	226 960	244 749	72.3
Q3	38 742	144 857	27 613	37 405	248 617	17 917	230 700	248 617	73.7
Q4	39 255	146 393	28 552	36 765	250 965	17 967	232 998	250 965	73.7
2006 Q1	39 621	148 360	29 907	38 158	256 046	18 187	237 859	256 046	74.2
Q2	40 019	150 003	28 835	37 995	256 852	18 423	238 429	256 852	72.7
Q3	40 502	151 554	29 242	38 507	259 805	19 040	240 765	259 805	72.7

1 This sector includes households and non-profit institutions serving households

2 The balance of gross primary incomes of the households and non-profit institutions serving households sector as a percentage of gross national income.

Source: Office for National Statistics: 020 7533 6031

1.6 Households sector[1]: secondary distribution of income account

£ million

	RESOURCES					USES						
	Gross balance of primary incomes	Social contributi-ons	Social benefits other than social transfers in kind	Other current transfers	Total resources	Current taxes on incomes etc.	Social contributi-ons	Social benefits other than social transfers in kind	Other current transfers	Gross disposable income	Total uses	Real households' disposable income
	QWMJ	RVFH	QWML	QWMO	QWMP	QWMS	QWMY	QWMZ	QWNC	QWND	QWMP	RVGK
1996	564 345	429	144 508	39 370	748 652	89 008	104 548	899	27 831	526 366	748 652	602 417
1997	600 498	410	150 844	34 521	786 273	91 208	110 848	880	23 396	559 941	786 273	625 184
1998	639 963	478	154 438	36 405	831 284	106 566	116 012	950	24 966	582 790	831 284	634 508
1999	676 240	450	157 647	38 154	872 491	115 186	123 516	922	23 879	608 988	872 491	652 060
2000	719 804	476	162 833	43 670	926 783	124 726	130 679	948	27 015	643 415	926 783	681 249
2001	766 026	502	171 814	44 687	983 029	133 054	135 998	977	26 688	686 312	983 029	710 531
2002	783 785	530	182 673	50 218	1 017 206	134 959	143 558	1 006	28 635	709 048	1 017 206	722 823
2003	821 127	505	193 596	49 511	1 064 739	138 261	158 348	987	26 754	740 389	1 064 739	740 389
2004	857 734	495	202 074	51 778	1 112 081	147 134	170 437	984	27 843	765 683	1 112 081	752 890
2005	914 063	500	214 114	56 222	1 184 899	159 423	187 600	994	31 894	804 756	1 184 899	772 033
Unadjusted												
2001 Q4	193 609	130	45 714	11 613	251 066	28 400	34 004	249	6 624	181 789	251 066	187 370
2002 Q1	195 652	132	44 006	13 175	252 965	44 162	36 616	251	6 948	164 988	252 965	169 321
Q2	195 888	132	44 904	12 086	253 010	28 065	35 491	251	6 790	182 413	253 010	185 919
Q3	193 486	133	45 665	12 604	251 888	33 621	35 297	252	7 395	175 323	251 888	178 554
Q4	198 759	133	48 098	12 353	259 343	29 111	36 154	252	7 502	186 324	259 343	189 029
2003 Q1	202 799	129	46 491	12 406	261 825	44 396	39 244	249	6 649	171 287	261 825	172 922
Q2	205 624	128	47 224	12 360	265 336	29 351	38 188	248	6 728	190 821	265 336	191 102
Q3	203 929	125	48 544	12 254	264 852	34 429	40 199	246	7 100	182 878	264 852	182 136
Q4	208 775	123	51 337	12 491	272 726	30 085	40 717	244	6 277	195 403	272 726	194 229
2004 Q1	214 837	123	48 626	12 786	276 372	46 462	44 785	245	6 955	177 925	276 372	176 256
Q2	212 183	124	49 977	12 954	275 238	30 540	41 067	246	7 209	196 176	275 238	193 050
Q3	212 290	124	50 522	13 553	276 489	37 117	41 171	246	7 089	190 866	276 489	187 382
Q4	218 424	124	52 949	12 485	283 982	33 015	43 414	247	6 590	200 716	283 982	196 202
2005 Q1	228 410	125	50 466	14 154	293 155	50 716	47 937	248	8 165	185 438	293 155	180 031
Q2	226 337	125	52 536	14 808	293 806	34 076	45 598	248	8 675	205 665	293 806	197 825
Q3	227 242	125	54 184	14 111	295 662	40 423	46 290	249	8 256	200 577	295 662	191 380
Q4	232 074	125	56 928	13 149	302 276	34 208	47 775	249	6 798	213 076	302 276	202 797
2006 Q1	244 258	127	53 429	13 000	310 814	53 844	54 875	251	7 810	193 492	310 814	183 058
Q2	237 543	127	54 993	14 860	307 523	36 041	..	251	8 865	213 529	307 523	200 484
Q3	237 074	127	57 044	14 705	308 950	43 124	..	251	8 228	208 688	308 950	194 625
Seasonally adjusted												
	ROYS		RPHL	RPHM	RPHP	RPHR	RPHU	RPIA	RPIB	RPHQ	RPHP	NRJR
2001 Q4	193 063	130	43 739	11 927	248 859	33 315	34 805	249	6 938	173 552	248 859	178 680
2002 Q1	193 169	132	45 060	12 939	251 300	33 921	35 252	251	6 712	175 164	251 300	179 363
Q2	195 577	132	45 510	12 050	253 269	33 463	35 635	251	6 754	177 166	253 269	180 917
Q3	196 670	133	45 831	12 471	255 105	33 804	35 961	252	7 262	177 826	255 105	181 266
Q4	198 369	133	46 272	12 758	257 532	33 771	36 710	252	7 907	178 892	257 532	181 277
2003 Q1	200 594	129	47 778	12 121	260 622	33 808	37 125	249	6 364	183 076	260 622	184 156
Q2	204 863	128	47 783	12 237	265 011	34 897	38 697	248	6 605	184 564	265 011	185 216
Q3	206 221	125	48 613	12 152	267 111	34 263	41 102	246	6 998	184 502	267 111	184 087
Q4	209 449	123	49 422	13 001	271 995	35 293	41 424	244	6 787	188 247	271 995	186 930
2004 Q1	211 012	123	50 048	12 565	273 748	34 979	42 135	245	6 734	189 655	273 748	187 493
Q2	212 202	124	50 428	12 794	275 548	36 312	41 825	246	7 049	190 116	275 548	187 472
Q3	215 156	124	50 578	13 399	279 257	37 285	42 176	246	6 935	192 615	279 257	189 038
Q4	219 364	124	51 020	13 020	283 528	38 558	44 301	247	7 125	193 297	283 528	188 887
2005 Q1	223 405	125	51 664	14 047	289 241	39 376	44 983	248	8 058	196 576	289 241	190 600
Q2	226 960	125	53 345	14 409	294 839	39 912	46 369	248	8 276	200 034	294 839	192 798
Q3	230 700	125	54 479	13 927	299 231	40 551	47 886	249	8 072	202 473	299 231	193 492
Q4	232 998	125	54 626	13 839	301 588	39 584	48 594	249	7 488	205 673	301 588	195 143
2006 Q1	237 859	127	54 655	12 955	305 596	42 119	50 689	251	7 765	204 772	305 596	193 692
Q2	238 429	127	55 784	14 399	308 739	41 956	50 228	251	8 404	207 900	308 739	195 629
Q3	240 765	127	56 768	14 546	312 206	43 253	50 606	251	8 069	210 027	312 206	195 990

1 This sector includes households and non-profit institutions serving house-holds.

Source: Office for National Statistics: 020 7533 6031

1.7 Households sector[1]: use of disposable income account

£ million

	RESOURCES			USES			
	Gross disposable income	Adjustment for the change in net equity of households in pension funds	Total resources	Individual consumption expenditure	Gross saving	Total uses	Households' saving ratio[2]
	QWND	NSSE	NSSF	NSSG	NSSH	NSSF	RVGL
1996	526 366	15 657	542 023	490 840	51 183	542 023	9.4
1997	559 941	15 111	575 052	520 662	54 390	575 052	9.5
1998	582 790	14 044	596 834	554 990	41 844	596 834	7.0
1999	608 988	14 016	623 004	589 868	33 136	623 004	5.3
2000	643 415	14 164	657 579	623 995	33 584	657 579	5.1
2001	686 312	16 041	702 353	657 216	45 137	702 353	6.4
2002	709 048	17 783	726 831	690 530	36 301	726 831	5.0
2003	740 389	21 377	761 766	724 345	37 421	761 766	4.9
2004	765 683	25 108	790 791	761 484	29 307	790 791	3.7
2005	804 756	30 510	835 266	791 302	43 964	835 266	5.3

Unadjusted

2001 Q4	181 789	3 803	185 592	172 740	12 852	185 592	6.9
2002 Q1	164 988	4 550	169 538	164 176	5 362	169 538	3.2
Q2	182 413	4 774	187 187	170 605	16 582	187 187	8.9
Q3	175 323	4 242	179 565	174 798	4 767	179 565	2.7
Q4	186 324	4 217	190 541	180 951	9 590	190 541	5.0
2003 Q1	171 287	5 998	177 285	171 107	6 178	177 285	3.5
Q2	190 821	4 228	195 049	178 557	16 492	195 049	8.5
Q3	182 878	5 570	188 448	183 989	4 459	188 448	2.4
Q4	195 403	5 581	200 984	190 692	10 292	200 984	5.1
2004 Q1	177 925	7 516	185 441	180 439	5 002	185 441	2.7
Q2	196 176	5 766	201 942	188 114	13 828	201 942	6.8
Q3	190 866	5 346	196 212	192 829	3 383	196 212	1.7
Q4	200 716	6 480	207 196	200 102	7 094	207 196	3.4
2005 Q1	185 438	8 957	194 395	187 732	6 663	194 395	3.4
Q2	205 665	6 421	212 086	194 992	17 094	212 086	8.1
Q3	200 577	6 805	207 382	200 053	7 329	207 382	3.5
Q4	213 076	8 327	221 403	208 525	12 878	221 403	5.8
2006 Q1	193 492	13 124	206 616	195 424	11 192	206 616	5.4
Q2	213 529	8 390	221 919	203 289	18 630	221 919	8.4
Q3	208 688	7 569	216 257	209 151	7 106	216 257	3.3

Seasonally adjusted

	RPHQ	RPQJ	RPQK	RPQM	RPQL	RPQK	NRJS
2001 Q4	173 552	4 501	178 053	167 326	10 727	178 053	6.0
2002 Q1	175 164	4 144	179 308	170 261	9 047	179 308	5.0
Q2	177 166	4 126	181 292	171 913	9 379	181 292	5.2
Q3	177 826	4 706	182 532	173 151	9 381	182 532	5.1
Q4	178 892	4 807	183 699	175 205	8 494	183 699	4.6
2003 Q1	183 076	5 107	188 183	177 616	10 567	188 183	5.6
Q2	184 564	4 035	188 599	180 286	8 313	188 599	4.4
Q3	184 502	6 086	190 588	182 339	8 249	190 588	4.3
Q4	188 247	6 149	194 396	184 104	10 292	194 396	5.3
2004 Q1	189 655	6 273	195 928	187 158	8 770	195 928	4.5
Q2	190 116	5 788	195 904	189 804	6 100	195 904	3.1
Q3	192 615	5 892	198 507	191 410	7 097	198 507	3.6
Q4	193 297	7 155	200 452	193 112	7 340	200 452	3.7
2005 Q1	196 576	6 941	203 517	195 279	8 238	203 517	4.0
Q2	200 034	6 913	206 947	196 346	10 601	206 947	5.1
Q3	202 473	7 977	210 450	198 567	11 883	210 450	5.6
Q4	205 673	8 679	214 352	201 110	13 242	214 352	6.2
2006 Q1	204 772	10 144	214 916	202 362	12 554	214 916	5.8
Q2	207 900	9 027	216 927	205 205	11 722	216 927	5.4
Q3	210 027	8 928	218 955	207 721	11 234	218 955	5.1

1 This sector includes households and non-profit institutions serving households.

2 Households' and non-profit institutions serving households' gross saving as a percentage of total resources.

Source: Office for National Statistics: 020 7533 6031

1.8 Household final consumption expenditure[1]

£ million

			UK National[2]												
						UK Domestic[3]									
		Net tourism	Total	Food & Drink	Alcohol & tobacco[4]	Clothing & footwear	Housing	Household goods & services	Health	Transport[4]	Communic-ation	Recreati-on & culture	Education	Restaura-nts & hotels[4]	Miscella-neous
	Total														
COICOP	-	-	0	01	02	03	04	05	06	07	08	09	10	11	12

At current prices

	ABPB	ABTE	ABQI	ABZV	ADFL	ADFP	ADFS	ADFY	ADGP	ADGT	ADGX	ADGY	ADIE	ADIF	ADII
1996	472 711	339	472 372	53 025	20 439	29 535	87 700	27 758	7 432	70 380	9 359	53 575	6 565	55 071	51 533
1997	501 290	905	500 385	53 787	21 553	30 901	91 977	29 492	7 757	77 204	9 984	58 012	7 440	57 164	55 114
1998	534 153	2 369	531 784	55 162	22 459	31 947	98 114	31 002	8 306	82 506	10 902	63 246	7 814	61 807	58 519
1999	567 994	5 378	562 616	57 040	24 458	33 375	103 193	32 846	8 775	87 237	12 005	67 481	8 943	64 387	62 876
2000	600 826	6 941	593 885	58 628	24 617	35 479	108 050	35 675	9 208	93 052	13 356	70 154	9 534	68 557	67 575
2001	632 496	9 524	622 972	59 804	25 158	36 822	115 905	37 974	9 976	96 435	14 157	73 452	9 409	71 620	72 260
2002	664 562	10 563	653 999	61 310	25 966	39 092	121 238	40 448	10 778	100 147	14 675	79 122	9 381	76 426	75 416
2003	697 160	12 158	685 002	63 174	27 297	41 155	129 051	42 466	11 335	104 569	15 654	84 386	9 610	78 902	77 403
2004	732 531	12 041	720 490	65 521	27 713	42 792	138 040	44 029	11 932	109 213	16 448	91 057	9 990	83 595	80 160
2005	759 714	12 049	747 665	67 101	27 997	43 764	146 935	43 662	12 025	113 001	16 746	94 258	10 409	88 860	82 907

Percentage change, year on previous year

1996	7.2		7.2	6.7	8.9	5.5	5.5	5.6	6.2	9.8	3.2	8.7	5.9	9.3	5.7
1997	6.0		5.9	1.4	5.5	4.6	4.9	6.2	4.4	9.7	6.7	8.3	13.3	3.8	6.9
1998	6.6		6.3	2.6	4.2	3.4	6.7	5.1	7.1	6.9	9.2	9.0	5.0	8.1	6.2
1999	6.3		5.8	3.4	8.9	4.5	5.2	5.9	5.6	5.7	10.1	6.7	14.4	4.2	7.4
2000	5.8		5.6	2.8	0.7	6.3	4.7	8.6	4.9	6.7	11.3	4.0	6.6	6.5	7.5
2001	5.3		4.9	2.0	2.2	3.8	7.3	6.4	8.3	3.6	6.0	4.7	−1.3	4.5	6.9
2002	5.1		5.0	2.5	3.2	6.2	4.6	6.5	8.0	3.8	3.7	7.7	−0.3	6.7	4.4
2003	4.9		4.7	3.0	5.1	5.3	6.4	5.0	5.2	4.4	6.7	6.7	2.4	3.2	2.6
2004	5.1		5.2	3.7	1.5	4.0	7.0	3.7	5.3	4.4	5.1	7.9	4.0	5.9	3.6
2005	3.7		3.8	2.4	1.0	2.3	6.4	−0.8	0.8	3.5	1.8	3.5	4.2	6.3	3.4

Not seasonally adjusted

2003 Q3	177 152	4 629	172 523	15 352	6 760	9 864	31 234	10 316	2 811	29 056	3 914	20 463	2 410	21 499	18 844
Q4	183 754	1 787	181 967	16 471	7 618	12 922	33 896	11 750	3 007	23 806	4 210	24 113	2 451	20 754	20 969
2004 Q1	173 338	2 364	170 974	16 171	6 496	9 075	35 251	10 292	2 881	26 485	4 070	20 878	2 470	18 299	18 606
Q2	180 921	3 214	177 707	16 298	6 832	10 242	33 609	10 871	2 990	26 904	4 024	22 666	2 487	20 980	19 804
Q3	185 550	4 680	180 870	15 665	6 745	10 224	33 148	10 933	2 973	30 377	4 089	22 034	2 501	22 492	19 689
Q4	192 722	1 783	190 939	17 387	7 640	13 251	36 032	11 933	3 088	25 447	4 265	25 479	2 532	21 824	22 061
2005 Q1	179 986	2 481	177 505	16 420	6 538	9 283	37 094	10 454	2 911	27 497	4 136	21 895	2 558	19 513	19 206
Q2	187 144	3 126	184 018	16 937	6 873	10 396	35 663	10 729	2 921	28 337	4 115	22 840	2 592	22 436	20 179
Q3	192 103	4 902	187 201	16 099	6 803	10 384	35 450	10 175	3 005	31 210	4 092	22 760	2 617	23 857	20 749
Q4	200 481	1 540	198 941	17 645	7 783	13 701	38 728	12 304	3 188	25 957	4 403	26 763	2 642	23 054	22 773
2006 Q1	187 196	2 517	184 679	16 852	6 699	9 356	39 942	10 492	3 107	28 512	4 226	22 520	2 672	20 127	20 174
Q2	194 966	3 228	191 738	17 478	7 109	10 844	38 246	11 052	3 156	28 932	4 160	24 110	2 711	23 215	20 725
Q3	200 751	4 517	196 234	17 297	7 105	10 968	37 704	10 926	3 145	32 460	4 152	23 793	2 751	24 853	21 080

Seasonally adjusted

	ABJQ	ABTF	ZAKV	ZWUM	ZAKX	ZAKZ	ZAVN	ZAVV	ZAWB	ZAWL	ZAWV	ZAWZ	ZWUS	ZAXR	ZAYF
2003 Q3	175 502	3 083	172 419	15 906	6 923	10 359	32 551	10 619	2 838	26 172	3 988	21 349	2 410	20 059	19 245
Q4	177 166	2 923	174 243	15 761	6 941	10 399	33 227	10 840	2 903	26 282	4 028	21 738	2 451	20 170	19 503
2004 Q1	180 057	2 975	177 082	16 381	6 926	10 617	33 870	10 671	2 925	26 806	4 101	22 150	2 470	20 527	19 638
Q2	182 611	2 905	179 706	16 211	6 925	10 736	34 287	11 083	3 030	26 991	4 083	23 052	2 487	20 827	19 994
Q3	184 131	3 121	181 010	16 285	6 918	10 734	34 668	11 294	2 995	27 407	4 166	22 936	2 501	20 999	20 107
Q4	185 732	3 040	182 692	16 644	6 944	10 705	35 215	10 981	2 982	28 009	4 098	22 919	2 532	21 242	20 421
2005 Q1	187 533	3 071	184 462	16 686	6 957	10 924	35 657	11 050	2 956	28 072	4 159	23 306	2 558	21 956	20 181
Q2	188 498	2 880	185 618	16 799	6 975	10 921	36 423	10 777	2 952	28 251	4 193	23 144	2 592	22 180	20 411
Q3	190 617	3 196	187 421	16 694	6 990	10 902	37 009	10 766	3 035	28 175	4 168	23 678	2 617	22 254	21 133
Q4	193 066	2 902	190 164	16 922	7 075	11 017	37 846	11 069	3 082	28 503	4 226	24 130	2 642	22 470	21 182
2006 Q1	194 134	3 085	191 049	17 134	7 132	11 024	38 293	11 039	3 150	28 844	4 241	23 913	2 672	22 634	20 973
Q2	196 882	3 000	193 882	17 331	7 221	11 356	39 167	11 339	3 185	29 047	4 229	24 403	2 711	22 909	20 984
Q3	199 321	2 869	196 452	17 892	7 304	11 515	39 606	11 419	3 177	29 331	4 231	24 751	2 751	23 134	21 341

1.8 Household final consumption expenditure[1]

continued

	Total	Net tourism	Total	Food & Drink	Alcohol & tobacco[4]	Clothing & footwear	Housing	Household goods & services	Health	Transport[4]	Communic-ation	Recreati-on & culture	Education	Restaura-nts & hotels[4]	Miscella-neous
							UK National[2]								
							UK Domestic[3]								
COICOP	-	-	0	01	02	03	04	05	06	07	08	09	10	11	12

Chained volume measures

	ABPF	ABTG	ABQJ	ADIP	ADIS	ADIW	ADIZ	ADJF	ADJM	ADJQ	ADJU	ADJV	ADMJ	ADMK	ADMN
1996	539 138	–3 165	542 297	56 292	26 798	24 777	119 956	28 793	10 672	82 239	7 945	49 099	9 819	71 200	62 811
1997	558 064	–319	558 122	57 261	27 125	25 696	121 226	30 248	10 531	85 325	8 698	52 550	10 582	71 229	64 600
1998	579 342	2 074	576 994	58 058	26 829	26 736	122 959	31 443	10 472	89 008	9 644	57 871	10 530	73 811	65 059
1999	606 648	5 868	600 627	59 904	27 623	28 689	123 662	33 130	10 362	92 969	10 948	63 601	11 394	74 191	67 867
2000	633 662	8 151	625 437	61 944	26 704	31 744	125 299	36 305	10 421	96 209	12 698	68 038	11 489	76 252	70 524
2001	653 326	10 733	642 595	61 048	26 497	34 485	126 749	38 310	10 697	98 485	14 452	72 552	10 692	76 434	73 239
2002	676 833	12 084	664 790	62 143	26 884	38 499	127 979	40 552	10 980	101 621	14 796	77 597	10 091	78 303	75 715
2003	697 160	12 158	685 002	63 174	27 297	41 155	129 051	42 466	11 335	104 569	15 654	84 386	9 610	78 902	77 403
2004	721 434	12 770	708 664	65 181	27 444	44 087	131 490	43 577	11 609	106 610	16 361	92 889	9 541	81 796	78 079
2005	730 994	11 593	719 401	65 806	27 245	46 145	131 952	42 946	11 552	107 365	17 093	98 707	9 474	83 815	77 301

Percentage change, year on previous year

1996	3.9		3.9	3.3	4.4	6.1	2.1	2.7	–0.2	5.7	5.1	6.0	1.7	5.4	1.7
1997	3.5		2.9	1.7	1.2	3.7	1.1	5.1	–1.3	3.8	9.5	7.0	7.8	–	2.8
1998	3.8		3.4	1.4	–1.1	4.0	1.4	4.0	–0.6	4.3	10.9	10.1	–0.5	3.6	0.7
1999	4.7		4.1	3.2	3.0	7.3	0.6	5.4	–1.1	4.5	13.5	9.9	8.2	0.5	4.3
2000	4.5		4.1	3.4	–3.3	10.6	1.3	9.6	0.6	3.5	16.0	7.0	0.8	2.8	3.9
2001	3.1		2.7	–1.4	–0.8	8.6	1.2	5.5	2.6	2.4	13.8	6.6	–6.9	0.2	3.8
2002	3.6		3.5	1.8	1.5	11.6	1.0	5.9	2.6	3.2	2.4	7.0	–5.6	2.4	3.4
2003	3.0		3.0	1.7	1.5	6.9	0.8	4.7	3.2	2.9	5.8	8.7	–4.8	0.8	2.2
2004	3.5		3.5	3.2	0.5	7.1	1.9	2.6	2.4	2.0	4.5	10.1	–0.7	3.7	0.9
2005	1.3		1.5	1.0	–0.7	4.7	0.4	–1.4	–0.5	0.7	4.5	6.3	–0.7	2.5	–1.0

Not seasonally adjusted

2003 Q3	176 448	4 523	172 065	15 331	6 692	9 965	30 904	10 326	2 799	28 724	3 895	20 478	2 401	21 520	19 051
Q4	182 769	1 741	180 953	16 328	7 561	12 871	33 187	11 696	2 951	23 639	4 197	24 368	2 411	20 673	20 951
2004 Q1	171 913	2 630	169 283	16 006	6 451	9 303	34 090	10 271	2 831	26 231	4 034	21 221	2 401	18 117	18 327
Q2	178 308	3 388	174 920	16 177	6 776	10 469	32 195	10 700	2 908	26 500	3 949	22 943	2 389	20 596	19 318
Q3	182 480	4 911	177 569	15 721	6 648	10 696	31 512	10 886	2 883	29 355	4 084	22 420	2 380	21 950	19 034
Q4	188 733	1 841	186 892	17 277	7 569	13 619	33 693	11 720	2 987	24 524	4 294	26 305	2 371	21 133	21 400
2005 Q1	175 209	2 593	172 616	16 105	6 408	9 812	34 054	10 438	2 807	26 689	4 186	22 679	2 372	18 714	18 352
Q2	180 537	3 044	177 493	16 548	6 694	10 880	32 260	10 539	2 806	27 329	4 186	23 534	2 371	21 244	19 102
Q3	183 843	4 498	179 345	15 874	6 595	11 095	31 711	10 015	2 880	29 252	4 191	23 826	2 367	22 370	19 169
Q4	191 405	1 458	189 947	17 279	7 548	14 358	33 927	11 954	3 059	24 095	4 530	28 668	2 364	21 487	20 678
2006 Q1	177 692	2 406	175 286	16 437	6 422	10 080	34 378	10 345	2 959	26 836	4 277	23 828	2 367	18 613	18 744
Q2	183 679	2 956	180 723	16 935	6 725	11 451	32 585	10 765	2 989	26 820	4 254	25 349	2 377	21 253	19 220
Q3	187 851	4 268	183 583	16 524	6 663	11 812	31 919	10 640	2 965	29 584	4 290	25 206	2 378	22 504	19 098

Seasonally adjusted

	ABJR	ABTH	ZAKW	ZWUN	ZAKY	ZALA	ZAVO	ZAVW	ZAWC	ZAWM	ZAWW	ZAXA	ZWUT	ZAXS	ZAYG
2003 Q3	175 140	3 019	172 120	15 797	6 868	10 339	32 229	10 590	2 828	26 169	3 974	21 450	2 401	20 060	19 399
Q4	176 046	2 803	173 236	15 590	6 870	10 507	32 491	10 841	2 848	26 142	4 020	21 894	2 411	20 085	19 514
2004 Q1	178 197	3 141	175 056	16 262	6 869	10 769	32 750	10 587	2 870	26 324	4 065	22 500	2 401	20 321	19 338
Q2	180 362	3 165	177 197	16 153	6 877	11 047	32 902	10 950	2 950	26 391	4 008	23 490	2 389	20 460	19 580
Q3	181 032	3 310	177 722	16 239	6 837	11 108	32 881	11 207	2 908	26 738	4 162	23 396	2 380	20 464	19 402
Q4	181 843	3 154	178 689	16 527	6 861	11 163	32 957	10 833	2 881	27 157	4 126	23 503	2 371	20 551	19 759
2005 Q1	182 294	3 079	179 215	16 389	6 814	11 420	32 786	10 952	2 845	26 961	4 224	24 129	2 372	21 039	19 284
Q2	182 222	2 817	179 405	16 462	6 808	11 501	32 968	10 632	2 840	27 092	4 269	24 113	2 371	21 004	19 345
Q3	182 723	2 968	179 755	16 426	6 800	11 526	32 980	10 557	2 914	26 571	4 270	24 946	2 367	20 868	19 530
Q4	183 755	2 729	181 026	16 529	6 823	11 698	33 218	10 805	2 953	26 741	4 330	25 519	2 364	20 904	19 142
2006 Q1	184 218	2 797	181 421	16 738	6 844	11 746	33 108	10 812	2 995	26 810	4 315	25 270	2 367	20 890	19 526
Q2	185 902	2 740	183 162	16 825	6 849	12 066	33 246	11 088	3 023	26 777	4 339	26 100	2 377	20 958	19 514
Q3	186 632	2 766	183 866	17 065	6 886	12 275	33 086	11 107	2 999	26 877	4 376	26 464	2 378	20 965	19 388

1 Until September 2001, Household Expenditure was published and broken down into 13 main headings according to existing UK National Accounts convention. From September 2001 it has been reclassified so as to conform to the European System of Accounts 1995 (ESA 95) COICOP (Classification Of Individual Consumption by Purpose).
2 Final consumption expenditure by UK households in the UK and abroad.
3 Final expenditure consumption in the UK by UK and foreign households.

4 Following reclassification to COICOP, alcohol consumed on the premises has been transferred from the "alcohol and tobacco" heading to "restaurants and hotels". Similarly, under reclassification, transport now includes purchase of bicycles.

Source: Office for National Statistics: 020 7533 6031

1.9 Change in inventories at chained volume measures

£ million[1]

	Mining and quarrying	Manufacturing industries				Elect-ricity, gas and water supply	Distributive trades		Other industries[3]	Change in inventories
		Materials and fuel	Work in progress	Finished goods	Total		Wholesale[2]	Retail[2]		
Value of stocks held at end-December 2005	984	16 220	16 052	19 787	52 059	1 833	28 486	24 786	54 056	162 204
	FADO	FBID	FBIE	FBIF	DHBH	FADP	FAJM	FBYH	DLWV	ABMQ
1996	−77	−199	−351	−3	−553	−21	783	801	452	1 231
1997	72	254	−1 413	295	−864	54	1 703	979	1 713	3 394
1998	367	537	−703	317	151	−163	666	1 186	2 636	4 291
1999	−325	503	−259	−430	−430	−167	1 743	1 722	3 464	5 803
2000	−263	543	358	418	1 319	202	1 939	1 480	−283	4 648
2001	87	−513	369	160	16	16	887	1 113	3 458	5 577
2002	−37	−496	−149	−372	−1 017	−132	788	1 716	971	2 289
2003	−66	−198	−650	−138	−986	−13	407	1 241	3 399	3 982
2004	−46	7	−614	−296	−903	8	304	1 000	4 234	4 597
2005	−47	−179	863	56	740	586	978	−412	1 766	3 611

Unadjusted

2002 Q2	−29	−97	12	−297	−382	157	635	1 302	−1 610	73
Q3	−16	212	577	−489	300	4	1 093	664	597	2 642
Q4	−31	−495	−715	−1 048	−2 258	−187	−1 145	−348	1 368	−2 601
2003 Q1	−33	342	124	822	1 288	35	508	−409	1 255	2 644
Q2	56	86	519	−363	242	−4	−614	778	−1 705	−1 247
Q3	−89	−313	−55	113	−255	22	786	1 054	2 275	3 793
Q4	−	−313	−1 238	−710	−2 261	−66	−273	−182	1 574	−1 208
2004 Q1	−28	−512	625	−61	52	117	629	691	1 787	3 248
Q2	12	−7	−345	160	−192	−116	325	−349	166	−154
Q3	−29	551	−70	190	671	105	−177	305	1 510	2 385
Q4	−1	−25	−824	−585	−1 434	−98	−473	353	771	−882
2005 Q1	−	207	248	763	1 218	−150	−696	−966	2 528	1 934
Q2	−29	−300	372	−434	−362	228	1 100	−883	−427	−373
Q3	−11	36	219	7	262	198	666	1 525	592	3 232
Q4	−7	−122	24	−280	−378	310	−92	−88	−927	−1 182
2006 Q1	−78	−43	846	469	1 272	−299	−1 270	−367	752	7
Q2	18	−168	427	−566	−307	226	618	−543	−460	−448
Q3	−17	197	213	−282	128	272	736	1 687	1 423	4 231

Seasonally adjusted

	FAEA	FBNF	FBNG	FBNH	DHBM	FAEB	FAJX	FBYN	DLWX	CAFU
2002 Q2	−32	−108	−195	−130	−433	132	854	1 136	−1 272	385
Q3	−22	−141	305	−265	−101	−74	475	−50	283	511
Q4	−29	−339	−259	−590	−1 188	−119	−598	−68	2 348	346
2003 Q1	−28	482	−29	−236	217	77	108	−156	−789	−571
Q2	55	−8	306	−31	267	−33	−370	894	−1 457	−644
Q3	−99	−557	−243	273	−527	−44	291	445	2 198	2 264
Q4	6	−115	−684	−144	−943	−13	378	58	3 448	2 934
2004 Q1	−27	−435	420	−1 177	−1 192	159	270	927	−518	−381
Q2	12	−76	−547	580	−43	−145	436	−128	918	1 050
Q3	−35	355	−199	283	439	39	−582	−362	1 526	1 025
Q4	4	163	−288	18	−107	−45	180	563	2 308	2 903
2005 Q1	4	246	197	57	500	−106	110	−352	1 598	1 754
Q2	−28	−186	151	−125	−160	188	496	−631	312	177
Q3	−19	−219	103	7	−109	133	157	712	−39	835
Q4	−4	−20	412	117	509	371	215	−141	−105	845
2006 Q1	−69	−73	428	55	410	−250	−651	270	1 875	1 585
Q2	15	8	135	−91	52	180	338	−127	874	1 332
Q3	−22	−22	−1	−326	−349	188	313	453	929	1 512

1 Estimates are given to the nearest £ million but cannot be regarded as accurate to this degree.
2 Wholesaling and retailing estimates exclude the motor trades.
3 Quarterly alignment adjustment included in this series. For description see notes.

Source: Office for National Statistics: 0207 533 5934

1.10 Gross fixed capital formation by sector and type of asset

£ million

		Analysis by sector						Analysis by asset					
		Public corporations[1]		Private sector									
			Transfer costs of non-prod-uced assets		Transfer costs on non-prod-uced assets				Other machinery and equipment		Other buildings and struct-ures	Intang-ible fixed assets	
	Business invest-ment[2]	General govern-ment[3]	Dwellings		Dwellings		Total	Transport equipment		Dwellings			Total

At current prices

	NPEM	NNBF	DEER	DLXQ	DFDF	EQBY	NPQX	DLWZ	DLXI	DFDK	EQEC	DLXP	NPQX
1996	86 662	11 859	1 995	−860	20 205	6 732	126 593	12 519	50 102	22 516	37 320	4 136	126 593
1997	92 801	10 487	1 623	−1 009	22 017	7 701	133 620	12 580	51 465	23 928	41 398	4 249	133 620
1998	107 882	11 910	1 632	−1 162	23 317	7 504	151 083	16 113	58 915	25 222	46 286	4 547	151 083
1999	110 417	12 599	1 529	−1 906	23 921	9 784	156 344	14 683	60 670	25 700	50 646	4 645	156 344
2000	113 213	12 227	1 421	−2 171	25 604	11 174	161 468	13 577	63 535	27 394	51 996	4 966	161 468
2001	112 024	13 533	2 387	−2 254	27 085	12 697	165 472	14 656	60 929	29 806	55 065	5 016	165 472
2002	111 146	15 452	2 837	−2 764	31 455	15 399	173 525	16 314	57 152	34 499	59 972	5 588	173 525
2003	109 218	20 509	3 509	−5 674	34 804	16 385	178 751	15 592	54 441	38 462	64 355	5 901	178 751
2004	111 811	23 206	3 235	−5 440	40 927	20 752	194 491	14 939	57 053	44 299	71 805	6 395	194 491
2005	132 000	8 173	3 574	−2 671	44 618	20 942	206 636	14 950	57 760	48 375	78 776	6 775	206 636

Unadjusted

2002 Q2	26 879	2 899	365	−517	7 873	3 419	40 918	4 577	13 586	8 290	13 066	1 399	40 918
Q3	26 951	3 732	522	−648	8 058	4 380	42 995	4 092	14 036	8 656	14 796	1 415	42 995
Q4	29 640	3 539	890	−791	9 064	4 315	46 657	3 619	15 365	10 014	16 186	1 473	46 657
2003 Q1	27 135	7 017	1 478	−2 121	7 457	4 450	45 416	3 953	14 245	8 955	16 824	1 439	45 416
Q2	26 717	3 658	479	−1 123	8 724	3 462	41 917	4 134	12 519	9 231	14 578	1 455	41 917
Q3	26 515	4 591	721	−1 124	8 835	4 181	43 719	3 941	12 818	9 608	15 866	1 486	43 719
Q4	28 851	5 243	831	−1 306	9 788	4 292	47 699	3 564	14 859	10 668	17 087	1 521	47 699
2004 Q1	27 823	7 250	1 157	−1 923	8 957	5 361	48 625	3 727	14 144	10 140	19 063	1 551	48 625
Q2	26 832	4 710	520	−1 149	10 404	4 876	46 193	4 202	13 123	10 952	16 332	1 584	46 193
Q3	28 242	5 233	708	−1 211	10 247	5 740	48 959	3 721	14 446	11 005	18 172	1 615	48 959
Q4	28 914	6 013	850	−1 157	11 319	4 775	50 714	3 289	15 340	12 202	18 238	1 645	50 714
2005 Q1	28 567	8 071	1 237	−1 080	9 600	4 765	51 160	3 578	14 544	10 893	20 482	1 663	51 160
Q2	44 031	−11 473	525	−578	11 178	4 796	48 479	4 112	13 380	11 743	17 561	1 683	48 479
Q3	29 180	5 719	878	−498	11 434	5 816	52 529	3 902	14 147	12 356	20 422	1 702	52 529
Q4	30 222	5 856	934	−515	12 406	5 565	54 468	3 358	15 689	13 383	20 311	1 727	54 468
2006 Q1	29 584	8 300	1 632	−947	10 736	5 907	55 212	3 264	15 358	12 408	22 429	1 753	55 212
Q2	29 963	4 861	587	−455	13 398	4 752	53 106	4 392	13 791	14 040	19 101	1 782	53 106
Q3	31 363	5 927	938	−542	13 398	4 906	55 990	3 934	15 090	14 389	20 745	1 832	55 990

Seasonally adjusted

	NPEK	RPZG	DKQG	TLNI	GGAG	TLOP	NPQS	TLPX	TLPW	GGAE	EQED	TLPK	NPQS
2002 Q2	27 667	3 850	592	−666	7 754	3 735	42 932	4 084	14 668	8 404	14 377	1 399	42 932
Q3	27 426	4 112	647	−819	8 183	4 230	43 779	4 165	14 393	8 910	14 896	1 415	43 779
Q4	28 796	3 710	915	−844	8 500	4 297	45 374	4 040	14 197	9 470	16 194	1 473	45 374
2003 Q1	27 054	5 291	985	−1 685	8 150	4 178	43 973	4 059	14 146	9 155	15 174	1 439	43 973
Q2	27 439	4 658	802	−1 303	8 586	3 918	44 100	3 641	13 478	9 416	16 110	1 455	44 100
Q3	26 840	5 073	885	−1 308	8 896	3 918	44 304	3 923	13 188	9 833	15 874	1 486	44 304
Q4	27 885	5 487	837	−1 378	9 172	4 371	46 374	3 969	13 629	10 058	17 197	1 521	46 374
2004 Q1	27 739	5 345	757	−1 453	9 805	4 919	47 112	3 847	13 989	10 588	17 137	1 551	47 112
Q2	27 555	5 771	840	−1 353	10 231	5 575	48 619	3 755	14 063	11 099	18 118	1 584	48 619
Q3	28 439	5 811	818	−1 392	10 278	5 409	49 363	3 683	14 688	11 146	18 231	1 615	49 363
Q4	28 078	6 279	820	−1 242	10 613	4 849	49 397	3 654	14 313	11 466	18 319	1 645	49 397
2005 Q1	28 791	6 084	783	−735	10 728	4 350	50 001	3 713	14 480	11 567	18 578	1 663	50 001
Q2	44 622	−10 316	869	−776	10 873	5 288	50 560	3 601	14 296	11 782	19 198	1 683	50 560
Q3	29 215	6 180	994	−628	11 377	5 620	52 758	3 844	14 345	12 415	20 452	1 702	52 758
Q4	29 372	6 225	928	−532	11 640	5 684	53 317	3 792	14 639	12 611	20 548	1 727	53 317
2006 Q1	29 900	6 207	1 050	−622	12 011	5 612	54 158	3 487	15 255	13 101	20 562	1 753	54 158
Q2	30 511	6 231	976	−643	13 028	5 383	55 486	3 820	14 790	14 059	21 035	1 782	55 486
Q3	31 316	6 388	1 048	−666	13 299	5 362	56 747	3 858	15 254	14 400	21 403	1 832	56 747

1.10 Gross fixed capital formation by sector and type of asset

continued

£ million

			Analysis by sector						Analysis by asset				
			Public corporations[1]		Private sector						Other new		
	Business invest-ment[2]	General govern-ment[3]	Dwellings	Transfer costs of non-prod-uced assets	Dwellings	Transfer costs of non-prod-uced assets	Total	Transport equipment	Other machinery and equipment	Dwellings	buildings and struct-ures	Intang-ible fixed assets	Total

Chained volume measures

	NPEN	EQDN	DEEW	EQDF	DFDP	EQCY	NPQR	DLWJ	DLWM	DFDV	DLWQ	EQDT	NPQR
1996	76 053	12 333	2 542	−2 172	29 852	17 765	130 555	12 608	35 715	32 839	49 484	4 796	130 555
1997	83 481	11 140	2 032	−2 215	31 610	18 197	139 064	12 960	38 200	33 942	54 092	4 950	139 064
1998	100 225	12 218	1 974	−2 284	31 971	15 614	158 525	16 279	47 919	34 201	58 200	4 982	158 525
1999	104 205	13 059	1 747	−3 141	30 928	16 821	163 039	14 602	51 650	32 863	59 956	4 956	163 039
2000	108 933	12 665	1 552	−3 093	31 041	16 293	167 486	13 489	55 766	32 888	58 736	5 172	167 486
2001	110 390	13 980	2 521	−2 825	31 318	16 173	171 639	14 698	56 779	34 172	59 527	5 129	171 639
2002	111 678	15 740	2 898	−3 092	33 748	17 369	178 066	16 414	55 968	36 839	62 088	5 676	178 066
2003	109 218	20 509	3 509	−5 674	34 804	16 385	178 751	15 592	54 441	38 462	64 355	5 901	178 751
2004	111 765	22 266	3 161	−5 561	38 245	19 616	189 492	14 706	58 817	41 541	68 135	6 294	189 492
2005	130 960	7 029	3 577	−2 891	39 102	18 290	195 913	14 906	59 626	42 701	72 108	6 573	195 913

Unadjusted

2002 Q2	27 035	2 921	374	−604	8 357	4 085	42 040	4 574	13 336	8 752	13 652	1 427	42 040
Q3	27 116	3 790	531	−694	8 547	4 759	43 845	4 121	13 750	9 135	15 142	1 433	43 845
Q4	29 863	3 609	898	−791	9 543	4 330	47 463	3 642	15 098	10 502	16 422	1 488	47 463
2003 Q1	27 246	7 542	1 491	−2 284	7 834	4 859	46 880	4 101	13 996	9 377	17 816	1 449	46 880
Q2	26 764	3 530	479	−1 177	8 639	3 770	42 110	4 089	12 255	9 127	15 069	1 458	42 110
Q3	26 453	4 426	717	−1 038	8 724	3 893	43 129	3 900	13 063	9 486	15 203	1 483	43 129
Q4	28 755	5 011	822	−1 175	9 607	3 863	46 632	3 502	15 127	10 472	16 267	1 511	46 632
2004 Q1	27 727	7 430	1 140	−2 187	8 761	5 730	48 601	3 755	14 580	9 927	18 805	1 534	48 601
Q2	26 939	4 545	509	−904	9 584	4 414	45 087	4 151	13 633	10 121	15 618	1 563	45 087
Q3	28 259	4 925	689	−857	9 476	4 841	47 333	3 663	14 888	10 214	16 982	1 587	47 333
Q4	28 840	5 366	823	−1 613	10 424	4 631	48 471	3 137	15 716	11 279	16 730	1 610	48 471
2005 Q1	28 407	7 595	1 197	−651	8 794	3 912	49 254	3 512	14 881	10 045	19 195	1 622	49 254
Q2	43 541	−11 244	521	−624	9 658	4 340	46 174	4 142	13 937	10 200	16 258	1 637	46 174
Q3	28 920	5 281	871	−744	9 954	5 032	49 281	3 910	14 523	10 834	18 364	1 649	49 281
Q4	30 092	5 397	988	−872	10 696	5 006	51 204	3 342	16 285	11 622	18 291	1 665	51 204
2006 Q1	29 448	7 723	1 505	−625	9 250	4 677	52 016	3 327	15 981	10 831	20 191	1 687	52 016
Q2	29 831	4 512	551	−365	10 737	3 905	49 170	4 372	14 539	11 340	17 208	1 711	49 170
Q3	31 389	5 245	881	−526	10 737	4 079	51 805	3 910	15 867	11 668	18 609	1 751	51 805

Seasonally adjusted

	NPEL	DLWF	DKQH	DLWH	DFEA	DLWI	NPQT	DLWL	DLWO	DFEG	DLWT	EQDO	NPQT
2002 Q2	27 677	3 846	608	−780	8 304	4 405	43 981	4 178	14 378	8 958	14 950	1 426	43 981
Q3	27 574	4 259	660	−894	8 669	4 613	44 765	4 269	14 253	9 400	15 363	1 433	44 765
Q4	28 980	3 875	925	−863	8 954	4 582	46 393	4 213	14 201	9 943	16 569	1 492	46 393
2003 Q1	27 111	5 673	994	−1 833	8 452	4 517	44 934	4 049	13 815	9 467	16 148	1 450	44 934
Q2	27 395	4 507	804	−1 378	8 695	4 145	44 161	3 726	13 165	9 536	16 287	1 463	44 161
Q3	26 712	4 999	882	−1 243	8 812	3 772	43 924	3 896	13 392	9 752	15 405	1 482	43 924
Q4	28 000	5 330	829	−1 220	8 845	3 951	45 732	3 921	14 069	9 707	16 515	1 506	45 732
2004 Q1	27 166	5 970	746	−1 598	9 421	5 551	47 256	3 771	14 083	10 193	17 675	1 534	47 256
Q2	27 757	5 360	824	−1 174	9 578	4 757	47 102	3 760	14 627	10 430	16 722	1 563	47 102
Q3	28 634	5 311	797	−1 186	9 524	4 733	47 813	3 635	15 299	10 370	16 922	1 587	47 813
Q4	28 208	5 625	794	−1 603	9 722	4 575	47 321	3 540	14 808	10 548	16 816	1 610	47 321
2005 Q1	28 653	5 613	758	−116	9 573	3 690	48 171	3 644	14 686	10 385	17 835	1 622	48 171
Q2	44 137	−10 232	835	−899	9 610	4 711	48 162	3 685	14 919	10 484	17 436	1 637	48 162
Q3	28 983	5 784	949	−965	9 950	4 962	49 663	3 828	14 821	10 941	18 424	1 649	49 663
Q4	29 187	5 864	881	−911	9 969	4 927	49 917	3 749	15 200	10 891	18 413	1 665	49 917
2006 Q1	29 807	5 682	994	−88	10 103	4 421	50 919	3 477	15 729	11 135	18 892	1 687	50 919
Q2	30 404	5 604	918	−649	10 674	4 222	51 173	3 854	15 578	11 644	18 386	1 711	51 173
Q3	31 359	5 772	984	−740	10 724	4 017	52 116	3 803	16 140	11 758	18 664	1 751	52 116

1 Remaining investment by public corporations included within business in-vestment.

2 Not including dwellings and purchases less sales of land and existing build-ings.

3 Please note that the data in the second quarter of 2005 is due to the transfer of nuclear reactors. In April 2005 British Nuclear Fuels (BNFL) transferred to the Nuclear Decommissioning Authority (NDA) nuclear reactors that were reaching the ends of their productive lives. BNFL is classified as a public corporation in the National Accounts and the NDA as central government. This transfer does not affect whole economy gross fixed capital formation (GFCF) since it is an acquisition by one sector and a disposal by another. The value of the transfer was -£15.6 billion. The negative value reflects the fact that the reactors are at the end of their productive lives and have large decommissioning and clean-up liabilities.

Source: Office for National Statistics: 0207 533 5934

1.11 Business Investment[1] by Industry, Chained volume measures

Reference year 2003, £ million[2]

| | Manufacturing | | | Non-manufacturing | | | | | | | Total Business Investment |
| | | | | Private Sector [3] | | | | Public Corporations | Total | |
	Private [3] Sector	Public Corporations [5]	Total	Other [4] Production	Construction	Distribution Services	Other Services			
2002	13 434	378	13 810	12 713	3 256	12 889	65 599	3 408	97 874	111 678
2003	13 138	309	13 447	11 853	3 296	12 109	64 800	3 713	95 771	109 218
2004	12 222	262	12 484	11 783	3 646	12 088	68 628	3 136	99 281	111 765
2005	13 976	15 335	29 311	10 780	2 621	13 468	70 792	3 988	101 649	130 960

Not seasonally adjusted

	INKL	APIA	APIL	IOCQ	KWOC	IOYO	JZKH	APII	APIP	NPEN
2002 Q4	3 710	94	3 805	3 210	866	3 712	16 640	1 613	26 048	29 863
2003 Q1	3 233	92	3 325	3 214	816	3 096	15 263	1 527	23 919	27 246
Q2	2 999	68	3 067	2 872	553	2 828	16 727	723	23 695	26 764
Q3	3 138	67	3 205	2 891	880	2 800	15 936	741	23 251	26 453
Q4	3 768	82	3 850	2 876	1 047	3 385	16 874	722	24 906	28 755
2004 Q1	2 541	77	2 618	3 033	845	3 177	17 234	820	25 109	27 727
Q2	3 007	55	3 062	2 811	1 001	2 402	16 917	746	23 877	26 939
Q3	3 034	55	3 089	2 998	880	3 149	17 378	765	25 170	28 259
Q4	3 640	75	3 715	2 941	920	3 360	17 099	805	25 125	28 840
2005 Q1	2 996	74	3 070	2 871	563	3 822	17 129	952	25 337	28 407
Q2	3 400	15 256	18 656	2 667	636	2 892	17 704	986	24 885	43 541
Q3	3 525	2	3 527	2 679	711	3 303	17 653	1 047	25 393	28 920
Q4	4 055	3	4 058	2 563	711	3 451	18 306	1 003	26 034	30 092
2006 Q1	3 210	2	3 212	3 192	764	3 128	18 113	1 039	26 236	29 448
Q2	3 167	4	3 171	3 121	811	2 821	19 001	906	26 660	29 831
Q3	3 428	2	3 430	3 616	714	3 325	19 294	1 010	27 959	31 389

Percentage change, latest quarter on previous quarter

2006 Q3	8.2	−50.0	8.2	15.9	−12.0	17.9	1.5	11.5	4.9	5.2

Percentage change, latest quarter on corresponding quarter of previous year

2006 Q3	−2.8	−	−2.8	35.0	0.4	0.7	9.3	−3.5	10.1	8.5

Seasonally adjusted

	INLN	APIE	APIN	IOCR	KWOE	IOYQ	JZKI	APIK	APIT	NPEL
2002 Q4	3 275	73	3 349	3 224	831	3 270	16 693	1 600	25 640	28 980
2003 Q1	3 535	62	3 595	3 063	833	3 036	15 298	1 341	23 576	27 111
Q2	3 116	77	3 194	2 988	598	3 213	16 520	844	24 135	27 395
Q3	3 099	85	3 186	2 897	887	2 842	16 004	812	23 466	26 712
Q4	3 388	85	3 472	2 905	978	3 018	16 978	716	24 594	28 000
2004 Q1	2 788	40	2 828	2 886	892	3 047	16 777	736	24 338	27 166
Q2	3 130	63	3 193	2 897	1 053	2 799	16 981	834	24 564	27 757
Q3	3 029	77	3 106	3 010	861	3 178	17 678	801	25 528	28 634
Q4	3 275	82	3 357	2 990	840	3 064	17 192	765	24 851	28 208
2005 Q1	3 280	65	3 345	2 752	610	3 692	17 407	847	25 308	28 653
Q2	3 586	15 259	18 845	2 765	648	3 377	17 440	1 062	25 292	44 137
Q3	3 517	7	3 524	2 676	712	3 280	17 718	1 073	25 459	28 983
Q4	3 593	4	3 597	2 587	651	3 119	18 227	1 006	25 590	29 187
2006 Q1	3 508	4	3 512	3 054	846	3 028	18 418	949	26 295	29 807
Q2	3 345	11	3 356	3 258	820	3 275	18 734	961	27 048	30 404
Q3	3 434	9	3 443	3 609	714	3 284	19 286	1 023	27 916	31 359

Percentage change, latest quarter on previous quarter

2006 Q3	2.7	−18.2	2.6	10.8	−12.9	0.3	2.9	6.5	3.2	3.1

Percentage change, latest quarter on corresponding quarter of previous year

2006 Q3	−2.4	28.6	−2.3	34.9	0.3	0.1	8.8	−4.7	9.7	8.2

1 All figures are exclusive of expenditure on land and existing buildings.
2 Estimates are shown to the nearest £ million but should not be regarded as accurate to this degree.
3 All private sector figures are exclusive of expenditure on dwellings.
4 Includes Agricultural Contractors.
5 Please note that the data in the second quarter of 2005 is due to the transfer of nuclear reactors. In April 2005 British Nuclear Fuels (BNFL) transferred to the Nuclear Decommissioning Authority (NDA) nuclear reactors that were reaching the end of their productive lives. BNFL is classified as a public corporation in the National Accounts and the NDA as central government. This transfer does not affect whole economy gross fixed capital formation (GFCF) since it is an acquisition by one sector and a disposal by another. The value of the transfer was -£15.6 billion. The negative value reflects the fact that the reactors are at the end of their productive lives and have large decommissioning and clean-up liabilities.

Source: Office for National Statistics: 020 7533 5934

1.12 Business Investment[1] by Industry at Current Prices

£ million[2]

| | Manufacturing | | | Non-manufacturing | | | | | | | Total |
| | | | | Private Sector [3] | | | | | | | Business |
	Private Sector [3]	Public Corporations [5]	Total	Other Production [4]	Construction	Distribution Services	Other Services	Public Corporations	Total		Investment
2002	13 441	366	13 807	12 386	3 309	12 972	65 281	3 391	97 339		111 146
2003	13 138	309	13 447	11 853	3 296	12 109	64 800	3 713	95 771		109 218
2004	12 141	275	12 416	11 854	3 633	12 145	68 573	3 190	99 395		111 811
2005	13 953	15 667	29 620	11 692	2 653	13 699	70 143	4 193	102 380		132 000

Not seasonally adjusted

	INJJ	APGG	APGZ	IOCP	KWOD	IOYP	JZKF	APGS	APHR		NPEM
2002 Q4	3 712	92	3 804	3 156	877	3 720	16 462	1 621	25 836		29 640
2003 Q1	3 221	90	3 311	3 178	816	3 089	15 215	1 526	23 824		27 135
Q2	2 994	68	3 062	2 875	552	2 827	16 678	723	23 655		26 717
Q3	3 144	67	3 211	2 907	881	2 802	15 974	740	23 304		26 515
Q4	3 779	84	3 863	2 893	1 047	3 391	16 933	724	24 988		28 851
2004 Q1	2 538	79	2 617	3 019	844	3 189	17 332	822	25 206		27 823
Q2	2 974	58	3 032	2 805	994	2 392	16 850	759	23 800		26 832
Q3	3 013	58	3 071	3 017	875	3 162	17 337	780	25 171		28 242
Q4	3 616	80	3 696	3 013	920	3 402	17 054	829	25 218		28 914
2005 Q1	2 978	79	3 057	3 012	560	3 871	17 077	990	25 510		28 567
Q2	3 389	15 583	18 972	2 893	643	2 935	17 552	1 036	25 059		44 031
Q3	3 520	2	3 522	2 953	723	3 367	17 515	1 100	25 658		29 180
Q4	4 066	3	4 069	2 834	727	3 526	17 999	1 067	26 153		30 222
2006 Q1	3 230	2	3 232	3 455	781	3 197	17 812	1 107	26 352		29 584
Q2	3 190	4	3 194	3 349	830	2 882	18 726	982	26 769		29 963
Q3	3 431	2	3 433	3 782	729	3 391	18 953	1 075	27 930		31 363

Percentage change, latest quarter on previous quarter

2006 Q3	7.6	−50.0	7.5	12.9	−12.2	17.7	1.2	9.5	4.3		4.7

Percentage change, latest quarter on corresponding quarter of previous year

2006 Q3	−2.5	–	−2.5	28.1	0.8	0.7	8.2	−2.3	8.9		7.5

Seasonally adjusted

	IOBN	APID	APIF	IOBM	IOYV	IOYW	JZKG	APIJ	APIO		NPEK
2002 Q4	3 273	84	3 357	3 180	839	3 278	16 536	1 606	25 439		28 796
2003 Q1	3 523	64	3 587	3 037	832	3 023	15 233	1 342	23 467		27 054
Q2	3 124	78	3 202	2 985	598	3 219	16 592	843	24 237		27 439
Q3	3 094	86	3 180	2 909	891	2 846	16 203	811	23 660		26 840
Q4	3 397	81	3 478	2 922	975	3 021	16 772	717	24 407		27 885
2004 Q1	2 786	42	2 828	2 887	890	3 052	17 343	739	24 911		27 739
Q2	3 125	64	3 189	2 894	1 048	2 799	16 777	848	24 366		27 555
Q3	2 991	81	3 072	3 021	858	3 195	17 476	817	25 367		28 439
Q4	3 239	88	3 327	3 052	837	3 099	16 977	786	24 751		28 078
2005 Q1	3 266	48	3 314	2 880	608	3 769	17 344	876	25 477		28 791
Q2	3 579	15 598	19 177	2 974	655	3 405	17 299	1 112	25 445		44 622
Q3	3 514	13	3 527	2 944	724	3 344	17 536	1 140	25 688		29 215
Q4	3 594	8	3 602	2 894	666	3 181	17 964	1 065	25 770		29 372
2006 Q1	3 538	8	3 546	3 285	867	3 090	18 114	998	26 354		29 900
Q2	3 365	10	3 375	3 471	840	3 345	18 441	1 039	27 136		30 511
Q3	3 439	11	3 450	3 771	729	3 353	18 902	1 111	27 866		31 316

Percentage change, latest quarter on previous quarter

2006 Q3	2.2	10.0	2.2	8.6	−13.2	0.2	2.5	6.9	2.7		2.6

Percentage change, latest quarter on corresponding quarter of previous year

2006 Q3	−2.1	−15.4	−2.2	28.1	0.7	0.3	7.8	−2.5	8.5		7.2

1 All figures are exclusive of expenditure on land and existing buildings.
2 Estimates are shown to the nearest £ million but should not be regarded as accurate to this degree.
3 All private sector figures are exclusive of expenditure on dwellings.
4 Includes Agricultural Contractors.
5 The data in the second quarter of 2005 is due to the transfer of nuclear reactors. In April 2005 British Nuclear Fuels (BNFL) transferred to the Nuclear Decommissioning Authority (NDA) nuclear reactors that were reaching the ends of their productive lives. BNFL is classified as a public corporation in the National Accounts and the NDA as central government. This transfer does not affect whole economy gross fixed capital formation (GFCF) since it is an acquisition by one sector and a disposal by another. The value of the transfer was -£15.6 billion. The negative value reflects the fact that the reactors are at the end of their productive lives and have large decommissioning and clean-up liabilities.

Source: Office for National Statistics: 020 7533 5934

1.13 Private Sector[1] Manufacturing Business Investment[2] by Industry, Chained volume measures

Reference year 2003, £ million[3]

	Analysis by industry group							
	Solid & nuclear fuels, oil refining	Metals & metal goods	Chemicals and man made fibres	Engineering and vehicles	Food, drink and tobacco	Textiles, clothing, leather and footwear	Other manufactur-ing	Total all manufacturing
2002	375	1 167	2 130	3 585	2 214	330	3 623	13 434
2003	418	1 159	1 885	3 155	2 357	243	3 921	13 138
2004	355	1 114	1 912	3 523	1 994	148	3 176	12 222
2005	341	1 263	1 793	4 061	2 233	188	4 097	13 976

Not seasonally adjusted

	INJX	INKA	INJY	INJO	INJT	INJU	JZKL	INKL
2002 Q4	135	328	695	922	603	66	957	3 710
2003 Q1	90	296	450	813	554	68	963	3 233
Q2	87	262	415	722	577	45	893	2 999
Q3	119	276	485	780	546	57	873	3 138
Q4	122	325	535	840	680	73	1 192	3 768
2004 Q1	65	211	395	664	430	40	736	2 541
Q2	70	247	443	967	498	40	742	3 007
Q3	88	327	505	781	542	37	754	3 034
Q4	132	329	569	1 111	524	31	944	3 640
2005 Q1	68	258	408	839	491	56	876	2 996
Q2	71	275	421	1 074	534	48	977	3 400
Q3	78	307	438	1 006	580	44	1 072	3 525
Q4	124	423	526	1 142	628	40	1 172	4 055
2006 Q1	119	319	375	928	527	34	908	3 210
Q2	134	362	366	761	574	43	927	3 167
Q3	137	322	429	944	554	25	1 017	3 428

Percentage change, latest quarter on previous quarter

2006 Q3	2.2	−11.0	17.2	24.0	−3.5	−41.9	9.7	8.2

Percentage change, latest quarter on corresponding quarter of previous year

2006 Q3	75.6	4.9	−2.1	−6.2	−4.5	−43.2	−5.1	−2.8

Seasonally adjusted

	INKZ	INLC	INLA	INKQ	INKV	INKW	JZKM	INLN
2002 Q4	103	300	563	791	561	66	889	3 275
2003 Q1	107	314	542	891	599	68	1 013	3 535
Q2	102	278	427	745	589	45	931	3 116
Q3	117	274	483	777	528	57	862	3 099
Q4	92	293	433	742	641	73	1 115	3 388
2004 Q1	78	228	467	742	461	40	772	2 788
Q2	85	268	468	983	507	40	779	3 130
Q3	91	325	505	798	525	37	748	3 029
Q4	101	293	472	1 000	501	31	877	3 275
2005 Q1	82	290	472	925	535	56	920	3 280
Q2	82	293	458	1 130	541	48	1 034	3 586
Q3	81	317	428	1 008	571	44	1 068	3 517
Q4	96	363	435	998	586	40	1 075	3 593
2006 Q1	140	362	428	1 017	573	34	954	3 508
Q2	152	384	400	796	582	43	988	3 345
Q3	145	335	417	952	546	25	1 014	3 434

Percentage change, latest quarter on previous quarter

2006 Q3	−4.6	−12.8	4.3	19.6	−6.2	−41.9	2.6	2.7

Percentage change, latest quarter on corresponding quarter of previous year

2006 Q3	79.0	5.7	−2.6	−5.6	−4.4	−43.2	−5.1	−2.4

1 All private sector figures are exclusive of expenditure on dwellings.
2 All figures are exclusive of expenditure on land and existing buildings.
3 Estimates are shown to the nearest £ million but should not be regarded as accurate to this degree.

Source: Office for National Statistics: 020 7533 5934

1.14 Private Sector[1] Manufacturing Business Investment[2] by Industry at Current Prices

£ million[3]

	Analysis by industry group							
	Solid & nuclear fuels, oil refining	Metals & metal goods	Chemicals and man made fibres	Engineering and vehicles	Food, drink and tobacco	Textiles, clothing, leather and footwear	Other manu-factur-ing	Total all manufacturing
2002	374	1 177	2 149	3 614	2 170	332	3 625	13 441
2003	418	1 159	1 885	3 155	2 357	243	3 921	13 138
2004	360	1 108	1 907	3 501	1 955	150	3 160	12 141
2005	341	1 268	1 774	4 055	2 223	189	4 103	13 953
Not seasonally adjusted								
	INIV	INIY	INIW	INIM	INIR	INIS	JZKJ	INJJ
2002 Q4	135	330	702	926	596	66	957	3 712
2003 Q1	90	295	450	810	547	68	961	3 221
Q2	87	261	414	718	576	45	893	2 994
Q3	119	277	485	784	550	57	872	3 144
Q4	122	326	536	843	684	73	1 195	3 779
2004 Q1	65	210	397	664	426	40	736	2 538
Q2	71	244	441	957	483	42	736	2 974
Q3	90	325	504	776	530	37	751	3 013
Q4	134	329	565	1 104	516	31	937	3 616
2005 Q1	68	257	403	834	486	56	874	2 978
Q2	71	275	417	1 074	527	48	977	3 389
Q3	78	308	432	1 006	580	44	1 072	3 520
Q4	124	428	522	1 141	630	41	1 180	4 066
2006 Q1	119	323	373	931	533	34	917	3 230
Q2	134	369	363	761	582	45	936	3 190
Q3	137	326	425	940	561	25	1 017	3 431
Percentage change, latest quarter on previous quarter								
2006 Q3	2.2	−11.7	17.1	23.5	−3.6	−44.4	8.7	7.6
Percentage change, latest quarter on corresponding quarter of previous year								
2006 Q3	75.6	5.8	−1.6	−6.6	−3.3	−43.2	−5.1	−2.5
Seasonally adjusted								
	IOAZ	IOBC	IOBA	IOAQ	IOAV	IOAW	JZKK	IOBN
2002 Q4	103	301	569	790	553	66	891	3 273
2003 Q1	107	314	543	887	591	68	1 013	3 523
Q2	103	277	426	748	593	45	932	3 124
Q3	116	275	482	780	530	57	854	3 094
Q4	92	293	434	740	643	73	1 122	3 397
2004 Q1	78	227	469	744	456	40	772	2 786
Q2	86	265	466	994	498	42	774	3 125
Q3	92	323	503	787	510	37	739	2 991
Q4	104	293	469	976	491	31	875	3 239
2005 Q1	82	290	466	919	535	56	918	3 266
Q2	82	294	454	1 131	537	48	1 033	3 579
Q3	81	319	423	1 008	571	44	1 068	3 514
Q4	96	365	431	997	580	41	1 084	3 594
2006 Q1	141	368	426	1 020	587	34	962	3 538
Q2	152	391	397	795	591	45	994	3 365
Q3	145	340	414	948	554	25	1 013	3 439
Percentage change, latest quarter on previous quarter								
2006 Q3	−4.6	−13.0	4.3	19.2	−6.3	−44.4	1.9	2.2
Percentage change, latest quarter on corresponding quarter of previous year								
2006 Q3	79.0	6.6	−2.1	−6.0	−3.0	−43.2	−5.1	−2.1

1 All private sector figures are exclusive of expenditure on dwellings.
2 All figures are exclusive of expenditure on land and existing buildings.
3 Estimates are shown to the nearest £ million but should not be regarded as accurate to this degree.

Source: Office for National Statistics: 020 7533 5934

1.15 Private Sector[1] Manufacturing Business Investment[2] by Asset

£ million[3]

	Chained volume measures, reference year 2003				Current prices			
	Analysis by asset				Analysis by asset			
	New Building Work	Vehicles	Other Capital Equipment	Total all manufacturing	New Building Work	Vehicles	Other Capital Equipment	Total all manufacturing
2002	1 394	535	11 417	13 434	1 376	535	11 530	13 441
2003	1 471	551	11 116	13 138	1 471	551	11 116	13 138
2004	1 272	480	10 470	12 222	1 304	481	10 356	12 141
2005	1 635	643	11 698	13 976	1 676	647	11 630	13 953
Not seasonally adjusted								
	IMGV	IMSG	INDR	INKL	IMDA	IMOL	IMZW	INJJ
2002 Q4	373	109	3 186	3 710	372	109	3 231	3 712
2003 Q1	371	162	2 742	3 233	371	162	2 688	3 221
Q2	348	131	2 526	2 999	347	131	2 516	2 994
Q3	324	140	2 657	3 138	324	140	2 680	3 144
Q4	428	118	3 191	3 768	429	118	3 232	3 779
2004 Q1	242	118	2 181	2 541	247	118	2 173	2 538
Q2	312	113	2 582	3 007	322	113	2 539	2 974
Q3	325	122	2 587	3 034	332	123	2 558	3 013
Q4	393	127	3 120	3 640	403	127	3 086	3 616
2005 Q1	303	139	2 554	2 996	309	139	2 530	2 978
Q2	379	214	2 807	3 400	389	216	2 784	3 389
Q3	469	139	2 917	3 525	481	140	2 899	3 520
Q4	484	151	3 420	4 055	497	152	3 417	4 066
2006 Q1	405	110	2 695	3 210	413	111	2 706	3 230
Q2	392	121	2 654	3 167	408	123	2 659	3 190
Q3	395	154	2 879	3 428	405	155	2 871	3 431
Percentage change, latest quarter on previous quarter								
2006 Q3	0.8	27.3	8.5	8.2	−0.7	26.0	8.0	7.6
Percentage change, latest quarter on corresponding quarter of previous year								
2006 Q3	−15.8	10.8	−1.3	−2.8	−15.8	10.7	−1.0	−2.5
Seasonally adjusted								
	IMKQ	IMWB	INHM	INLN	INSA	INVV	INZQ	IOBN
2002 Q4	353	112	2 754	3 275	336	112	2 825	3 273
2003 Q1	413	162	3 003	3 535	413	161	2 949	3 523
Q2	339	135	2 646	3 116	352	134	2 638	3 124
Q3	295	133	2 652	3 099	314	134	2 646	3 094
Q4	424	121	2 815	3 388	392	122	2 883	3 397
2004 Q1	273	121	2 394	2 788	280	120	2 386	2 786
Q2	303	114	2 713	3 130	328	113	2 684	3 125
Q3	308	116	2 605	3 029	326	118	2 547	2 991
Q4	388	129	2 758	3 275	370	130	2 739	3 239
2005 Q1	335	143	2 802	3 280	352	144	2 770	3 266
Q2	389	220	2 977	3 586	399	221	2 959	3 579
Q3	456	126	2 935	3 517	476	128	2 910	3 514
Q4	455	154	2 984	3 593	449	154	2 991	3 594
2006 Q1	449	115	2 944	3 508	469	117	2 952	3 538
Q2	406	123	2 816	3 345	420	124	2 821	3 365
Q3	388	139	2 907	3 434	402	140	2 897	3 439
Percentage change, latest quarter on previous quarter								
2006 Q3	−4.4	13.0	3.2	2.7	−4.3	12.9	2.7	2.2
Percentage change, latest quarter on corresponding quarter of previous year								
2006 Q3	−14.9	10.3	−1.0	−2.4	−15.5	9.4	−0.4	−2.1

1 All private sector figures are exclusive of expenditure on dwellings.
2 All figures are exclusive of expenditure on land and existing buildings.
3 Estimates are shown to the nearest £ million but shown not be regarded as accurate to this degree.

Source: Office for National Statistics: 020 7533 5934

2 Population and vital statistics

2.1 Mid-year estimates of resident population

Thousands

	England and Wales			Scotland			Northern Ireland			United Kingdom		
	Males	Females	Persons	Males	Females	Persons	Males	Females	Persons	Males	Females	Persons
	BBAE	BBAF	BBAD	BBAH	BBAI	BBAG	BBAK	BBAL	BBAJ	BBAB	BBAC	DYAY
1984	24 185	25 528	49 713	2 475	2 664	5 139	761	796	1 557	27 421	28 989	56 409
1985	24 254	25 606	49 861	2 470	2 658	5 128	765	800	1 565	27 489	29 065	56 554
1986	24 311	25 687	49 999	2 462	2 649	5 112	768	805	1 574	27 542	29 142	56 684
1987	24 371	25 752	50 123	2 455	2 644	5 099	773	809	1 582	27 599	29 205	56 804
1988	24 434	25 820	50 254	2 444	2 633	5 077	774	812	1 585	27 652	29 265	56 916
1989	24 510	25 898	50 408	2 443	2 635	5 078	776	814	1 590	27 729	29 348	57 076
1990	24 597	25 964	50 561	2 444	2 637	5 081	778	818	1 596	27 819	29 419	57 237
1991	24 681	26 067	50 748	2 445	2 639	5 083	783	824	1 607	27 909	29 530	57 439
1992	24 739	26 136	50 876	2 445	2 640	5 086	792	831	1 623	27 977	29 608	57 585
1993	24 793	26 193	50 986	2 448	2 644	5 092	798	837	1 636	28 039	29 675	57 714
1994	24 853	26 263	51 116	2 453	2 649	5 102	802	842	1 644	28 108	29 754	57 862
1995	24 946	26 326	51 272	2 453	2 650	5 104	804	845	1 649	28 204	29 821	58 025
1996	25 030	26 381	51 410	2 447	2 645	5 092	810	851	1 662	28 287	29 877	58 164
1997	25 113	26 446	51 560	2 442	2 641	5 083	816	856	1 671	28 371	29 943	58 314
1998	25 201	26 519	51 720	2 439	2 638	5 077	819	859	1 678	28 458	30 017	58 475
1999	25 323	26 610	51 933	2 437	2 635	5 072	818	861	1 679	28 578	30 106	58 684
2000	25 438	26 702	52 140	2 432	2 631	5 063	820	862	1 683	28 690	30 196	58 886
2001	25 574	26 786	52 360	2 434	2 630	5 064	824	865	1 689	28 832	30 281	59 113
2002	25 702	26 868	52 570	2 432	2 623	5 055	829	868	1 697	28 963	30 359	59 322
2003	25 841	26 953	52 794	2 435	2 623	5 057	833	870	1 703	29 108	30 446	59 554
2004	25 988	27 057	53 046	2 446	2 632	5 078	836	874	1 710	29 271	30 563	59 834
2005	26 179	27 211	53 390	2 456	2 639	5 095	844	880	1 724	29 479	30 730	60 209

Sources: Office for National Statistics;
General Register Office (Scotland);
Northern Ireland Statistics and Research Agency

2.2 Age distribution of estimated resident population at 30 June 2005

Thousands

	Resident population											
	England and Wales[1]		Wales		Scotland		Northern Ireland		United Kingdom[1]			
	Males	Females	Males	Females	Males	Females	Males	Females	Males	Females	Persons	
0-4	1 564	1 489	82	77	136	130	57	54	1 756	1 672	3 428	
5-9	1 616	1 541	90	85	146	139	61	58	1 823	1 737	3 560	
10-14	1 736	1 644	99	94	161	154	64	62	1 962	1 859	3 821	
15-19	1 802	1 703	103	97	168	159	68	64	2 038	1 927	3 965	
20-24	1 751	1 703	96	94	167	164	62	59	1 980	1 926	3 906	
25-29	1 664	1 663	79	80	149	149	54	55	1 867	1 867	3 735	
30-34	1 848	1 864	86	92	159	171	59	61	2 066	2 096	4 162	
35-39	2 054	2 071	100	107	187	202	64	66	2 305	2 338	4 643	
40-44	2 016	2 054	105	110	195	209	63	66	2 274	2 329	4 602	
45-49	1 765	1 793	95	99	180	190	56	58	2 001	2 041	4 042	
50-54	1 602	1 637	92	96	163	168	50	50	1 815	1 855	3 670	
55-59	1 715	1 760	103	105	168	173	47	49	1 929	1 982	3 912	
60-64	1 347	1 410	84	88	131	142	41	43	1 519	1 595	3 114	
65-69	1 158	1 238	71	75	115	131	33	37	1 306	1 406	2 711	
70-74	963	1 104	59	67	94	118	26	32	1 083	1 254	2 337	
75-79	751	981	45	59	69	99	20	29	839	1 108	1 947	
80-84	508	817	31	50	43	75	13	22	564	914	1 477	
85-89	225	456	13	28	18	41	5	11	248	508	756	
90 and over	94	285	5	16	7	25	2	7	103	316	419	
0-14	4 916	4 673	270	257	443	422	182	173	5 541	5 269	10 809	
15-64	17 564	17 658	942	968	1 668	1 729	563	570	19 795	19 956	39 751	
65 and over	3 699	4 880	225	296	345	488	99	137	4 143	5 505	9 649	
All ages	26 179	27 211	1 438	1 521	2 456	2 639	844	880	29 479	30 730	60 209	

Sources: Office for National Statistics;
General Register Office (Scotland);
Northern Ireland Statistics and Research Agency

2.3 Births[1] and marriages

Thousands

	Live births[2]						Marriages					
	England and Wales				Northern Ireland[3]	United Kingdom[3]	England and Wales				Northern Ireland	United Kingdom
	Total	England	Wales	Scotland			Total	England	Wales	Scotland		
	BBCB	G8ZT	BBCC	BBCD	BBCE	BBCA	BBCG	G8ZU	BBCH	BBCI	BBCJ	BBCF
2001	594.6	563.7	30.6	52.5	22.0	669.1	249.2	236.2	13.0	29.6	7.3	286.1
2002	596.1	565.7	30.2	51.3	21.4	668.8	255.6	242.1	13.5	29.8	7.6	293.0
2003	621.5	589.9	31.4	52.4	21.6	695.6	270.1	255.6	14.5	30.8	7.8	308.6
2004	639.7	607.2	32.3	54.0	22.3	716.0	270.7	255.9	14.8	32.2	8.3	311.2
2005	645.8	613.0	32.6	54.4	22.3	722.6	..	..	..	30.9	..	..
2002 Q4	150.6	142.9	7.7	13.1	5.2	168.9	46.9	46.5	2.4	6.2	1.3	54.4
2003 Q1	147.4	139.9	7.5	12.8	5.4	165.6	34.0	32.3	1.7	3.7	0.8	38.2
Q2	155.1	147.3	7.8	12.9	5.4	173.4	75.2	71.2	4.0	8.4	2.2	85.9
Q3	162.9	154.4	8.3	13.8	5.6	182.2	111.9	105.6	6.2	12.3	3.3	127.0
Q4	156.0	148.2	7.8	13.0	5.3	174.3	49.1	46.5	2.6	6.3	1.4	56.1
2004 Q1	155.2	147.3	7.8	13.5	5.7	174.3	35.0	33.3[4]	1.7	3.9	0.8	39.7
Q2	157.4	149.6	7.8	13.3	5.4	176.2	74.6	70.6[4]	4.0	8.7	2.4	85.7
Q3	165.4	156.9	8.4	13.8	5.8	185.1	112.6	106.2[4]	6.4	12.7	3.5	128.8
Q4	161.7	153.3	8.3	13.3	5.4	180.4	49.5	46.8[4]	2.7	6.8	1.6	57.9
2005 Q1	154.3	146.4	7.8	13.4	5.5[4]	173.2[4]	..	..	..	3.8	0.9	..
Q2	159.8	151.8	7.9	13.6	5.7[4]	179.0[4]	..	..	..	8.6	2.2	..
Q3	170.2	161.4	8.7	14.2	5.9[4]	190.3[4]	..	..	..	12.3	3.5	..
Q4	161.7	153.4	8.2	13.2	5.2[4]	180.1[4]	..	..	..	6.1	1.4	..
2006 Q1	159.5[4]	151.3[4]	8.1[4]	13.6[4]	5.8[4]	178.9[4]	..	..	..	3.5[4]	..	..
Q2	165.3[4]	157.0[4]	8.3[4]	14.0[4]	5.8[4]	185.1[4]	..	..	..	8.3[4]	..	..

Note: Figures may not add exactly due to rounding.
1 Excluding stillbirths.
2 Birth figures for England and also for Wales each exclude events for persons usually resident outside England and Wales. These events are however, included in the totals for England and Wales combined, and for the United Kingdom.

3 For England and Wales, figures relate to numbers occurring in a period; for Scotland and Northern Ireland, figures relate to those registered in a period.
4 Provisional.

Sources: Office for National Statistics;
General Register Office for Scotland;;
Northern Ireland Statistics & Research Agency.

2.4 Deaths registered

Thousands

	Total					Infants aged under one year				
	England and Wales		Scotland	Northern Ireland	United Kingdom	England and Wales		Scotland	Northern Ireland	United Kingdom
	Total	Wales				Total	Wales			
	BBDB	BBDC	BBDD	BBDE	BBDA	BBDG	BBDH	BBDI	BBDJ	BBDF
2001	532.5	33.2	57.4	14.5	604.4	3.27	0.17	0.29	0.13	3.69
2002	535.4	33.3	58.1	14.6	608.0	3.20	0.10	0.30	0.10	3.50
2003	539.2	33.8	58.4	14.5	612.0	3.30	0.13	0.27	0.12	3.69
2004	514.3	32.3	56.2	14.4	584.8	3.27	0.16	0.27	0.12	3.66
2005	513.0	32.2	55.7	14.2	583.0	3.25	0.14	0.28	0.14	3.67
2002 Q4	137.7	8.5	15.2	3.7	156.5	0.83	0.04	0.07	0.03	0.92
2003 Q1	143.1	9.0	15.7	3.9	162.6	0.83	0.04	0.07	0.03	0.93
Q2	129.2	8.3	14.1	3.4	146.7	0.79	0.03	0.06	0.02	0.87
Q3	124.3	7.7	13.3	3.5	141.0	0.80	0.04	0.07	0.04	0.91
Q4	142.6	8.9	15.3	3.7	161.7	0.87	0.02	0.07	0.03	0.97
2004 Q1	142.0	8.9	15.3	3.9	161.1	0.86	0.05	0.06	0.03	0.96
Q2	122.5	7.7	13.6	3.6	139.7	0.78	0.04	0.07	0.03	0.88
Q3	119.0	7.5	13.1	3.4	135.5	0.81	0.05	0.07	0.04	0.92
Q4	130.6	8.2	14.2	3.5	148.3	0.82	0.03	0.06	0.02	0.91
2005 Q1	145.3	9.2	15.6	3.8	164.7	0.82	0.03	0.07	0.03	0.91
Q2	125.9	8.0	13.7	3.7	143.3	0.83	0.04	0.07	0.04	0.94
Q3	115.4	7.2	12.8	3.4	131.6	0.80	0.03	0.08	0.04	0.92
Q4	126.2	7.8	13.6	3.4	143.3	0.80	0.04	0.07	0.03	0.90
2006 Q1[1]	140.8	8.7	14.9	4.0	159.7	0.82	0.03	0.05	0.03	0.90
Q2[1]	123.6	7.6	13.9	3.6	141.1	0.84	0.03	0.07	0.03	0.93

1 Provisional.

Sources: Office for National Statistics;
General Register Office (Scotland);
Northern Ireland Statistics and Research Agency

3 Labour market

3.1 Labour market activity
United Kingdom

Thousands, seasonally adjusted[1]

| | Employment categories | | | | | | | | | Employment rate: aged 16 - 59/64[2] % |
	Employees	Self employed	Unpaid family workers	Government training and employment programmes	Total employment	Unemployment	Total economically active	Economically inactive	Total aged 16 and over	
	MGRN	MGRQ	MGRT	MGRW	MGRZ	MGSC	MGSF	MGSI	MGSL	MGSU
2003 Q3	24 353	3 645	108	107	28 212	1 504	29 716	17 382	47 098	74.6
Q4	24 402	3 655	99	107	28 263	1 453	29 716	17 467	47 183	74.6
2004 Q1	24 558	3 623	104	116	28 402	1 432	29 834	17 434	47 268	74.8
Q2	24 514	3 676	98	123	28 412	1 433	29 844	17 508	47 352	74.7
Q3	24 649	3 583	89	129	28 450	1 400	29 850	17 593	47 443	74.7
Q4	24 738	3 637	97	125	28 597	1 411	30 008	17 538	47 547	74.9
2005 Q1	24 823	3 622	105	126	28 676	1 411	30 087	17 563	47 650	74.9
Q2	24 848	3 630	101	114	28 693	1 433	30 126	17 628	47 753	74.7
Q3	24 936	3 661	90	107	28 794	1 447	30 242	17 611	47 853	74.8
Q4	24 861	3 699	90	108	28 758	1 554	30 312	17 634	47 946	74.5
2006 Q1	24 966	3 740	88	93	28 887	1 599	30 486	17 552	48 038	74.6
Q2	25 023	3 719	93	94	28 930	1 683	30 613	17 518	48 131	74.6
Q3	25 026	3 759	104	97	28 986	1 711	30 696	17 527	48 224	74.5

1 Seasonally adjusted estimates are subject to periodic revision.
2 The employment rate equals those in employment aged 16-64 (male) and 16-59 (female), as a percentage of all in these age groups.

Source: Labour Force Survey, Office for National Statistics: 020 7533 6094

3.2 Distribution of the workforce[1,2]
United Kingdom

Thousands

| | Not seasonally adjusted | | | | | | Seasonally adjusted | | |
| | | Employee jobs | | | Self-employment jobs (with or without employees)[3] | HM Forces[4] | Workforce jobs | Employee jobs | Self-employment jobs |
	Workforce jobs	Males	Females	Total					
At June									
	DYDA	BCAE	BCAF	BCAD	BCAG	BCAH	DYDC	BCAJ	DYZN
2002	29 974	13 080	13 005	26 085	3 588	214	29 985	26 107	3 573
2003	30 264	13 172	12 974	26 146	3 807	223	30 283	26 175	3 793
2004	30 543	13 195	13 148	26 343	3 878	218	30 572	26 381	3 866
2005	30 776	13 341	13 267	26 608	3 866	210	30 810	26 650	3 855
2006	31 022†	13 476†	13 306†	26 782†	3 976†	204	31 064†	26 819†	3 976†
2003 Q4	30 597	13 315	13 093	26 408	3 865	222	30 489	26 284	3 883
2004 Q1	30 420	13 109	13 123	26 232	3 863	220	30 524	26 334	3 869
Q2	30 543	13 195	13 148	26 343	3 878	218	30 572	26 381	3 866
Q3	30 565	13 246	13 152	26 398	3 850	215	30 558	26 396	3 843
Q4	30 863	13 449	13 252	26 701	3 845	215	30 747	26 569	3 863
2005 Q1	30 735	13 325	13 244	26 569	3 850	213	30 832	26 663	3 857
Q2	30 776	13 341	13 267	26 608	3 866	210	30 810	26 650	3 855
Q3	30 828	13 399	13 241	26 640	3 886	207	30 827	26 647	3 878
Q4	31 053	13 479	13 336	26 815	3 942	206	30 926	26 683	3 950
2006 Q1	30 886	13 384	13 231	26 615	3 985	206	30 993	26 718	3 991
Q2	31 022†	13 476†	13 306†	26 782†	3 976†	204	31 064†	26 819†	3 976†
Q3	31 110	13 548	13 266	26 815	4 040	202	31 118	26 824	4 037

1 The data in this table include revised figures for self-employment to reflect the results of the 2001 Census.
2 Estimates for employee jobs and workforce jobs for Great Britain now use the Annual Business Inquiry as a benchmark on which the quarterly movements are based. For further information see Labour Market Statistics First Release April 2001.

3 Estimates of the self-employed are based on the results of the Labour Force Survey. The estimates given in the table are unadjusted.
4 HM Forces figures, provided by the Ministry of Defence, represent the total number of UK service personnel, male and female, in HM Regular Forces wherever serving and including those on release leave. The numbers are not subject to seasonal adjustment.

Sources: Office for National Statistics;
Department of Economic Development (Northern Ireland)

3.3 Employee jobs: all industries[1,2]
Great Britain

Not seasonally adjusted

Thousands

	All employee jobs	Employee jobs		Manufact-uring indus-tries; all jobs	Production indus-tries; all jobs	Production and constru-ction; all jobs	Production and constru-ction; male	Production and constru-ction; female	Service industries; all jobs
		male	female						
SIC 1992 Divisions or Classes	*A-O*			*D*	*C-E*	*C-F*			*G-O*
	LMAB	DYCA	DYCB	LMAD	LMAF	LMAH	LMBL	LMBM	LMAJ
2005	25 916	13 006	12 911	3 045	3 199	4 354	3 352	1 003	21 336
2006	26 086†	13 138†	12 947†	2 958	3 121	4 347	3 350	997	21 525†
2005 Q2	25 916	13 006	12 911	3 045	3 199	4 354	3 352	1 003	21 336
Q3	25 951	13 065	12 886	3 021	3 179	4 375	3 384	991	21 357
Q4	26 116	13 141	12 975	2 991	3 149	4 343	3 358	984	21 559
2006 Q1	25 918	13 046	12 871	2 965	3 123	4 319	3 332	987	21 389
Q2	26 086†	13 138†	12 947†	2 958	3 121	4 347	3 350	997	21 525†
Q3	26 113	13 209	12 905	2 952†	3 118†	4 367	3 384	983	21 517
2005 Dec	..	..	..	2 991	3 149	..	..	..	..
2006 Jan	..	..	..	2 972	3 129	..	..	..	..
Feb	..	..	..	2 970	3 128	..	..	..	..
Mar	..	..	..	2 965	3 123	..	..	..	..
Apr	..	..	..	2 960	3 119	..	..	..	..
May	..	..	..	2 955	3 116	..	..	..	..
Jun	..	..	..	2 958	3 121	..	..	..	..
Jul	..	..	..	2 956†	3 119†	..	..	..	..
Aug	..	..	..	2 953	3 117	..	..	..	..
Sep	..	..	..	2 952	3 118	..	..	..	..
Oct	..	..	..	2 950	3 116	..	..	..	..
Nov	..	..	..	2 946	3 112	..	..	..	..

3.3
continued

Employee jobs: all industries[1,2]
Great Britain
Not seasonally adjusted

Thousands

	Agriculture, hunting, forestry and fishing	Mining and quarrying, electricity, gas and water supply	Food products, beverages and tobacco	Manufacture of clothing, textiles and leather production	Wood and wood products	Paper, pulp, printing, publishing and recording media	Chemicals, chemical products and man-made fibres	Rubber and plastic products	Non-metallic mineral products, metal and metal products	Machinery and equipment, nec	Electrical and optical equipment	Transport equipment	Coke, nuclear fuel and other manufacturing, nec
SIC 1992 Divisions or Classes	A,B 01-05	C,E 10-12, 40-41	DA 15-16	DB/DC 17-19	DD 20	DE 21-22	DG 24	DH 25	DI/DJ 26-28	DK 29	DL 30-33	DM 34-35	DF, DN 23, 36-37
	LMAL	LMAM	LMAN	LMAO	LMAP	LMAQ	LMAR	LMAS	LMAT	LMAU	LMAV	LMAW	LMAX
2005	226	154	414	136	79	397	198	195	501	280	326	315	204
2006	214	163	409	130	76	384	193	182	493	274	313	306	197
2005 Q2	226	154	414	136	79	397	198	195	501	280	326	315	204
Q3	219	158	418	135	77	393	196	189	501	278	324	310	200
Q4	214	158	417	133	76	391	194	185	493	277	321	307	196
2006 Q1	210	158	408	132	75	387	195	184	491	275	316	307	196
Q2	214	163	409	130	76	384	193	182	493	274	313	306	197
Q3	229	166†	411†	129	78	381	192†	180	491†	276†	310	304	199†
2005 Dec	..	158	417	133	76	391	194	185	493	277	321	307	196
2006 Jan	..	157	411	132	74	390	194	185	490	276	319	306	195
Feb	..	157	410	132	75	389	195	184	491	275	317	306	196
Mar	..	158	408	132	75	387	195	184	491	275	316	307	196
Apr	..	160	408	131	75	385	194	184	490	275	315	307	197
May	..	160	407	131	75	385	194	183	490	275	313	306	196
Jun	..	163	409	130	76	384	193	182	493	274	313	306	197
Jul	..	163	411†	130	76	383†	193	182	491†	276	312	305†	198†
Aug	..	164	411	130	78†	381	193	181	490	276	311†	304	198
Sep	..	166†	411	129	78	381	192†	180	491	276†	310	304	199
Oct	..	166	412	129	78	380	192	180	491	277	309	304	198
Nov	..	166	412	129	78	379	192	179	489	278	309	303	199

	Construction	Wholesale and retail trade and repairs	Hotels and restaurants	Transport and storage	Post and telecommunications	Financial intermediation	Real estate	Renting, research, computer and other business activities	Public administration and defence, compulsory social security	Education	Health and Social work activities	Other community social and personal activities
SIC 1992 Divisions or Classes	F 45	G 50-52	H 55	I 60-63	64	J 65-67	K 70	71-74	L 75	M 80	N 85	O 90-93
	LMAY	LMAZ	LMBA	LMBB	LMBC	LMBD	LMBE	LMBF	LMBG	LMBH	LOJV	LMBK
2005	1 156	4 484	1 801	1 069	484	1 061	422	3 724	1 477	2 255	3 181	1 378
2006	1 225	4 492	1 774	1 076†	479	1 064	442	3 781	1 477	2 295	3 239†	1 405
2005 Q2	1 156	4 484	1 801	1 069	484	1 061	422	3 724	1 477	2 255	3 181	1 378
Q3	1 196	4 484	1 781	1 078	482	1 061	424	3 762	1 473	2 245	3 196	1 370
Q4	1 194	4 620	1 758	1 069	493	1 062	428	3 782	1 478	2 289	3 216	1 364
2006 Q1	1 196	4 482	1 743	1 068	482	1 064	433	3 745	1 477	2 298	3 231	1 367
Q2	1 225	4 492	1 774	1 076†	479	1 064	442	3 781	1 477	2 295	3 239†	1 405
Q3	1 249	4 488	1 756	1 076	478	1 065	443	3 823	1 479	2 271	3 242	1 396

1 The data in this table have not been adjusted to reflect the 2001 Census population data.

2 Estimates of employee jobs and workforce jobs for Great Britain now use the Annual Business Inquiry as a benchmark on which quarterly movements are based. For further information see Labour Market Statistics First Release April 2001.

Source: Office for National Statistics

3.4 Civil Service employment by department[1]

Full-time equivalents, Great Britain, not seasonally adjusted

		2006 Q2	2006 Q3
Attorney General's Departments	GB3F	9 430	9 510
Cabinet Office	BBGD	1 620	1 590
Other Cabinet Office Agencies[2]	GB3G	780	780
HM Treasury	GB3H	1 160	1 130
Chancellor's other departments	GB3I	5 390	5 240
Charity Commission	GB3J	520	520
Communities and Local Government[2]	YEGA	5 610	5 450
Constitutional Affairs	GB3K	34 100	33 780
Culture, Media and Sport	DMTC	620	640
Defence	BCDW	87 260	85 690
Education and Skills	LNFW	4 170	4 060
Environment, Food and Rural Affairs	LNFX	13 100	12 730
Export Credits Guarantee Department	GB3L	260	250
Foreign and Commonwealth	BCDK	6 160	6 070
Health[3]	BAKR	6 100	6 020
HM Revenue and Customs	GB3M	97 360	95 780
Home Office[2]	BCDL	72 990	73 560
International Development	DMUA	1 780	1 740
Northern Ireland Office	BBGG	140	140
Office for Standards in Education	GB3N	2 280	2 300
Security and Intelligence Services	GB3O	4 890	4 890
Trade and Industry[2]	BCDQ	10 110	10 080
Transport	BCDR	19 120	18 950
Work and Pensions	LNGA	117 180	116 660
Central Governments Departments Total	GB3P	502 100	497 540
Scottish Executive	GB3Q	16 100	15 900
Welsh Assembly	GB3R	5 960	6 460
TOTAL	BCDX	524 160	519 900

1 Numbers are rounded to the nearest ten. Data not available are represented by "-".

2 The Office of the Third Sector (OTS) was established in May 2006. Employee numbers for OTS will not be available until Q4. For this publication, the employees are therefore still included with their former departments.

3 The Wine Standards Board (former NDPB) merged with the Food Standards Agency on 1 July 2006.

Source: Office for National Statistics

3.5 Intake and outflow of UK Regular Armed Forces Personnel

	Intake[1]						Outflow[2]					
	Financial Year					12 months to 1 October 2006	Financial Year					12 months to 1 October 2006
	2001-02	2002-03	2003-04	2004-05	2005-06		2001-02	2002-03	2003-04	2004-05	2005-06	
All Services[3]												
Total	23.7	26.3	23.5	17.6	18.1	19.5	24.7	24.1	23.4	23.4	23.3	24.0
Male	21.0	23.1	20.8	15.7	16.4	17.6	22.4	21.8	21.2	21.3	21.3	22.0
Female	2.7	3.3	2.7	1.9	1.7	2.0	2.4	2.3	2.2	2.1	2.0	2.0
Naval Service												
Total	5.0	5.2	4.1	3.7	3.9	3.9	5.8	5.3	4.8	4.6	4.5	4.4
Male	4.3	4.4	3.5	3.2	3.5	3.5	5.1	4.7	4.2	4.1	4.0	3.9
Female	0.7	0.8	0.6	0.5	0.5	0.5	0.7	0.6	0.5	0.5	0.5	0.5
Army												
Total	14.9	16.7	15.3	11.7	12.7	13.8	14.4	14.6	14.6	15.1	14.2	14.5
Male	13.7	15.1	14.0	10.8	11.7	12.7	13.3	13.4	13.5	14.0	13.2	13.5
Female	1.2	1.6	1.3	0.9	1.0	1.1	1.1	1.1	1.1	1.1	1.0	0.9
RAF[3]												
Total	3.8	4.4	4.2	2.2	1.5	1.7	4.5	4.2	4.0	3.7	4.6	5.2
Male	3.1	3.6	3.3	1.6	1.2	1.4	4.0	3.7	3.5	3.2	4.1	4.6
Female	0.7	0.9	0.9	0.5	0.3	0.4	0.6	0.6	0.6	0.5	0.5	0.6

1 Intake from civilian life, includes re-enlistments and rejoined reservists.

2 Outflow includes recalled reservists on release and outflow to the Home Service battallions of the Royal Irish Regiment.

3 Denotes provisional. Due to the introduction of a new Personnel Administration System for RAF, all data from 1st May 2006 and onwards are provisional and subject to review.

Figures are for UK Regular Forces (including both Trained and Untrained personnel), and therefore exclude Gurkhas, Full Time Reserve Service personnel, the Home Service battalions of the Royal Irish Regiment, mobilised reservists and Naval Activated Reservists.

Source: Defence Analytical Services Agency: 0207 218 4439

3.6 UK armed forces full-time strengths[1]

	1 Apr 2002	1 Apr 2003	1 Apr 2004	1 Apr 2005	1 Apr 2006	1 Oct 2006
All Services[2]						
Trained						
Total	187 110	188 520	190 190	188 050	183 180	179 420
UK regulars	181 680	182 780	184 590	182 840	178 300	174 590
Full Time Reserve Service	1 980	2 360	2 220	1 690	1 540	1 560
Gurkhas	3 450	3 380	3 390	3 520	3 330	3 280
Untrained						
Total	23 350	24 520	22 770	18 430	17 880	18 710
UK regulars	23 000	24 160	22 430	18 260	17 550	18 400
Gurkhas	350	380	330	170	330	310
Naval Service						
Trained						
Total	37 490	37 610	37 510	36 400	35 620	35 260
UK regulars	36 770	36 610	36 420	35 500	34 890	34 590
Full Time Reserve Service	720	1 010	1 090	900	720	670
Untrained						
Total	4 850	4 940	4 460	4 440	4 500	4 360
UK regulars	4 850	4 940	4 460	4 440	4 500	4 360
Army						
Trained						
Total	100 410	102 000	103 560	102 440	100 620	99 570
UK regulars	96 020	97 640	99 420	98 490	96 790	95 760
Full Time Reserve Service	940	990	750	430	490	540
Gurkhas	3 450	3 380	3 390	3 520	3 330	3 280
Untrained						
Total	14 380	14 880	13 650	10 970	11 260	12 060
UK regulars	14 030	14 490	13 320	10 800	10 940	11 750
Gurkhas	350	380	330	170	330	310
RAF[2]						
Trained						
Total	49 200	48 900	49 120	49 210	46 940	44 590
UK regulars	48 880	48 540	48 740	48 850	46 620	44 240
Full Time Reserve Service	320	360	380	360	330	350
Untrained						
Total	4 120	4 700	4 650	3 020	2 110	2 300
UK regulars	4 120	4 700	4 650	3 020	2 110	2 300

1 The full-time strength includes UK Regular Forces, Gurkhas and FTRS (Full Time Reserve Service) personnel.

2 Denotes provisional. Due to the introduction of a new Personnel Administration System for RAF, all data from 1st May 2006 and onwards are provisional and subject to review.

Due to the rounding methods used, totals may not always equal the sum of the parts. When rounding to the nearest 10, numbers ending in 5 have been rounded to the nearest multiple of 20 to prevent systematic bias.

Source: Defence Analytical Services Agency: 0207 218 4439

3.7 Number of workers employed in agriculture[1]

Thousands

	Regular workers					Seasonal or casual workers			All workers		
	Whole-time		Part-time								
	Male	Female	Male	Female	Total	Male	Female	Total	Male	Female	Total
	BAMY	BAMZ	BANA	BANB	BANC	BAND	BANE	BANF	BANG	BANH	BANI
1991 Jun	104.7	15.0	29.7	27.4	176.9	53.8	32.8	86.6	188.2	75.3	263.5
1992 Jun	99.9	14.8	29.1	26.1	169.9	54.4	31.9	86.2	183.3	72.8	256.2
1993 Jun	96.5	13.7	29.8	25.3	165.3	55.0	30.4	85.4	181.3	69.4	250.7
1994 Jun	93.6	13.2	30.0	24.2	161.0	53.9	28.4	82.2	177.5	65.7	243.2
1995 Jun	90.4	13.0	30.0	24.1	157.4	56.5	27.2	83.7	176.8	64.3	241.2
1996 Jun	89.2	12.6	31.2	23.4	156.4	55.6	25.8	81.5	176.0	61.9	237.9
1997 Jun	87.5	12.6	31.2	23.1	154.4	55.3	25.5	80.9	174.0	61.2	235.2
1998 Jun[2]	88.0	13.1	29.7	24.7	155.6	55.6	23.8	79.5	172.8	62.2	235.0
1999 Jun	82.7	11.9	27.5	22.6	144.7	51.8	21.2	73.0	162.0	55.6	217.7
2000 Jun	73.4	10.3	24.6	20.6	128.9	45.9	18.5	64.4	143.9	49.4	193.3
2001 Jun[3]	69.0	10.9	22.0	18.9	120.8	44.6	18.6	63.2	135.6	48.5	184.0
	70.3	11.2	22.5	19.4	123.5	45.4	18.8	64.1	138.2	49.4	187.6
2002 Jun	64.7	11.5	21.7	18.4	116.3	46.2	18.0	64.2	132.6	47.9	180.6
2003 Jun	60.4	10.0	21.0	17.0	108.4	44.8	17.8	62.6	126.2	44.8	170.9
2004 Jun	58.1	9.8	23.5	17.4	108.8	49.6	18.6	68.3	131.2	45.8	177.0
2005 Jun	57.2	10.3	24.5	17.2	109.2	46.4	18.7	65.1	128.1	46.2	174.3

1 Figures exclude farmers, partners, directors and their spouses, salaried managers, school children and most trainees. Includes estimates for minor holdings.

2 In 1998, fundamental changes were introduced to the labour questions on the June Agricultural and Horticultural Census in England, Wales and Scotland. It appears that this change in questions may have led to the recording of additional Labour who were not previously included in the returns. The change in questions has also led to a redistribution of labour between the various categories. We therefore advise caution when comparing the results from 1998 onwards with previous years.

3 Due to an English register improvement only the top figure for 2001 is directly comparable with June 2000, while the bottom figure for 2001 is only comparable with data from June 2002.

Source: Department for Environment, Food and Rural Affairs: 01904 455095

3.8 Unemployment in United Kingdom
Analysis by duration[1]

Thousands, seasonally adjusted[1]

	Males				Females			
	Up to 26 weeks	Over 26 and up to 52 weeks	Over 52 weeks	Total	Up to 26 weeks	Over 26 and up to 52 weeks	Over 52 weeks	Total
	MGYK	MGYM	MGYO	MGSD	MGYL	MGYN	MGYP	MGSE
1994 Q4	579	264	806	1 650	424	152	294	870
1995 Q1	571	247	798	1 617	446	145	286	877
Q2	551	254	776	1 581	449	143	269	861
Q3	580	229	766	1 575	462	149	257	869
Q4	559	245	708	1 512	456	154	230	840
1996 Q1	581	265	693	1 539	437	132	221	790
Q2	591	246	680	1 517	458	143	220	820
Q3	579	227	664	1 471	458	133	219	809
Q4	543	205	653	1 400	471	133	223	827
1997 Q1	512	189	597	1 297	443	126	216	785
Q2	550	178	538	1 265	459	121	203	783
Q3	533	173	491	1 197	450	116	185	750
Q4	523	175	443	1 141	439	109	171	719
1998 Q1	512	171	418	1 101	448	98	163	709
Q2	518	160	396	1 074	462	93	160	715
Q3	550	164	377	1 092	446	102	144	692
Q4	556	166	356	1 079	445	92	147	684
1999 Q1	561	165	353	1 080	450	99	147	695
Q2	536	165	359	1 059	450	102	131	684
Q3	513	160	353	1 026	438	100	138	676
Q4	504	144	355	1 002	442	108	132	682
2000 Q1	515	142	326	984	458	108	126	691
Q2	500	140	317	957	420	100	122	642
Q3	476	137	297	910	429	92	114	635
Q4	486	136	289	912	415	87	107	609
2001 Q1	475	128	283	885	400	85	101	586
Q2	481	131	269	882	396	87	107	591
Q3	504	132	261	897	401	82	104	586
Q4	529	132	253	914	432	76	101	609
2002 Q1	533	147	239	919	419	77	97	593
Q2	536	147	227	910	441	71	94	606
Q3	571	141	234	945	439	78	98	615
Q4	526	148	218	891	443	85	95	623
2003 Q1	557	130	239	926	425	83	89	597
Q2	537	131	218	886	410	74	94	578
Q3	531	145	223	899	434	75	95	604
Q4	502	146	227	876	414	74	89	577
2004 Q1	489	141	212	841	412	82	97	591
Q2	503	141	197	841	419	86	86	592
Q3	492	133	195	821	420	75	84	579
Q4	494	143	192	829	418	77	87	583
2005 Q1	490	137	205	832	415	75	89	579
Q2	487	137	209	833	424	77	98	599
Q3	500	146	212	859	409	95	84	588
Q4	535	143	238	916	439	105	94	638
2006 Q1	531	166	235	932	454	112	101	667
Q2	546	177	252	975	477	126	105	708
Q3	545	176	273	994	479	117	121	716

1 Seasonally adjusted estimates are subject to periodic revision.

Source: Labour Force Survey, Office for National Statistics: 020 7533 6094

3.9 Claimant count in United Kingdom
Analysis of claimant by duration - computerised claims only

Thousands, seasonally adjusted

	Males				Females			
	Up to 26 weeks	Over 26 and up to 52 weeks	Over 52 weeks	Total[1]	Up to 26 weeks	Over 26 and up to 52 weeks	Over 52 weeks	Total[1]
	AGXK	ELNP	ELON	AGNG	JLGK	JLGJ	JLGL	JLGI
2001	449.4	125.4	158.8	733.6	160.4	35.1	32.5	227.9
2002	457.4	124.2	126.7	708.3	163.6	35.5	27.6	226.8
2003	451.2	127.1	114.7	693.0	166.3	37.2	26.5	230.1
2004	408.9	113.7	108.3	630.9	153.1	34.9	26.7	214.7
2005	423.4	113.3	98.3	635.1	159.3	35.6	25.1	220.0
2006	438.7	136.3	118.9	693.9	171.3	43.6	30.8	245.7
2002 Nov	455.5	125.4	119.8	700.7	163.5	35.3	26.8	225.6
Dec	453.1	125.1	118.8	697.0	165.3	35.4	26.8	227.5
2003 Jan	456.4	125.1	117.6	699.1	166.9	35.5	26.7	229.1
Feb	459.0	125.8	116.4	701.2	167.9	35.5	26.2	229.6
Mar	457.1	126.5	115.7	699.3	168.2	36.1	26.2	230.5
Apr	456.0	124.8	113.9	694.7	169.9	35.4	25.7	231.0
May	462.0	127.1	114.6	703.7	169.8	36.6	26.4	232.8
Jun	461.4	128.2	114.2	703.8	169.1	37.4	26.3	232.8
Jul	455.7	128.6	113.9	698.2	167.7	38.1	26.4	232.2
Aug	449.7	128.9	114.3	692.9	166.0	38.8	26.7	231.5
Sep	446.8	128.5	114.0	689.3	165.2	38.7	26.7	230.6
Oct	440.9	129.1	114.1	684.1	163.5	38.9	26.9	229.3
Nov	436.9	126.9	113.6	677.4	161.6	38.2	26.9	226.7
Dec	432.2	126.1	113.7	672.0	160.4	37.7	27.2	225.3
2004 Jan	427.6	124.2	113.1	664.9	158.6	37.3	27.3	223.2
Feb	424.6	121.2	112.4	658.2	158.0	36.5	27.2	221.7
Mar	419.2	120.6	111.9	651.7	156.6	36.5	27.1	220.2
Apr	417.3	118.2	111.2	646.7	153.9	36.5	27.6	218.0
May	407.3	116.9	110.7	634.9	152.8	35.7	27.2	215.7
Jun	402.3	114.2	109.9	626.4	151.4	35.1	27.0	213.5
Jul	400.4	111.2	108.2	619.8	148.9	34.4	26.8	210.1
Aug	399.6	110.5	107.0	617.1	150.4	34.1	26.5	211.0
Sep	401.5	109.3	106.0	616.8	151.2	33.8	26.4	211.4
Oct	404.8	106.3	104.5	615.6	151.7	33.0	25.9	210.6
Nov	401.7	106.4	103.1	611.2	151.7	33.2	25.9	210.8
Dec	400.2	105.3	101.6	607.1	152.0	32.9	25.4	210.3
2005 Jan	399.1	103.2	99.1	601.4	152.1	32.7	25.3	210.1
Feb	401.3	102.3	97.9	601.5	152.2	32.6	25.1	209.9
Mar	410.6	103.5	97.7	611.8	154.4	32.9	24.9	212.2
Apr	415.2	105.9	96.5	617.6	156.8	33.2	24.3	214.3
May	426.9	107.0	96.2	630.1	158.4	33.8	24.5	216.7
Jun	430.9	109.4	96.7	637.0	160.2	34.6	24.5	219.3
Jul	429.7	113.0	96.6	639.3	160.1	35.4	24.7	220.2
Aug	429.1	115.8	96.7	641.6	160.2	36.2	24.8	221.2
Sep	430.4	120.5	98.0	648.9	161.4	37.5	25.1	224.0
Oct	433.9	123.6	99.8	657.3	163.7	38.4	25.6	227.7
Nov	436.3	127.3	101.4	665.0	165.4	39.4	26.3	231.1
Dec	437.8	128.6	103.0	669.4	166.3	40.1	26.7	233.1
2006 Jan	433.5	128.1	104.6	666.2	165.8	40.5	27.2	233.5
Feb	441.9	131.9	107.7	681.5	168.8	41.4	27.9	238.1
Mar	447.4	133.6	110.7	691.7	170.8	41.5	28.4	240.7
Apr	446.2	137.2	113.8	697.2	169.7	43.4	29.6	242.7
May	443.9	139.8	116.9	700.6	171.5	43.9	30.1	245.5
Jun	441.9	141.7	119.5	703.1	171.4	44.9	30.8	247.1
Jul	437.4	141.8	121.8	701.0	171.2	45.6	31.7	248.5
Aug	434.5	140.0	124.0	698.5	171.5	45.2	32.2	248.9
Sep	436.6	139.9	126.6	703.1	174.2	45.4	32.9	252.5
Oct	433.8	138.4	126.9	699.1	174.6	45.0	33.0	252.6
Nov	433.7[†]	133.2[†]	127.1[†]	694.0[†]	173.7[†]	43.7[†]	33.0[†]	250.4[†]
Dec	433.3	130.3	126.7	690.3	172.4	43.2	32.8	248.4

1 Total computerised claims only.

Sources: Jobcentre Plus Administrative;
Labour Market Statistics;
Helpline: 020 7533 6094

3.10 Claimant count

Thousands

	United Kingdom						Great Britain	
	Not seasonally adjusted		Seasonally adjusted[1]				Seasonally adjusted[1]	
	Total	Percentage rate[2]	Males	Females	Total	Percentage rate[2]	Total	Percentage rate[2]
	BCJA	BCJB	DPAE	DPAF	BCJD	BCJE	DPAG	DPAJ
2001	983.0	3.2	739.6	230.3	969.9	3.2	930.5	3.1
2002	958.8	3.1	717.1	229.6	946.6	3.1	910.2	3.0
2003	945.9	3.0	700.3	232.8	933.0	3.0	898.5	3.0
2004	866.1	2.8	636.4	217.1	853.5	2.7	822.7	2.7
2005	874.4	2.8	639.8	222.0	861.8	2.7	833.2	2.7
2006	956.7	3.1	697.1	247.3	944.4	3.0	916.5	3.0
2003 Nov	884.6	2.8	683.5	229.5	913.0	2.9	878.9	2.9
Dec	889.7	2.9	677.9	228.1	906.0	2.9	872.0	2.9
2004 Jan	952.4	3.0	671.2	226.0	897.2	2.9	863.7	2.8
Feb	957.0	3.0	664.2	224.5	888.7	2.8	855.9	2.8
Mar	932.0	3.0	657.6	222.9	880.5	2.8	848.3	2.8
Apr	905.2	2.9	651.6	220.3	871.9	2.8	840.1	2.8
May	869.7	2.8	640.2	217.9	858.1	2.7	826.9	2.7
Jun	840.5	2.7	631.9	215.8	847.7	2.7	817.1	2.7
Jul	841.5	2.7	625.1	212.0	837.1	2.7	807.3	2.6
Aug	847.6	2.7	622.3	213.2	835.5	2.7	805.6	2.6
Sep	827.8	2.6	622.1	213.6	835.7	2.7	805.9	2.6
Oct	806.8	2.6	621.1	213.1	834.2	2.7	804.5	2.6
Nov	803.0	2.6	616.8	213.2	830.0	2.6	800.5	2.6
Dec	810.2	2.6	612.9	213.0	825.9	2.6	796.8	2.6
2005 Jan	872.1	2.8	607.1	212.5	819.6	2.6	790.5	2.6
Feb	885.0	2.8	607.0	212.0	819.0	2.6	789.9	2.6
Mar	882.3	2.8	617.0	214.4	831.4	2.6	802.5	2.6
Apr	871.8	2.8	622.7	216.5	839.2	2.7	810.4	2.6
May	867.6	2.7	635.2	219.0	854.2	2.7	825.5	2.7
Jun	858.2	2.7	641.9	221.4	863.3	2.7	834.6	2.7
Jul	871.0	2.8	643.9	222.2	866.1	2.7	837.9	2.7
Aug	880.7	2.8	646.1	223.2	869.3	2.7	841.1	2.7
Sep	871.5	2.8	653.3	226.0	879.3	2.8	851.2	2.8
Oct	864.8	2.7	661.5	229.7	891.2	2.8	863.0	2.8
Nov	875.3	2.8	668.6	232.7	901.3	2.8	872.7	2.8
Dec	892.7	2.8	673.3	234.6	907.9	2.9	879.5	2.9
2006 Jan	955.3	3.0	669.9	235.2	905.1	2.9	876.9	2.8
Feb	984.7	3.1	685.2	239.8	925.0	2.9	896.7	2.9
Mar	989.1	3.1	695.2	242.6	937.8	3.0	909.5	3.0
Apr	981.2	3.1	700.7	244.4	945.1	3.0	916.8	3.0
May	965.7	3.1	703.8	246.9	950.7	3.0	922.4	3.0
Jun	952.9	3.0	706.4	248.6	955.0	3.0	927.1	3.0
Jul	960.8	3.0	704.3	249.7	954.0	3.0	926.4	3.0
Aug	958.9	3.0	701.5	250.3	951.8	3.0	924.2	3.0
Sep	952.9	3.0	706.1	254.0	960.1	3.0	932.4	3.0
Oct	933.7	3.0	702.4	254.1	956.5	3.0	928.7	3.0
Nov	922.1	2.9	696.9†	251.7†	948.6†	3.0	921.3†	3.0
Dec	923.5	2.9	693.4	249.7	943.1	3.0	916.2	3.0

1 The seasonally adjusted series relate only to claimants aged 18 or over in order to maintain the consistent series, available back to 1971 (1974 for the regions - see p.608 of the December 1990 *Employment Gazette* and pS16 of the April 1994 issue for the list of discontinuities taken into account). It also takes into account the effect of the change in benefit eligibility rules introduced with Jobseeker's Allowance (see p.219-24, Labour Market Trends, May 2000). The latest national and regional seasonally adjusted claimant count figures are provisional and are subject to revision mainly in the following month.

2 Percentage rates have been calculated by expressing the number of claimants as a percentage of the estimated total workforce (the sum of claimants, employees jobs, self-employed, HM Forces and participants on work related government training programmes) at mid-2005 estimates for 2005 and 2006 figures and at the corresponding mid-year estimates for earlier years.

Sources: Jobcentre Plus Administrative;
Labour Market Statistics;
Helpline: 020 7533 6094

3.11 Unemployed
Analysis by Government Office Regions

Thousands, seasonally adjusted

	North East	North West	Yorkshire and the Humber	East Midlands	West Midlands	East	London	South East	South West	England	Wales	Scotland	Great Britain	Northern Ireland	United Kingdom[1]
	YCMP	YCMQ	YCMR	YCMS	YCMT	YCMU	YCMV	YCMW	YCMX	YCMY	YCMZ	YCNA	YCNB	ZSFA	MGSC
2000 Q1	103	193	153	107	156	106	282	148	105	1 354	87	186	1 627	48	1 675
Q2	105	171	148	101	156	100	271	137	105	1 295	80	178	1 552	49	1 599
Q3	105	172	143	100	145	101	254	130	101	1 250	88	167	1 505	41	1 545
Q4	90	170	148	97	152	101	250	140	95	1 243	76	157	1 476	45	1 521
2001 Q1	88	167	129	97	143	97	242	140	96	1 200	79	147	1 426	46	1 472
Q2	86	171	130	105	140	100	234	133	88	1 188	80	159	1 427	46	1 472
Q3	82	163	127	97	138	110	249	143	90	1 198	72	167	1 437	46	1 483
Q4	84	172	122	96	144	109	278	143	89	1 236	76	168	1 481	44	1 523
2002 Q1	86	172	122	99	145	103	262	151	86	1 227	74	166	1 467	46	1 511
Q2	75	179	128	97	148	102	258	162	92	1 241	75	159	1 475	43	1 515
Q3	72	179	135	100	155	108	266	169	100	1 284	70	161	1 515	47	1 561
Q4	84	161	121	102	148	113	250	168	101	1 247	69	156	1 472	43	1 514
2003 Q1	77	161	129	85	158	129	265	163	96	1 263	67	151	1 481	42	1 523
Q2	70	163	124	91	147	108	274	166	85	1 229	63	134	1 426	41	1 464
Q3	76	164	120	99	153	112	276	163	82	1 245	66	150	1 461	43	1 504
Q4	74	154	121	96	147	98	263	162	76	1 191	67	148	1 406	48	1 453
2004 Q1	67	148	119	101	144	98	266	161	75	1 179	65	149	1 393	40	1 432
Q2	64	146	112	90	144	108	269	154	93	1 181	59	154	1 394	39	1 433
Q3	70	148	113	88	130	102	274	154	83	1 161	67	136	1 364	38	1 400
Q4	76	154	114	91	125	107	272	148	83	1 171	58	148	1 377	36	1 411
2005 Q1	69	158	108	93	124	109	257	156	92	1 167	64	144	1 374	37	1 411
Q2	83	147	118	92	123	110	275	162	82	1 192	63	140	1 395	39	1 433
Q3	80	150	113	98	123	119	260	171	95	1 209	65	142	1 416	34	1 447
Q4	79	162	137	103	139	130	288	180	101	1 318	68	136	1 521	35	1 554
2006 Q1	82	165	136	112	139	138	304	192	92	1 360	67	139	1 566	35	1 599
Q2	75	178	145	121	154	144	315	202	96	1 430	81	141	1 652	34	1 683
Q3	85	190	152	119	164	143	321	195	100	1 469	77	131	1 677	38	1 711

Unemployment rate[2]

	YCNC	YCND	YCNE	YCNF	YCNG	YCNH	YCNI	YCNJ	YCNK	YCNL	YCNM	YCNN	YCNO	ZSFB	MGSX
2006 Q3	6.9	5.6	6.0	5.3	6.1	5.0	8.0	4.5	3.9	5.7	5.4	5.0	5.6	4.7	5.6

1 Due to slight methodological differences between the way the national and regional LFS estimates have been interim adjusted for the 2001 Census, there may be small differences between the UK totals and the sum of the regional components.

2 Unemployed as a percentage of total economically active (the sum of unemployed and those in employment).

Source: Labour Force Survey, Office for National Statistics: 020 7533 6094

3.12 Claimant count[1]
Analysis by Government Office Regions

Thousands, seasonally adjusted

	North East	North West	Yorkshire and the Humber	East Midlands	West Midlands	East	London	South East	South West	Wales	Scotland	Northern Ireland
	DPDG	IBWA	DPAX	DPAY	DPBC	DPDJ	DPDK	DPDL	DPBB	DPBE	DPBF	DPBG
1999	79.9	153.8	123.0	76.2	119.7	76.5	203.1	95.3	75.3	64.1	130.4	50.7
2000	72.2	136.9	107.0	69.4	108.0	64.1	174.5	78.9	61.8	57.3	116.3	42.1
2001	62.7	123.5	96.0	63.6	99.0	55.0	154.9	66.6	52.7	51.2	105.2	39.5
2002	57.9	118.1	88.8	58.7	93.7	56.6	166.0	71.2	50.1	47.1	102.0	36.4
2003	52.8	111.7	83.7	58.9	94.7	58.1	170.7	75.6	48.3	44.6	99.5	34.6
2004	46.3	99.2	73.4	52.5	88.3	55.4	162.8	70.7	41.9	40.2	92.0	30.8
2005	46.0	101.3	76.0	54.1	93.9	58.1	163.0	71.6	42.2	41.2	85.9	28.6
2006	50.2	115.5	87.4	62.0	108.4	65.4	166.6	81.6	48.0	44.3	87.2	27.9
2003 Jul	52.6	112.5	84.2	59.9	94.9	58.7	171.7	76.4	49.1	45.0	100.6	34.6
Aug	52.2	110.9	83.0	59.6	95.0	58.3	171.0	76.4	48.4	44.3	99.4	34.8
Sep	51.8	110.0	82.6	59.4	94.6	57.8	170.4	76.4	48.0	43.6	99.2	34.9
Oct	51.1	108.7	81.6	58.8	94.2	57.3	170.2	76.0	47.4	43.1	98.9	34.7
Nov	50.7	107.0	80.0	57.8	93.5	57.3	169.3	75.8	46.5	42.7	98.3	34.1
Dec	50.1	106.0	78.7	57.2	93.0	56.9	168.9	75.3	45.9	42.2	97.8	34.0
2004 Jan	49.6	104.2	78.1	56.1	92.9	56.6	167.8	74.8	45.0	41.9	96.7	33.5
Feb	48.3	103.6	77.2	55.0	92.4	56.6	166.6	74.0	44.2	41.8	96.2	32.8
Mar	47.5	102.8	76.4	54.5	91.7	56.4	165.0	73.3	43.7	41.5	95.5	32.2
Apr	47.4	101.3	75.4	53.9	90.0	56.2	165.0	71.9	42.9	41.7	94.4	31.8
May	46.6	99.6	73.9	52.5	88.5	55.3	164.8	70.9	41.9	40.5	92.4	31.2
Jun	45.7	98.4	72.7	51.8	87.7	54.7	163.1	70.2	41.3	39.9	91.6	30.6
Jul	45.4	96.7	71.7	51.1	86.8	54.4	162.2	68.8	40.5	39.5	90.2	29.8
Aug	45.1	97.0	71.9	50.9	86.1	54.7	161.0	68.8	40.6	39.3	90.2	29.9
Sep	45.0	97.2	71.4	51.1	86.0	54.9	160.5	69.0	40.7	39.5	90.6	29.8
Oct	45.4	97.0	71.4	51.1	85.9	55.1	159.3	69.6	40.7	39.2	89.8	29.7
Nov	44.9	96.5	70.6	51.4	85.7	55.1	159.3	68.8	40.6	38.8	88.8	29.5
Dec	44.7	96.2	69.7	50.9	85.8	55.3	158.9	67.9	40.7	38.9	87.8	29.1
2005 Jan	43.4	94.0	69.6	50.6	85.3	55.0	158.6	68.0	40.5	38.7	86.8	29.1
Feb	44.1	94.4	70.2	50.1	84.3	55.1	159.2	67.4	40.4	38.8	85.9	29.1
Mar	44.9	96.2	72.1	51.6	85.7	56.1	160.9	68.8	40.9	39.2	86.1	28.9
Apr	44.9	97.7	73.0	52.0	88.0	56.4	161.5	69.5	41.5	39.8	86.1	28.8
May	45.5	99.5	74.5	52.8	95.0	57.1	161.6	70.3	42.0	40.6	86.6	28.7
Jun	46.2	100.9	75.7	53.9	95.6	58.1	162.2	71.8	42.4	41.5	86.3	28.7
Jul	46.3	102.0	76.3	54.4	96.6	58.6	162.5	72.0	42.7	41.6	84.9	28.2
Aug	46.7	102.8	76.7	54.7	96.4	58.5	163.7	72.0	42.5	41.6	85.5	28.2
Sep	47.1	104.7	78.2	55.6	97.9	59.2	164.6	73.1	42.7	42.1	86.0	28.1
Oct	47.6	106.5	80.1	56.7	99.4	60.3	166.4	74.1	43.1	43.0	85.8	28.2
Nov	47.5	107.9	82.0	57.9	101.0	61.1	166.8	75.4	43.7	43.7	85.7	28.6
Dec	47.4	108.7	83.6	58.6	102.1	61.7	167.5	76.9	43.7	43.9	85.4	28.4
2006 Jan	46.5	108.3	83.4	58.4	102.0	61.9	167.8	77.6	43.5	43.4	84.1	28.2
Feb	48.7	111.9	85.2	59.9	105.0	63.2	167.6	79.9	44.9	44.4	86.0	28.3
Mar	49.4	113.7	86.3	60.9	107.3	64.4	167.3	81.2	46.4	45.1	87.5	28.3
Apr	49.8	114.9	86.8	61.9	108.6	65.1	167.1	81.3	48.0	45.3	88.0	28.3
May	50.4	115.8	87.7	62.1	108.7	65.3	167.5	82.5	48.7	45.1	88.6	28.3
Jun	50.8	116.8	88.2	62.5	108.9	65.4	168.3	83.4	49.1	44.8	88.9	27.9
Jul	50.4	116.5	88.6	62.9	109.3	65.1	167.8	83.7	49.2	44.3	88.6	27.6
Aug	50.6	116.0	88.8	63.5	109.7	65.1	165.6	82.8	49.5	44.4	88.2	27.6
Sep	51.1	118.4	89.1	63.6	110.3	66.9	167.5	83.8	49.6	44.2	87.9	27.7
Oct	51.3	118.2	89.0	63.2	110.1	67.7	166.2	82.6	49.1	43.9	87.4	27.8
Nov	51.5	117.7†	88.1†	62.6†	110.1†	67.5†	164.4†	81.0†	48.9†	43.2	86.3†	27.3†
Dec	52.1	117.9	87.6	62.2	110.3	66.9	162.4	79.6	48.9	42.9	85.4	26.9

Claimant count rate[2]												
	DPDM	IBWC	DPBI	DPBJ	DPBN	DPDP	DPDQ	DPDR	DPBM	DPBP	DPBQ	DPBR
2006 Dec	4.4	3.4	3.3	2.9	4.0	2.4	3.4	1.8	1.9	3.1	3.2	3.1

1 The seasonally adjusted series relate only to claimants aged 18 or over in order to maintain the consistent series, available back to 1971 (1974 for the regions - see p.608 of the December 1990 *Employment Gazette* and pS16 of the April 1994 issue for the list of discontinuities taken into account). It also takes into account the effect of the change in benefit eligibility rules introduced with Jobseeker's Allowance (see p.219-24, Labour Market Trends, May 2000). The latest national and regional seasonally adjusted claimant count figures are provisional and are subject to revision mainly in the following month.

2 Percentage rates have been calculated by expressing the number of claimants as a percentage of the estimated total workforce (the sum of claimants, employees jobs, self-employed, HM Forces and participants on work related government training programmes) at mid-2005 estimates for 2005 and 2006 figures and at the corresponding mid-year estimates for earlier years.

Sources: Jobcentre Plus Administrative;
Labour Market Statistics;
Helpline: 020 7533 6094

3.13 Vacancies at Jobcentres and career offices[1]
Analysis by Government Office Regions

Thousands

	North East	North West	Yorkshire and the Humber	East Midlands	West Midlands	East	London	South East	South West	Wales	Scotland	Great Britain	Northern Ireland[2]	United Kingdom
Total vacancies at Jobcentres: not seasonally adjusted[3]														
	DPCQ	IBWF	BCRG	BCRF	BCRE	DPCT	BCRB	DPCU	BCRD	BCRJ	BCRK	BCRL	BCRM	BCOM
1997	10.1	34.4	21.0	20.4	23.1	23.6	35.0	34.4	25.5	18.1	31.5	277.0	6.8	283.9
1998	11.0	41.0	22.6	20.6	30.5	24.1	28.2	34.8	26.1	17.9	31.0	287.7	8.9	296.6
1999	16.4	37.1	24.1	21.3	35.7	23.9	32.1	37.9	27.8	17.0	33.0	306.2	..	..
2000	19.7	41.2	32.8	22.3	35.9	24.4	36.4	43.6	34.6	19.0	40.1	349.9	–	–
2000 Apr	17.7	38.5	30.5	20.9	33.9	24.0	34.3	40.7	35.7	19.5	37.0	332.5	–	–
May	18.0	39.2	31.3	21.2	33.7	24.7	34.2	42.0	35.9	19.0	35.8	335.1	–	–
Jun	18.5	40.3	32.9	22.6	35.1	25.2	36.3	45.1	37.6	19.5	36.7	349.8	–	–
Jul	18.7	40.4	33.5	22.2	34.8	25.7	37.5	46.2	36.8	19.3	37.6	352.8	–	–
Aug	19.2	40.7	34.0	21.5	35.8	24.7	36.1	44.7	35.9	19.2	38.5	350.2	–	–
Sep	21.9	46.4	37.5	24.0	39.5	26.4	36.2	48.5	38.0	20.4	45.4	384.1	–	–
Oct	23.9	50.6	40.8	25.4	43.4	27.5	41.3	51.6	39.6	20.4	49.0	413.4	–	–
Nov	23.4	49.1	40.6	25.9	42.4	26.5	42.0	50.7	38.5	19.6	49.5	408.1	–	–
Dec	20.8	41.3	36.4	23.4	37.9	23.5	38.5	45.4	34.0	18.0	45.4	364.5	–	–
2001 Jan	20.3	40.0	35.3	22.0	36.1	21.6	36.6	41.0	33.1	18.1	45.3	349.4	–	–
Feb	20.6	40.9	34.6	22.3	35.6	21.8	33.8	42.6	32.5	18.0	42.7	345.5	–	–
Mar	22.9	43.0	36.2	22.9	37.0	23.2	33.9	44.2	34.0	19.4	43.9	360.6	–	–
Apr	23.6	44.5	38.7	22.1	37.2	24.9	30.1	42.6	35.9	20.1	42.7	362.5	–	–
Seasonally adjusted[3]														
	DPCL	IBWE	BCQG	BCQF	BCQE	DPCO	BCQB	DPCP	BCQD	BCQJ	BCQK	BCQL	BCQM	DPCB
2000 Apr	19.5	41.2	31.0	22.5	35.9	25.2	36.7	41.9	34.7	19.8	38.4	346.8	–	355.7
May	19.0	41.3	31.7	22.6	35.8	25.3	36.0	42.5	34.1	18.9	38.2	345.4	–	354.3
Jun	18.5	41.0	32.7	22.9	36.1	25.0	36.5	43.7	34.5	18.9	38.5	348.3	–	357.2
Jul	18.7	41.4	33.3	22.9	36.0	25.3	37.6	45.1	35.1	19.1	39.5	354.0	–	362.9
Aug	18.7	40.8	33.6	22.5	36.6	24.7	37.3	44.5	35.4	19.3	39.3	352.7	–	361.6
Sep	19.3	42.1	34.6	22.7	36.6	24.3	35.3	45.3	35.5	19.1	41.9	356.7	–	365.6
Oct	19.6	42.4	35.3	20.9	36.2	23.4	35.8	45.0	35.8	18.4	42.8	355.6	–	364.5
Nov	20.7	43.0	37.1	22.0	36.5	23.6	36.9	45.7	36.9	18.7	44.3	365.4	–	374.3
Dec	21.2	42.0	37.5	22.5	37.2	23.8	36.9	46.0	37.1	18.9	44.5	367.6	–	376.5
2001 Jan	22.4	44.0	39.5	23.5	39.7	24.5	39.0	47.1	39.6	19.8	47.7	386.8	–	395.7
Feb	23.8	45.0	38.8	24.7	39.0	24.9	36.4	48.0	37.3	19.6	45.3	382.7	–	391.6
Mar	25.6	46.3	39.3	25.3	39.8	25.4	35.7	47.0	36.3	20.2	45.1	386.0	–	394.9
Apr	25.2	46.7	39.4	23.9	39.4	26.4	32.6	44.8	35.9	20.6	44.2	378.9	–	387.8
Total vacancies at careers offices: not seasonally adjusted														
	DPCV	IBWJ	BCSG	BCSF	BCSE	DPCY	BCSB	DPCZ	BCSD	BCSJ	BCSK	BCSL	BCSM	BCSN
2001	0.3	2.0	2.4	1.0	1.8	1.9	3.5	3.6	1.4	0.4	1.4	19.8	..	..
2002	0.4	2.2	2.9	0.9	2.0	1.5	1.8	3.1	1.5	0.3	1.3	17.7	..	..
2003	0.3	2.2	2.4	0.8	1.2	1.4	1.5	2.7	2.4	0.3	1.4	16.6	..	..
2004	0.5	3.2	2.3	0.9	1.1	1.3	1.4	2.5	2.1	0.2	1.3	16.8	..	..
2004 Apr	0.4	2.7	2.2	0.9	1.7	1.2	1.3	2.4	2.3	0.2	1.5	16.9	..	..
May	0.5	3.9	2.2	0.8	0.9	1.4	1.4	1.6	2.4	0.2	1.4	16.8	..	..
Jun	0.5	3.2	2.3	1.1	0.8	1.5	1.6	2.8	2.5	0.3	1.5	18.0	..	..
Jul	0.6	4.2	2.8	1.1	1.1	1.7	1.6	3.0	2.2	0.2	1.6	20.1	..	..
Aug	0.6	4.2	2.6	1.1	1.0	1.6	1.7	3.0	2.4	0.2	1.5	20.0	..	..
Sep	0.6	4.0	2.5	1.0	1.1	1.5	1.4	2.7	2.3	0.2	1.5	18.8	..	..
Oct	0.6	3.7	2.4	0.9	0.9	1.4	1.4	2.6	2.2	0.3	1.6	18.0	..	..
Nov	0.5	3.5	2.1	0.9	0.9	1.2	1.3	2.8	1.5	0.2	1.2	16.1	..	..
Dec	0.4	3.4	1.9	0.8	0.8	1.1	1.2	2.6	1.5	0.2	1.1	15.1	..	..
2005 Jan	0.5	3.4	1.7	0.8	0.7	1.1	1.0	2.6	1.3	0.2	1.1	14.4	..	..
Feb	0.5	2.3	1.7	0.8	0.7	1.1	1.1	2.6	1.3	0.3	1.0	13.4	..	..
Mar	0.4	3.0	1.8	0.8	0.7	1.1	1.1	2.5	1.6	0.4	1.9	15.3	..	..
Apr	0.4	3.1	1.9	0.8	1.1	1.3	1.2	2.7	1.7	0.3	1.9	16.4	..	..

1 Only a proportion of all vacancies are notified to Jobcentres. These could include some that are suitable for young persons and similarly vacancies notified to careers offices could include some for adults. Because of possible duplication the two series should not be added together. The figures represent only the number of vacancies notified by employers and remaining unfilled on the day of the count.

2 The publication of the vacancy figures for Northern Ireland has been suspended since March 1999 as a result of a difficulty caused by the introduction of a new computer system for processing vacancies to Training and Employment Agency offices. For the purpose of the seasonally adjusted United Kingdom figures it has been assumed provisionally that the Northern Ireland figures have remained constant since February 1999.

3 Publication of the Jobcentre vacancy statistics has been deferred. Figures from May 2001 are affected by the introduction of Employer Direct. This major change involves transferring the vacancy taking process from local Jobcentres to regional Customer Service Centres, as part of Modernising the Employment Service. ONS and the Department of Work and Pensions will continue to monitor and review the data with the aim of publishing the series as soon it is possible to produce a consistent measure.

Source: Office for National Statistics: 020 7533 6094

3.14 Labour disputes[1]

Thousands

SIC 1992	Workers beginning involvement in period in any dispute	Total working days lost[2]						
		All industries and services	Manufacturing	Transport, storage and communication	Public administration and defence	Education	Health and social work	All other industries and services
		All classes	*15-37*	*60-64*	*75*	*80*	*85*	*All other classes*
	BBFV	BBFW	BBFX	BBFY	BBFZ	BBGA	BBGB	BBGC
2002	918	1 323	21	96	488	376	148	195
2003	123	499	63	126	138	131	15	25
2004	272	905	31	44	437	379	4	10
2005	92	157	16	33	23	43	–	43
2005 Jan	1	1	–	–	–	–	–	–
Feb	7	8	–	–	3	4	–	–
Mar	3	4	–	–	–	3	–	–
Apr	3	6	–	3	–	1	–	1
May	26	32	2	2	5	17	–	6
Jun	2	5	2	1	–	–	–	3
Jul	5	15	4	10	–	–	–	–
Aug	5	17	1	3	3	–	–	10
Sep	4	29	6	8	1	–	–	14
Oct	4	7	–	3	2	1	–	–
Nov	19	19	–	–	3	15	–	1
Dec	13	15	–	2	5	–	–	8
2006 Jan	45	77	–	5	69	–	–	3
Feb	2	14	–	10	1	1	–	2
Mar	577	482	–	2	461	17	–	1
Apr	2	3	–	–	–	1	–	–
May	49	83	2	2	70	4	–	6
Jun	2	6	2	3	–	–	–	–
Jul	8	11	1	2	4	1	2	1
Aug	2	6	1	1	1	–	1	3
Sep	5	23	–	1	16	2	2	1
Oct	7	13	3	1	2†	2	–	5
Nov	5	24	5	6	1	2	–	11

1 Excludes stoppages involving fewer than 10 workers or lasting less than one day except any in which the total number of working days lost are 100 or more. There may be some under-recording of small or short stoppages; this would have much more effect on the total stoppages than on working days lost.

2 The working days lost figures relate to the total working days lost within each of the periods shown as a result of stoppages *in progress* in that period, whether the stoppages began in that period or earlier.

Source: Office for National Statistics: 01633 819205

4 Social services

4.1 National Insurance and Child Benefit
Great Britain

Thousands

	National Insurance				Child Benefit[1,6]	
	Persons in receipt of Jobseeker's Allowance (contributions based)[2]	New claims Incapacity Benefit (Weekly averages)[5]	At end of period		Families receiving benefit	Children in families receiving benefits
			State Pension[3]	Widows/ Bereavement Benefit[4]		
	BDAD	BDAA	BDAE	BMCR	BDAG	BDAH
2005 Mar	..	13.4	..	..	..	..
Apr	..	13.4	..	..	..	..
May	145	12.5	10 526	183	7 073	12 573
Jun	..	13.7	..	..	..	..
Jul	..	12.2	..	..	..	..
Aug	145	12.6	10 547	178	7 135	12 698
Sep	..	13.3	..	..	..	..
Oct	..	12.5	..	..	..	..
Nov	141	13.7	10 557	172	6 986	12 323
Dec	..	10.7	..	..	..	..
2006 Jan	..	12.2	..	..	..	..
Feb	158	13.9	10 564	169	7 047	12 444
Mar	..	13.5	..	..	..	..
Apr	..	10.0	..	..	..	..
May	143	11.8	10 587	165	7 125	12 604
Jun	..	13.3	..	..	..	..
Jul	..	12.3	..	..	..	..
Aug	..	13.6	..	..	7 198	12 750
Sep	..	13.7	..	..	..	..
Oct	..	14.1	..	..	..	..

1 Child Benefit figures are taken from the Child Benefit Computer System 5% scan in the months shown. Figures exclude overseas cases.
2 Jobseeker's Allowance (JSA) figures have been derived by applying 5% proportions to 100% WPLS totals. Excludes recipients of contribution based and income based benefit.
3 Excluding pensioners in receipt of non-contributory State Pension awarded under National Insurance Acts 1970 and 1971. Also excludes overseas and Channel Islands.

4 Includes all Widow's Benefit and Bereavement Benefit except Widow's Payment and Bereavement Payment. Excludes overseas and Channel Island cases.
5 The figures for Incapacity Benefit are calculated from 100% counts but are provisional and therefore subject to amendment.
6 Child Benefit is the responsibility of HM Revenue and Customs.

Source: Department for Work and Pensions: 0191 225 9900

4.2 Child and Working Tax Credit[1]
United Kingdom

Thousands

	Families in work receiving credit:			No of children in these families
	All families	Two-adult families	One-adult families	
	WMPT	WMPU	WMPV	WMPW
2004 Apr	4 541	3 363	1 179	7 668
2004 Jul[2]	4 610	3 390	1 220	7 760
2004 Dec	4 518	3 310	1 209	7 455
2005 Apr	4 638	3 378	1 260	7 624
2005 Dec	4 538	3 261	1 277	7 324
2006 Apr	4 601	3 307	1 294	7 446
2006 Dec	4 526	3 237	1 289	7 292

1 For further information refer to Section 4 of the *Annual Supplement* in the January edition of *Monthly Digest*.

2 July figures rounded to nearest 10 thousand.

Source: Board of Inland Revenue: 020 7438 6275

4.3 Income Support/Pension Credit/Jobseeker's Allowance (income based)
Great Britain

In a week in the month shown, thousands

| | Income support[1] | | | | | Pension Credit[6] | Jobseeker's Allowance (income based)[7] | | |
	Aged 60 and over[2,6]	Incapacity benefits[3]	Lone parents[4]	Others[5]	All cases		With contribution based benefit	Without contribution based benefit	All cases
	BALZ	BAMD	BAME	BAMF	BAMG	A4EK	DMUB	DMUC	DMUD
2003 Feb	1 758	1 216	844	169	3 987	..	19	673	692
May	1 778	1 215	853	168	4 014	..	19	640	659
Aug	1 800	1 221	851	168	4 041	..	17	618	635
Nov	..	1 221	832	167	2 220	2 085	15	584	599
2004 Feb	..	1 221	830	166	2 217	2 282	16	625	641
May	..	1 205	823	164	2 193	2 491	14	576	590
Aug	..	1 207	818	166	2 191	2 593	12	576	588
Nov	..	1 206	797	165	2 167	2 630	13	553	566
2005 Feb	..	1 200	793	160	2 153	2 655	15	600	615
May	..	1 194	789	157	2 140	2 683	14	596	610
Aug	..	1 190	789	159	2 138	2 697	15	620	635
Nov	..	1 189	779	160	2 127	2 708	14	633	647
2006 Feb	..	1 188	777	158	2 123	2 709	15	702	717
May	..	1 183	775	157	2 115	2 717	14	690	704

1 IS claimants have been assigned to a statistical group according to a hierarchy. The order is as shown in the table, i.e. "Aged 60 and over", "Incapacity Benefits" etc. For example, lone parents with both Incapacity Benefits and Income Support will fall into the "Incapacity Benefits" category.

2 "Aged 60 and over" are benefit units where the claimant and/or partner is aged 60 or over.

3 "Incapacity Benefits" refer to claimants aged under 60 claiming Incapacity Benefit (IB) or Severe Disablement Allowance (SDA), including IB credits only cases.

4 "Lone Parents" are single recipients of Income Support aged under 60 with a child under 16 who are not in receipt of IB/SDA.

5 "Others" are recipients of Income Support not in one of the other categories.

6 Since 6th October 2003, Income Support for claimants aged 60 or over have been paid via the new Pension Credit. Pension Credit eligibility is also more generous than prior to 6th October 2003, increasing the numbers of pensioners in receipt.

7 JSA figures have been derived by applying 100% WPLS proportions to 5% totals.

Source: Department for Work and Pensions

4.4 Family health services

Thousands

| | England and Wales | | | | | | Scotland | | | |
| | Pharmaceutical services | Dental services | | | Ophthalmic services[1] | | Pharmaceutical services | Dental services | Ophthalmic services | |
	Number of prescriptions items dispensed by chemists etc[2]	Completed courses of treatment and cases of occasional treatment[3]	Personal dental services	Total courses of treatment[5]	Sight tests	Pairs of spectacles paid for by HAs under the Voucher Scheme	Number of prescription items dispensed[4]	Completed courses of adult treatment and cases of occasional treatment[3]	Sight tests paid for	Pairs of spectacles paid for by SHBs under the Voucher Scheme
	CKQJ	BDDB	F93Q	G92F	BDDC	BDDD	BDDE	BDDF	BDDG	BDDH
2004	740 389	26 604	1 616	28 220	..	..	74 335	2 942	943	462
2005	776 925	19 914	7 631	27 545	..	..	76 427	2 904	942	454
2006	..	..	..	..	..	..	..	2 920	..	..
2003 Q3	174 725	7 021	108	7 129	5 160	1 850	17 843	713	229	111
Q4	188 324	7 252	125	7 377	..	..	18 560	725	221	109
2004 Q1	177 162	6 795	175	6 970	5 331	1 921	17 936	731	243	117
Q2	181 577	6 885	249	7 134	..	..	18 545	745	237	118
Q3	184 288	6 411	461	6 872	5 466	1 979	18 599	728	236	116
Q4	197 362	6 513	731	7 244	..	..	19 255	738	227	111
2005 Q1	183 181	5 598	1 224	6 822	5 339	1 896	18 271	719	235	113
Q2	193 562	5 334	1 794	7 128	..	..	19 374	749	244	119
Q3	193 063	4 597	2 140	6 737	5 191	1 860	19 097	712	237	113
Q4	207 119	4 385	2 473	6 858	..	..	19 685	724	226	109
2006 Q1	196 783	4 103	2 532	6 635	5 163	1 819	..	727	252	117
Q2	200 015	..	..	4 819[6]	..	..	..	743	371	114
Q3	..	..	..	..	..	..	..	710	375	104
Q4	..	..	..	..	..	..	..	740	..	..

1 Data on Ophthalmic Services are collected six-monthly and presented against the second quarter covered.

2 The data covers all prescriptions dispensed by community pharmacists and appliance contractors, dispensing doctors, and prescriptions submitted by prescribing doctors for items personally administered.

3 Number scheduled for payment in the General Dental Service.

4 Includes prescriptions dispensed by Community Pharmacies, appliance suppliers, dispensing doctors and stock orders.

5 Some early information relates only to the General Dental Service.

6 A new NHS dental contract was introduced on 1 April 2006. Information based on the new contract are not comparable with information under the old contractual arrangements. Data cannot be separated between GDS and PDS contracts from 1 April 2006 and data only for England have been provided under the new contract.

Sources: The Health and Social Care Information Centre;;
NHS Scotland and the National Assembly for Wales

5 Law enforcement

5.1 Recorded crime statistics
England and Wales

Thousands

	Violence against the person	Sexual offences	Burglary	Robbery	Theft and handling stolen goods	Fraud and forgery	Drug offences	Criminal damage	Other	Total
1998/99	502.8	36.2	953.2	66.8	2 191.4	279.5	135.9	879.6	63.6	5 109.0
1999/2000	581.0	37.8	906.5	84.3	2 223.6	334.8	121.9	945.7	65.7	5 301.3
2000/2001	600.9	37.3	836.0	95.2	2 145.4	319.3	113.5	960.1	63.2	5 170.9
2001/2002[1]	650.3	41.4	878.5	121.4	2 267.0	314.9	121.4	1 064.5	65.7	5 525.1
2002/03[2,5]	845.1	49.2	890.1	110.3	2 411.6	331.1	143.3	1 120.6	73.7	5 975.0
2003/04[3,5]	967.2	52.7	820.0	103.7	2 312.9	319.6	143.5	1 218.5	75.5	6 013.6
2004/05[4,5]	1 048.2	62.1	681.1	90.7	2 069.4	280.5	145.5	1 198.2	64.9	5 640.6
2005/06[5]	1 059.9	62.1	645.1	98.2	2 019.3	233.0	178.5	1 184.7	75.7	5 556.5
	BEAB	BEAC	BEAD	BEAE	BEAF	BEAG	LQMO	BEAH	BEAI	BEAA
2003 Q1	207.8	11.9	216.7	27.9	583.7	81.1	36.2	291.8	18.1	1 475.2
Q2	236.3	12.8	221.6	27.6	605.8	84.6	34.2	303.7	19.1	1 545.7
Q3	253.0	14.1	208.7	26.2	591.7	81.2	36.2	281.2	19.7	1 512.0
Q4	236.9	12.6	196.4	24.6	560.7	76.2	37.4	307.7	18.1	1 470.6
2004 Q1	241.0	13.2	193.3	25.4	554.7	77.7	35.7	325.9	18.5	1 485.4
Q2	266.3	15.1	177.3	23.4	541.7	74.0	33.6	311.7	17.5	1 460.6
Q3	269.2	17.4	169.9	21.6	530.0	71.5	35.5	280.6	16.7	1 412.4
Q4	262.0	14.9	167.2	23.1	512.3	70.0	38.7	302.0	15.2	1 405.4
2005 Q1	250.7	14.7	166.6	22.5	485.4	65.0	37.7	308.8	15.6	1 367.0
Q2	277.6	16.7	162.0	24.2	516.1	63.1	41.5	302.1	19.2	1 422.5
Q3	278.2[†]	16.6[†]	160.7[†]	24.0[†]	514.5[†]	60.9[†]	42.0[†]	277.7[†]	19.8[†]	1 394.5[†]
Q4	262.4	15.0	164.1	24.6	503.4	55.7	46.7	305.2	18.8	1 395.9
2006 Q1	241.5	13.7	158.3	25.4	485.0	53.0	48.2	299.4	17.8	1 342.3
Q2	278.4	15.8	154.4	25.3	496.5	53.6	48.0	301.7	20.8	1 394.5
Q3	276.9	15.9	152.5	24.2	490.8	52.1	45.8	281.7	19.4	1 359.3

1 Some forces adopted the principles of the National Crime Recording Standard in advance of its national implementation on 1 April 2002. For 2001/02 as a whole it has been calculated that this in itself has inflated the total number of crimes recorded by 5%, although the impact differs for each offence group.

2 The National Crime Recording Standard (NCRS) was introduced across all police forces from April 2002, and this has increased the recorded crime figures significantly. For 2002/03 it has been estimated that the implementation of the Standard has inflated the total number of recorded crimes by 10%, although the impact differs for each offence group. Violence against the person was particularly affected. Due to the introduction of the NCRS in April 2002, figures before and after this date are not directly comparable.

3 Much of the increase in violence against the person in 2003/04 is likely to be due to the continuing impact of changes in recording.

4 The Sexual Offences Act 2003, introduced in May 2004, re-defined many sexual offences. This change in legislation could, in itself, account for much of the increase in recorded sexual offences and means figures for 2004/05 are not comparable with earlier years.

5 Figures for the British Transport Police have been added to these statistics as from April 2002 onwards.

Source: Home Office: 020 7035 0307

5.2 Crimes and offences recorded by the police[1,2]
Scotland

Thousands

	Non-sexual crimes of violence	Crimes of indecency	Crimes of dishonesty	Fire raising, vandalism etc	Other crimes	Motor vehicle offences	Miscellaneous offences	Total crimes and offences (annual)
	BEBC	BEBD	BEBE	BEBF	BEBG	BEBI	BEBH	BEBB
2001	15.1	6.0	239.8	94.9	65.2	362.1	162.5	945.8
2002	16.5	6.5	235.7	95.6	72.9	341.3	167.5	935.9
2003	15.2	6.6	210.8	100.1	74.3	409.4	176.7	993.1
2004	15.2	7.4	214.4	124.8	78.6	428.2	209.4	1 078.0
2005	13.6	6.7	191.8	128.3	79.4	381.4	216.9	1 018.2
2003 Q1	3.9	1.6	51.9	25.9	17.1	90.3	41.5	232.1
Q2	3.9	1.7	54.3	26.1	18.3	107.2	45.6	257.0
Q3	3.8	1.6	54.6	23.4	19.7	101.3	47.1	251.6
Q4	3.6	1.7	50.0	24.7	19.2	110.6	42.5	252.4
2004 Q1	3.8	1.8	52.1	29.6	20.3	107.5	45.8	260.8
Q2	4.2	2.0	55.4	32.9	20.2	107.9	57.5	280.1
Q3	3.8	1.8	56.0	32.0	19.6	110.0	56.2	279.4
Q4	3.4	1.8	50.9	30.3	18.5	102.8	49.9	257.7
2005 Q1	3.3	1.7	48.1	33.3	18.9	98.0	50.7	254.0
Q2	3.4	1.6	48.5	32.5	20.3	97.7	56.8	260.8
Q3	3.6	1.7	48.0	29.7	19.2	89.4	53.8	245.4
Q4	3.3	1.7	47.2	32.8	21.0	96.3	55.6	258.0
2006 Q1	3.5	1.6	44.1	32.9	21.4	90.9	53.3	247.4
Q2	3.7	1.8	46.2	33.4	21.9	93.7	59.0	259.8
Q3	3.6	1.9	47.3	31.6	21.0	92.2	62.7	260.2

1 Components may not add to totals due to separate rounding.
2 The introduction of the Scottish Crime recording Standard on 1 April 2004 has increased the number of minor crimes recorded, such as minor crimes of vandalism and petty thefts.

Sources: The Scottish Executive Justice Analytical Services Division;
Tel: 0131 244 2635

6 Agriculture, food, drinks and tobacco

6.1 Land use and crop areas[1]
Area at the June Census

Thousand hectares

		1999	2000	2001	2002	2003	2004	2005
Total agricultural area	BFAH	18 579	18 311	18 556	18 506	18 467	18 437	18 509
Crops	BFAA	4 709	4 665	4 455	4 573	4 478	4 593	4 443
Bare fallow	BFAB	33	37	43	33	29	29	140
All grasses	BFAC	6 675	6 589	6 789	6 761	6 884	6 866	6 904
Sole right rough grazing	BFAD	4 575	4 445	4 435	4 488	4 329	4 326	4 354
Set aside	DMNF	572	567	800	612	689	560	559
All other land on agricultural holdings, including woodland	BFAE	789	780	801	806	820	825	872
Total land on agricultural holdings	BFAF	17 352	17 083	17 323	17 271	17 230	17 200	17 273
Common rough grazing (estimated)	BFAG	1 227	1 228	1 232	1 234	1 236	1 237	1 236
Crops	BFAA	4 709	4 665	4 455	4 573	4 478	4 593	4 443
Cereals	BFAJ	3 141	3 348	3 014	3 245	3 059	3 133	2 925
Wheat	BFAK	1 847	2 086	1 635	1 996	1 837	1 990	1 868
Barley (winter and spring)	BFAL	1 179	1 128	1 245	1 101	1 078	1 010	942
Oats	BFAM	92	109	112	126	122	108	91
Mixed corn[2]	BFAN	2	2	3	4	4	..	..
Rye	BFAO	8	7	5	5	4	6	..
Triticale[2]	DMNH	13	16	14	14	15	..	..
Mixed corn and Triticale	C6GX	..	..	..	..	..	19	–
Mixed corn, Triticale and Rye	EFO7	..	..	..	..	..	25	24
Other arable crops (excluding potatoes)	DMNI	1 211	979	1 103	993	1 098	1 137	1 382
Oilseed rape[3]	BFAP	417	332	404	357	460	498	519
Sugar beet, not for stock feeding	BFAQ	183	173	177	169	162	154	148
Hops	DMNJ	3	2	2	2	2	2	..
Peas for harvesting dry and field beans	DMNK	202	208	275	249	235	242	239
Linseed	DMNL	209	71	31	12	32	30	..
Other crops	DMNM	197	192	214	204	207	211	306
Potatoes	BFAR	178	166	165	158	145	149	137
Horticultural	BFAV	179	172	173	176	176	175	170
Vegetables grown in the open	DMNN	126	119	120	124	125	125	121
Orchard fruits	BFBG	28	28	28	26	25	24	..
Soft fruit	DMNO	9	10	9	9	9	9	..
Orchard fruits and Soft fruit	EFO8	..	..	..	..	..	..	33
Ornamentals	DMNP	13	14	14	15	14	15	14
Glasshouse crops	DMNQ	2	2	2	2	2	2	2

1 Figures include estimates for minor holdings. For further information refer to Section 6 of the *Annual Supplement* in the January edition of *Monthly Digest*.

2 From 2004 onwards data for Mixed corn and Triticale amalgamated.
3 Area grown not on set-aside land.

Source: Department for Environment, Food and Rural Affairs: 01904 455095

6.2 Crops: yields and production[1]

		Yields per hectare (tonnes)						Production (thousand tonnes)				
		2001	2002	2003	2004	2005		2001	2002	2003	2004	2005
Agricultural crops												
Wheat	BFBJ	7.08	8.00	7.78	7.77	7.96	BADO	11 580	15 973	14 288	15 473	14 877
Barley (winter and spring)	BFBK	5.35	5.57	5.91	5.76	5.86	BADP	6 660	6 126	6 370	5 815	5 533
Oats	BFBO	5.53	6.00	6.16	5.83	5.84	BADQ	621	753	749	630	534
Sugar beet	BFBL	46.99	56.54	56.55	57.50	..	BADR	8 335	9 557	9 168	8 850	..
Potatoes	BFBM	40.21	43.96	40.74	42.47	..	BADS	6 649	6 966	5 918	6 316	..

		2000/01	2001/02	2002/03	2003/04	2004/05		2000/01	2001/02	2002/03	2003/04	2004/05
Horticultural crops[2]												
Field vegetables												
Brussels sprouts	BFBR	12.5	13.4	13.3	11.5	12.4	BADT	67.3	54.8	42.7	55.8	42.5
Cabbage, inc. savoy and spring greens	BFBS	86.8	88.5	88.6	80.3	94.7	BADU	273.2	295.4	255.2	245.6	290.9
Cauliflowers	BFBT	13.6	11.5	12.4	11.2	12.0	BADV	156.1	107.4	115.8	126.3	165.3
Carrots	BFBU	58.4	63.3	60.6	52.2	70.1	BADW	725.8	760.0	731.2	614.3	659.8
Turnips and swedes	BFBV	35.1	35.3	38.3	13.8	16.8	BADX	132.1	141.8	104.7	96.5	103.8
Beetroot	BFBW	35.5	36.4	33.5	34.9	41.8	BADY	67.1	68.6	56.3	58.8	53.9
Onions dry bulb	BFBX	41.9	35.5	41.8	40.7	40.8	BADZ	392.7	374.9	283.4	373.6	340.9
Peas green for market (in pod weight)	BFBY	7.4	6.7	6.8	6.5	7.9	BAEA	6.7	6.2	7.2	5.9	5.9
Peas green for processing (shelled weight)	BFBZ	4.7	4.4	4.5	4.0	3.7	BAEB	184.5	161.0	169.3	167.6	130.3
Lettuce	BFCA	21.5	23.3	23.2	22.4	18.1	BAEC	135.8	123.9	109.9	125.6	140.9
Protected crops												
Tomatoes	BFCB	422.6	412.6	425.5	437.2	416.0	BAED	113.0	109.1	100.9	75.6	78.5
Cucumbers	BFCC	447.5	431.4	435.8	557.8	472.0	BAEE	79.8	71.5	73.6	77.0	61.4
Lettuce	BFBP	35.4	43.8	33.9	39.3	37.3	BAEF	18.7	20.9	16.0	16.6	10.8
Fruit												
Dessert apples	BAEG	13.1	16.1	11.9	13.4	18.0	BFCD	101.3	104.4	84.0	69.0	96.3
Cooking apples	BAEH	18.5	23.7	17.1	18.8	27.5	BFCE	107.5	107.4	95.3	74.9	78.2
Soft fruit	BAEI	..	..	..	..	..	BFCF	65.4	64.6	66.7	79.6	85.9
Pears	BAEJ	14.4	15.3	17.1	17.0	13.6	BFBQ	26.6	38.5	34.2	29.6	22.7

1 For further information refer to Section 6 of the *Annual Supplement* in the January edition of *Monthly Digest*.

2 Yield data are marketed yield and production data are home production marketed.

Source: Department for Environment, Food and Rural Affairs

6.3 Livestock[1]

Thousands

		2000 Jun	2000 Dec	2001 Jun	2001 Dec	2002 Jun	2002 Dec	2003 Jun	2003 Dec	2004 Jun	2004 Dec	2005 Jun	2005 Dec
Total cattle and calves	BFCG	11 135	10 878	10 602	10 159	10 345	10 381	10 517	10 519	10 603	10 425	10 414	10 160
Dairy cows	BFCH	2 336	2 339	2 251	2 203	2 227	2 229	2 192	2 207	2 131	2 152	2 065	2 074
Beef cows	BFCI	1 842	1 783	1 708	1 673	1 657	1 694	1 700	1 702	1 739	1 733	1 768	1 657
Heifers in calf	BFCJ	718	684	701	691	728	684	680	678	691	680	639	676
Total sheep and lambs	BFCM	42 264	27 591	36 716	24 434	35 834	24 898	35 846	24 572	35 890	24 688	35 517	23 933
Ewes and shearlings	CKUQ	20 449	18 513	17 921	16 082	17 630	16 469	17 599	16 337	17 665	16 308	16 990	15 633
Lambs under one year old	BFCP	20 857	7 769	17 769	7 219	17 310	7 233	17 335	7 078	17 275	7 067	17 532	7 146
Total pigs	BFCQ	6 482	5 948	5 845	5 845	5 588	5 330	5 047	4 842	5 161	4 787	4 864	4 724
Sows in pig and other sows for breeding	CKUU	537	497	527	482	483	446	443	444	449	413	404	376
Gilts in pig	CKUR	73	81	71	65	74	74	73	70	66	63	67	65
Total fowls	CKUS	..	..	..	..	..	..	..	..	..	..	..	..
Total table chicken	CKUT	105 689	..	112 531	..	105 137	..	116 774	..	119 912	..	111 487	..
Birds in laying flock	CKUV	28 687	..	29 895	..	28 778	..	29 274	..	29 662	..	29 550	..
Growing pullets up to point of lay	CKUW	9 461	..	9 367	..	9 784	..	8 286	..	8 156	..	10 929	..

1 Figures include estimates for minor holdings. For further details refer to section 6 of the *Annual Supplement* in the January edition of *Monthly Digest*.

Source: Department for Environment, Food and Rural Affairs

6.4 Animals slaughtered and meat produced
Monthly averages or totals for four or five week periods

	Animals slaughtered (thousands)								Meat produced (thousand tonnes)					
	Steers, heifers and young bulls	Cows and adult bulls	Calves	Ewes and rams	Other sheep and lambs	Sows and boars	Other pigs	Poultry[2]	Beef and veal	Mutton and lamb	Pork	Poultry[2]	Offal[3]	Total
	BFHA	BFHB	BFHC	BFHD	BFHE	BFHF	BAKP	JYXD	BFHK	BFHL	BFHM	JYXE	BFHN	BFHJ
2001[1]	173	–	8	144	936	15	871	72 175	54.3	21.6	49.7	130.5	20.5	276.6
2002[1]	182	–	8	160	1 098	20	761	71 828	58.3	25.3	48.3	129.7	21.2	283.8
2003[1]	182	–	7	160	1 098	20	761	73 460	58.3	25.3	48.3	130.8	21.2	283.8
2004[1,4]	191	–	9	165	1 126	20	763	73 477	61.2	26.6	49.9	130.3	21.5	289.5
2005	190	1	9	183	1 174	17	748	75 218	63.5	27.6	48.9	131.8	21.7	293.4
2004 Dec[4]	230	–	12	209	1 446	22	890	66 120	73.6	33.3	59.5	117.5	22.8	306.7
2005 Jan	187	–	10	212	1 247	21	867	91 184	61.4	30.6	58.7	163.3	24.8	338.7
Feb	182	–	9	166	941	18	696	71 905	60.1	23.1	46.1	124.7	20.2	274.1
Mar	170	–	8	163	977	18	648	67 891	56.5	24.1	42.6	116.0	19.1	258.2
Apr	221	–	6	196	1 061	21	836	83 732	73.5	27.1	55.1	143.3	23.7	322.7
May	178	–	5	156	796	16	667	68 230	59.1	20.2	43.8	121.8	19.6	264.5
Jun	174	–	7	154	979	16	691	68 815	58.1	23.1	45.3	122.5	20.0	268.9
Jul	202	–	11	196	1 324	20	837	87 521	67.5	30.6	53.7	153.7	24.7	330.3
Aug	173	–	12	175	1 216	16	686	69 150	57.5	27.8	45.4	119.6	20.4	270.6
Sep	195	–	14	190	1 303	14	707	68 578	65.0	29.6	46.9	116.6	21.0	279.2
Oct	227	–	14	234	1 610	16	881	86 954	74.9	36.1	55.9	151.6	26.1	344.7
Nov	198	9	9	182	1 374	15	744	71 484	68.0	31.0	48.2	126.6	22.3	296.2
Dec	174	10	7	170	1 264	12	711	67 173	60.3	28.1	44.7	122.3	20.9	276.3
2006 Jan	217	18	7	221	1 341	17	818	80 586	77.7	31.7	53.8	145.0	25.0	333.3
Feb	175	19	6	173	1 052	14	685	68 195	63.8	25.1	45.7	116.1	20.3	271.1
Mar	176	21	5	172	1 039	15	681	68 295	65.0	25.4	45.5	114.6	20.4	270.9
Apr	202	24	3	199	1 090	17	783	85 971	74.3	27.1	50.3	144.7	24.1	320.5
May	179	25	2	159	831	14	665	69 034	67.5	20.3	43.4	119.8	20.3	271.3
Jun	170	26	3	163	965	14	666	67 676	64.7	22.6	42.9	120.9	20.4	271.5
Jul	202	38	5	212	1 309	20	832	83 915	78.8	30.0	53.4	149.0	25.3	336.5
Aug	152	31	4	184	1 216	15	686	67 886	59.4	27.5	44.8	118.8	20.4	270.8
Sep	177	37	4	192	1 302	17	726	68 308	69.0	29.3	48.7	117.2	21.5	285.7
Oct	217	53	4	221	1 587	21	915	84 799	86.5	35.7	58.4	153.3	27.2	361.0
Nov	183	47	3	179	1 221	17	749	69 364	74.2	28.0	48.6	124.4	22.4	297.6

1 Annual averages.
2 Includes chickens, turkeys, ducks and geese.
3 Includes poultry offal.
4 2004 is a 53 week statistical year; December is a 5 week statistical month rather than the usual 4 week statistical month.

Source: Department for Environment, Food and Rural Affairs: 01904 455097

6.5 Cereals and cereal products
Monthly averages or totals for four or five week periods. Stocks refer to the end of the period

Thousand tonnes

	Wheat and flour						Oats				Barley			
	Sales of home-grown wheat for food	Wheat milled		Stocks3 (including flour as wheat)	Flour produced	Flour disposals	Sales of home-grown oats for milling	Oats milled	Products of oat-milling	Stocks	Sales of home-grown barley for food1	Disposals for food and brewing	Stocks	Breakfast cereals:2 production
		Home-produced	Imported											
	BFDA	BFDB	BFDC	BFDD	BFDE	BFDF	BFDG	BFDH	BFDI	BFDJ	BFDK	BFDL	BFDM	BFDN
2001	407	399	73	718	374	373	24	24	14	46	222	219	1 315	29
2002	397	387	81	823	368	367	26	26	16	52	251	240	1 300	29
2003	402	394	70	738	364	363	27	27	16	44	260	256	1 096	27
2004	403	398	67	689	369	369	26	27	16	44	207	211	976	29
2005	385	379	88	652	366	367	28	28	17	33	230	218	880	32
2003 Q2	413	400	71	718	367	367	25	27	16	42	104	211	822	28
Q3	379	390	76	629	367	365	25	26	15	38	454	285	1 034	28
Q4	416	395	68	783	364	364	29	28	17	51	238	248	1 366	26
2004 Q1	425	411	64	819	377	381	28	26	15	54	173	219	1 149	27
Q2	395	392	61	687	362	361	26	26	16	45	102	193	768	29
Q3	386	403	61	560	368	368	22	26	16	34	354	238	833	30
Q4	407	385	82	692	367	366	30	29	17	41	201	195	1 153	29
2005 Q1	385	375	81	736	357	360	28	28	17	40	232	264	1 002	31
Q2	385	379	90	643	368	368	30	29	18	29	109	183	724	34
Q3	384	384	93	576	374	373	27	27	16	29	350	206	914	31
Q4	386	395	75	661	370	370	31	29	17	42	197	209	1 190	30
2006 Q1	379	390	68	654	363	364	30	29	17	40	158	198	889	32
2005 Oct	410	390	75	622	364	365	31	28	17	38	225	195	1 241	33
Nov	402	384	71	664	359	359	33	31	18	43	182	229	1 181	29
Dec	418	411	79	696	387	385	28	28	16	46	183	203	1 149	27
2006 Jan	368	361	66	690	338	343	27	26	15	42	169	212	1 018	33
Feb	373	361	62	665	334	331	32	30	17	40	151	194	887	29
Mar	447	447	77	607	417	417	32	31	18	38	154	189	763	33
Apr	378	367	65	874	343	342	25	30	17	29	103	176	806	31
May	382	373	62	887	345	342	32	31	18	28	99	173	708	30
Jun	441	434	80	912	405	403	37	35	20	26	84	178	588	35
Jul	358	369	73	856	347	349	22	28	16	27	148	156	641	32
Aug	352	366	72	453	347	345	31	31	18	24	384	189	785	31
Sep	438	434	82	554	410	408	41	37	21	44	434	214	1 170	36

1 Sales of UK grown barley to brewers, maltsters and distillers.
2 Other than oatmeal and oatmeal flakes.
3 Stocks held by wheat millers, feed compounders, cereal breakfast food manufacturers, brewers, maltsters and distillers, merchants and dealers.

Source: Department for Environment, Food and Rural Affairs: 01904 455076

6.6 Production of compound feedingstuffs
Monthly averages

Thousand tonnes

	Cattle feed	Calf feed	Pig feed	Poultry feed	Other compounds	Total
	BFFB	BFFC	BFFD	BFFE	BFFF	BFFA
1996	322.8	19.9	194.8	229.5	99.2	866.8
1997	284.9	16.5	200.5	234.0	92.7	834.9
1998	277.1	14.3	208.0	226.1	95.3	826.9
1999	305.2	13.0	185.9	219.3	108.6	837.6
2000	286.0	13.0	160.9	214.5	97.3	776.4
2001	304.2	12.3	150.0	226.9	94.7	793.2
2002	288.3	11.8	137.1	243.2	89.0	774.5
2003	306.3	12.9	116.9	230.7	95.2	766.9
2004	299.4	12.9	121.9	229.0	99.3	767.5
2005	280.3	12.1	119.4	219.0	95.5	731.3
2004 Q2	254.3	10.6	118.2	230.0	80.0	697.9
Q3	264.3	11.4	123.4	236.3	62.0	702.2
Q4	330.2	15.5	127.2	224.7	89.1	791.9
2005 Q1	311.1	14.7	116.3	211.1	150.8	808.7
Q2	243.3	10.7	116.4	219.7	80.5	675.4
Q3	254.0	10.3	119.0	230.1	60.7	679.2
Q4	312.7	12.9	125.9	214.9	90.2	761.7
2006 Q1	332.0	13.1	116.8	193.9	164.1	824.5
Q2	273.3	10.6	121.8	207.7	95.2	713.4
Q3	291.3	11.0	125.4	219.2	78.0	728.1

Source: Department for Environment, Food and Rural Affairs: 01904 455076

6.7 Potatoes and sugar[1]
Monthly averages, calendar months or totals for four or five week periods

Thousand tonnes

	Potatoes				Sugar (as refined)				
	Movement into human consumption in the United Kingdom				Production from home- grown sugar beet	Disposals			Glucose: production
						Total New Supply[5]	For food in the United Kingdom	Stocks	
	From home crop	Imports[2,3]	Exports[3]	Stocks[4]					
	BFGA	BFGB	BFGC	BFGD	BFGF	BFGG	BFGH	BFGI	BFGK
2002	417	150	30	3 386	1 430.0	2 197.0	..	..	60.6
2003	400	142	31	2 915	1 368.0	1 800.0	..	..	61.2
2004	391	148	26	2 820	1 390.0	1 874.0	..	..	61.4
2005	361[†]	124	27	2 509	1 300.0	1 859.0	..	..	57.9
2006	355	..	..	..	..	..	..	..	46.7
2005 Nov	440[†]	69	32	..	293.8	177.6	176.9	523.4	51.6
Dec	385	171	40	2 509	265.8	153.0	152.3	709.4	35.6
2006 Jan	368	..	..	..	134.4	137.9	131.8	779.2	45.8
Feb	343	..	..	..	0.1	144.1	143.4	731.9	44.5
Mar	377	..	..	..	2.9	151.6	151.3	627.5	50.7
Apr	301	..	..	..	3.3	165.0	164.0	525.6	47.4
May	305	..	..	..	3.1	151.9	151.3	428.9	49.7
Jun	281	..	..	..	..	..	..	..	44.8
Jul	277	..	..	..	..	..	..	..	47.6
Aug	390	..	..	..	..	..	..	..	47.8
Sep	444	..	..	..	..	..	..	..	45.7
Oct	521	..	..	..	..	..	..	..	47.7
Nov	420	..	..	..	..	..	..	..	47.0
Dec	368	..	..	..	..	..	..	..	41.2

*Note: The annual figures for sugar are calendar year totals, rather than 12 month averages.

1 For further information refer to Section 6 of the *Annual Supplement* in the January edition of *Monthly Digest*.

2 Includes Channel Isles exports to Great Britain.

3 Trade data provided by British Potato Council and Dept. of Agriculture and Rural Development in Northern Ireland. Figures currently unavailable for 2006.

4 Estimate of end - December stocks based on Potato Marketing returns.

5 Total New supply (including imports) for use by UK food and other industries (including sugar used in the chemical industry). For further information: http://statistics.defra.gov.uk/esg/publications/auk/2005/5-7.xls

Sources: Department for Environment, Food and Rural Affairs;
01904 455067 (glucose);
020 7238 3279 (sugar);
01904 455068 (potatoes)

6.8 Production of bacon, ham and canned meat and meat stocks in cold storage[1]
Monthly averages or totals for four or five week periods   Monthly averages or end of period stocks

Thousand tonnes

| | Bacon and ham | Meat stocks in cold storage[2,3] | | | | |
	Production	Beef and veal	Mutton and lamb	Pork	Offal	Total
	BAKQ	BFIF	BFIG	BFIH	BFII	BFIE
2001	16.9	..	..	..	..	..
2002	16.4	..	..	..	..	..
2003	16.9	..	..	..	..	..
2004	17.6	..	..	..	..	..
2005	17.8	..	..	..	..	..
2005 May	..	..	..	..	..	..
Jun	18.0	28.8	7.6	8.5	1.8	46.7
Jul	..	..	..	..	..	..
Aug	..	..	..	..	..	..
Sep	17.1	26.7	8.2	9.0	2.5	46.5
Oct	..	..	..	..	..	..
Nov	..	..	..	..	..	..
Dec	19.0	25.2	7.7	9.4	2.1	44.3
2006 Jan	..	..	..	..	..	..
Feb	..	..	..	..	..	..
Mar	16.0	26.6	6.6	10.8	1.4	45.4
Apr	..	..	..	..	..	..
May	..	..	..	..	..	..
Jun	17.0	..	..	..	..	..
Jul	..	..	..	..	..	..
Aug	..	..	..	..	..	..
Sep	16.6	..	..	..	..	..

1 Owing to change in methodology, the data are now collected on a quarterly basis, and consequently, cannot be provided for the intermediate months.

2 Stocks held in cold stores for private concerns or in undischarged cargos are not included.

3 The stocks held in Public Coldstores in the United Kingdom survey was discontinued in March 2006. As a result, stocks data is no longer published.

Source: Department for Environment, Food and Rural Affairs: 01904 455096

6.9 Fish, oils and fats
Monthly averages, calendar months or totals for four or five week periods; stocks: end of period

Thousand tonnes

	Fresh and frozen fish: UK landings	Oilseeds and nuts			Vegetable oil		Marine oil		Margar-ine: produc-tion	Solid cooking fat	Other table spreads
		Crushed	Crude oil produced	Stocks[5]: crude oil equiv-alent	Crude oil equivalent		Crude oil equivalent				
					Disposals[1]	Stocks[2,5]	Usage[3]	Stocks[4,5]			
	BFJA	BFJE	BFJF	BFJG	BFJJ	BFJK	BFJL	BFJM	BFJN	BFJO	BFJP
2001[6]	..	2 251.6	785.9	17.3	2 068.0	95.9	1.9	–	124.3	120.7	284.9
2002[6]	..	2 332.3	804.9	10.4	2 065.9	88.9	2.1	–	114.4	114.4	300.8
2003[6]	..	2 210.3	768.9	13.2	2 213.7	86.5	2.1	–	135.9	130.9	305.9
2004	..	2 128.2	747.5	23.1	2 058.2	170.8	1.7	..	114.5	130.5	316.5
2005	..	..	..	..	2 043.6	166.9	1.2	..	115.9	117.0	292.7
2005 Sep	17.4	–	–	–	–	–	–	–	–	–	–
Oct	42.0	–	–	–	–	–	–	–	–	–	–
Nov	13.1	–	–	–	–	–	–	–	–	–	–
Dec	11.1	–	–	–	–	–	–	–	–	–	–
2006 Jan	50.2	..	..	..	..	..	..	..	..	..	..
Feb	22.5	..	..	..	..	..	..	..	..	..	..
Mar	17.3	..	..	..	..	..	..	..	..	..	..
Apr	20.8	..	..	..	..	..	..	..	..	..	..
May	10.3	..	..	..	..	..	..	..	..	..	..
Jun	11.3	..	..	..	..	..	..	..	..	..	..
Jul	31.6	..	..	..	..	..	..	..	..	..	..
Aug	39.6	..	..	..	..	..	..	..	..	..	..
Sep	15.7	..	..	..	..	..	..	..	..	..	..
Oct	30.9	..	..	..	..	..	..	..	..	..	..
Nov	9.6	..	..	..	..	..	..	..	..	..	..

1 This series contains revisions following the incorporation of revised trade figures.
2 Comprising stocks of crude and refined oils held by seed crushers, oil refiners and manufacturers of margarine, solid cooking fat and other table spreads.
3 For the manufacture of margarine, solid cooking fat and other table spreads only.
4 Including quantities held by hardeners and refiners of oil and manufacturers of margarine.
5 Stocks are as at the end of December.
6 Figures for 2001 - 2003 are shown in actual annual totals.

Sources: Department for Environment, Food and Rural Affairs;
020 7238 5913 (fish landings);
01904 455067 (oils and fats)

6.10 Milk, milk products and eggs[1]
Monthly averages or calendar months; stocks: end of period

	Million litres				Thousand tonnes										Supply of hen eggs for human consump-tion (million dozen)[1,2]
					Condensed and evaporated milk		Milk powder				Butter		Cheese		
							Full-cream		Skimmed						
	Liquid milk[3,6]	Milk for manufac-ture[4,6]	Other dis-posals[5,6]	Total milk dis-posals	Pro-duction	Stocks	Pro-duction	Stocks	Pro-duction	Stocks	Pro-duction	Stocks	Pro-duction	Stocks	
	BFKB	BFKC	JYXF	BFKA	BFKH	BFKI	BFKJ	BFKK	BFKL	BFKM	BFKD	BFKE	BFKF	BFKG	BFKN
2001	563	563	43	1 168	13.4	9.7	7.3	6.2	5.9	12.4	10.5	18.4	32.3	15.0	70.20
2002	569	574	42	1 184	14.5	9.0	8.7	5.2	7.2	28.4	11.3	19.4	30.3	12.4	70.17
2003	563	595	44	1 202	13.2	7.2	8.5	3.4	9.6	50.8	10.9	17.5	28.7	7.1	71.74
2004	558	560	45	1 164	13.4	7.7	6.7	2.0	7.3	20.7	10.1	9.2	29.3	11.1	75.54
2005	555	541	59	1 155	11.9	4.8	4.5	1.9	5.8	10.1	10.8	2.9	32.0	3.3	73.94
2005 Jul	549	561	73	1 183	10.1	4.3	5.8	2.7	4.9	11.5	9.8	5.1	34.2	8.0	73.36
Aug	541	558	57	1 156	10.6	3.6	4.4	3.0	5.1	10.9	10.9	5.1	35.0	8.0	73.36
Sep	544	490	50	1 084	10.6	3.1	2.3	1.9	4.6	10.8	10.7	3.7	31.6	7.2	73.36
Oct	579	453	61	1 094	11.2	3.0	2.2	1.8	3.4	10.2	11.3	3.7	29.2	7.2	76.21
Nov	556	438	52	1 047	11.9	2.6	1.9	1.3	2.6	10.3	9.7	3.7	28.2	7.2	76.21
Dec	580	472	64	1 117	13.0	4.8	4.3	1.9	4.7	10.1	9.6	2.9	28.7	3.3	76.21
2006 Jan	570	513	70	1 153	..	3.2	..	1.5	..	9.2	10.9	2.9	32.7	3.3	75.30
Feb	528	481	53	1 062	..	3.4	..	2.3	..	3.7	9.2	2.8	30.0	3.3	75.30
Mar	589	567	40	1 195	..	1.4	..	2.2	..	2.5	10.9	3.4	34.9	2.6	75.30
Apr	555	579	75	1 208	..	2.4	..	3.0	..	3.1	9.6	4.6	35.3	2.6	72.39
May	585	650	62	1 297	..	1.0	..	2.6	..	4.3	10.4	6.6	37.9	2.6	72.39
Jun	556	582	59	1 198	..	0.9	..	2.8	..	6.0	9.7	6.9	34.8	2.6	72.39
Jul	561	548	65	1 174	..	1.3	..	5.2	..	5.3	9.1	6.9	34.6	2.6	73.44
Aug	556	511	55	1 123	..	1.2	..	4.6	..	4.4	9.5	6.8	33.5	2.6	73.44
Sep	544	472	61	1 077	..	1.4	..	2.4	..	2.2	8.8	6.7	32.4	2.6	73.44
Oct	565	456	67	1 089	..	1.4	..	0.9	..	2.4	9.2	6.7	30.8	2.6	76.29

1 Includes first and second quality eggs broken out.
2 This series has been revised as a result in survey methodology - see Explanatory Notes in the January edition of Monthly Digest.
3 Includes wholesale and direct sellers utilisation of milk for liquid milk.
4 Includes wholesale and direct sellers utilisation of milk for the manufacture of milk products.
5 Includes dairy wastage, stock changes and exports of raw milk.
6 Suckled milk, milk used on farm for farmhouse consumption, milk fed to livestock and farm waste are excluded. Utilisation of imported raw milk is included.

Source: Department for Environment, Food and Rural Affairs: 01904 455095

6.11 Beverages and confectionery
Monthly averages, calendar months or totals for four or five week periods; stocks: end of period

	Thousand tonnes					
	Chocolate and sugar confectionery		Tea		Raw coffee	
	Production	Disposals	Disposals[1]	Stocks	Disposals	Stocks
	BFLG	BFLH	BFLJ	BFLK	BFLL	BFLM
1999	72.68	88.45	11.1	38.2	9.6	6.8
2000	70.91	88.78	12.0	27.6	9.1	7.9
2001	69.60	88.82	11.0	31.2	8.7	12.5
2002	66.47	87.88	11.3	29.3	9.6	8.5
2003	65.02	88.79	10.0	24.3	9.1	8.9
2004	61.66	87.69	13.0	18.9	10.7	11.1
2005	57.08	84.87	9.0	19.0	8.7	9.2
2004 Oct	81.78	112.87	..	..	..	..
Nov	77.51	104.91	12.30	19.0	9.40	11.10
Dec	68.04	93.93	..	..	..	..
2005 Jan	44.68	68.45	..	..	..	..
Feb	56.52	78.47	8.40	18.4	7.80	6.70
Mar	71.54	93.41	..	..	..	..
Apr	51.84	75.14	..	..	..	..
May	56.59	80.40	10.40	19.7	12.00	3.60
Jun	55.84	82.25	..	..	..	..
Jul	47.13	71.51	..	..	..	..
Aug	50.54	81.59	7.30	19.0	7.70	3.30
Sep	67.76	105.37	..	..	..	..
Oct	72.47	107.66	..	..	..	..
Nov	64.37	99.10	6.60	19.0	9.90	3.20
Dec	45.51	75.45	..	..	..	..
2006 Jan	33.27	57.52	..	..	..	..
Feb	40.29	66.71	7.03	19.1	10.48	3.57
Mar	54.03	80.92	..	..	..	..

1 Excluding exports.

Source: Department for Environment, Food and Rural Affairs: 020 7270 8560

6.12 Tobacco products released for home consumption
Monthly averages or calendar months

	Million			Thousand kilogrammes			
	Cigarettes			Other tobacco products			Total tobacco products other than cigarettes
	Home-produced	Imported	Total	Cigars	Hand-rolling	Other[1]	
	LUQN	LUQO	LUQP	LUQQ	LUQR	LUQS	LUQT
2001	47 689	6 828	54 517	1 019	2 825	750	4 595
2002	49 574	6 514	56 088	969	2 864	688	4 522
2003	49 096	4 856	53 952	902	2 893	589	4 384
2004	48 166	4 454	52 620	826	3 052	549	4 428
2005	45 922	4 322	50 244	758	3 189	499	4 445
2005 Dec	4 590	379	4 969	67	285	44	396
2006 Jan	2 809	307	3 116	42	228	29	299
Feb	3 237	339	3 575	56	260	34	350
Mar	8 980	1 076	10 056	133	692	77	901
Apr	296	10	306	5	46	10	60
May	4 847	357	5 204	40	200	34	274
Jun	1 681	173	1 854	57	280	35	372
Jul	3 115	349	3 464	47	255	31	333
Aug	3 929	412	4 341	66	305	38	410
Sep	3 916	361	4 277	55	318	33	406
Oct	3 883	383	4 266	55	269	38	362
Nov	2 950	356	3 306	66	272	37	374

1 Excluding snuff.

Source: HM Revenue and Customs: 020 7147 0593

6.13 Alcoholic drink

			Thousand hectolitres							Thousand hectolitres of alcohol			Production of potable spirits[1]	
			Released for home consumption							Released for home consumption				
			Wine of fresh grapes				Made wine			Spirits				
			Still											
	Beer production[7]	Beer[7]	Not exceeding 15%[2]	Over 15%	Sparkling	Total	Coolers[3]	Other	Cider and perry	Home-produced whisky	Spirit-based coolers[4]	Other[5]	Home-produced whisky	Other
	BFNK	BAYL	BFNO	BFNP	BFNS	BFNT	BAYM	BAYN	BFNW	BFNX	YZUJ	BFNY	BAYO	BAYP
2001	56 802	58 234	9 533.7	287.0	515.0	10 335.6	3 712.3	363.7	5 910.9	320.8	..	647.2	3 691.6	676.2
2002	56 672	59 384	10 318.7	325.3	577.9	11 221.9	1 606.2	367.4	5 939.2	320.7	105.2	688.9	3 905.7	602.5
2003	58 014	60 301	10 646.9	296.4	640.1	11 583.5	423.2	339.2	5 876.1	318.3	124.4	744.3	3 936.6	616.0
2004	57 459	59 195	11 768.2	297.8	675.6	12 741.5	508.0	351.4	6 138.8	319.3	114.3	792.4	3 529.4	551.5
2005	56 255	57 572	12 117.1	305.7	720.6	13 143.4	597.5	333.9	6 376.9	300.7	84.2	821.7	3 758.6	606.0
2003 May	4 764	4 901	843.0	18.5	38.3	899.9	25.7	22.5	460.4	24.9	10.7	56.1	..	..
Jun	5 558	5 260	846.7	14.7	41.8	903.2	37.3	23.2	556.6	24.2	11.4	57.3	1 111.1	165.9
Jul	5 449	5 633	908.5	15.2	45.9	969.5	41.1	27.6	534.8	22.5	11.6	57.0	..	..
Aug	5 170	5 354	887.4	16.1	52.4	955.9	41.6	24.1	583.6	23.2	12.5	61.0	..	..
Sep	4 712	5 124	795.0	16.3	37.4	848.7	25.1	27.6	491.8	22.0	9.6	51.9	884.2	146.8
Oct	5 225	5 299	1 027.4	36.4	60.7	1 124.5	36.1	36.9	485.4	33.5	12.8	76.4	..	..
Nov	4 848	5 509	1 273.1	66.7	124.1	1 463.9	52.0	49.0	486.8	50.8	13.5	106.1	..	..
Dec	5 324	5 575	859.5	46.4	80.1	986.0	43.5	28.6	597.6	34.7	11.2	79.6	994.2	156.2
2004 Jan[6]	3 153	3 322	832.8	20.6	53.9	907.3	36.1	24.2	339.2	16.9	9.2	54.3	..	..
Feb	3 815	3 955	764.0	12.8	31.9	808.7	25.1	21.5	427.8	19.5	6.0	46.4	..	..
Mar	5 168	5 446	933.0	15.7	40.8	989.5	33.7	33.9	485.9	24.0	8.6	60.4	896.6	131.9
Apr	4 677	4 612	896.2	15.5	34.7	946.4	19.4	19.1	519.8	21.6	7.9	53.2	..	..
May	5 196	5 264	967.9	16.1	39.3	1 023.3	31.3	24.4	553.0	25.2	11.8	62.9	..	..
Jun	5 391	5 513	968.5	16.9	49.1	1 034.6	35.6	25.1	551.3	23.3	9.9	63.9	976.1	129.8
Jul	4 700	4 782	1 023.1	18.2	54.6	1 095.8	35.5	36.7	541.9	23.3	10.4	57.9	..	..
Aug	5 416	5 171	982.7	15.6	46.1	1 044.4	44.7	26.1	585.4	23.3	8.4	60.2	..	..
Sep	4 696	5 106	985.5	20.3	44.9	1 050.7	50.6	28.5	460.4	22.1	8.8	64.4	844.0	120.4
Oct	4 517	4 836	1 132.8	32.8	64.5	1 230.1	50.8	38.9	525.6	34.1	8.9	76.3	..	..
Nov	5 336	5 042	1 233.5	61.2	107.6	1 402.3	62.4	42.0	579.1	48.3	12.9	101.4	..	..
Dec	5 394	6 146	1 047.9	52.3	108.3	1 208.5	82.7	31.1	569.4	37.6	11.4	91.0	812.6	169.4
2005 Jan[6]	3 072	3 094	791.4	17.0	56.8	865.2	38.1	20.7	332.7	15.6	4.6	51.9	..	..
Feb	3 918	3 924	770.1	14.2	37.8	822.1	30.4	16.4	374.7	16.8	4.1	44.8	..	..
Mar	5 119	5 355	1 116.7	17.8	43.4	1 177.9	53.1	32.2	532.8	22.8	7.5	64.3	854.2	134.1
Apr	4 336	4 341	968.4	14.9	39.0	1 022.2	49.3	15.9	522.4	22.4	6.5	61.1	..	..
May	5 072	5 190	911.0	14.7	38.4	964.2	38.4	28.2	507.3	21.7	7.2	61.1	..	..
Jun	5 121	5 368	1 040.5	17.4	49.7	1 107.7	52.5	25.9	609.9	23.6	8.3	68.5	1 057.3	168.0
Jul	5 012	5 054	1 077.2	14.8	55.4	1 147.4	53.6	25.8	578.1	22.6	8.0	60.1	..	..
Aug	4 987	5 271	964.7	15.3	43.1	1 023.0	45.9	22.3	608.5	18.3	7.6	57.0	..	..
Sep	4 612	4 603	970.3	19.2	57.6	1 047.1	47.5	30.0	577.2	24.0	5.0	62.3	863.4	133.8
Oct	4 742	4 827	1 064.5	39.3	71.2	1 175.0	51.1	32.9	518.2	32.1	7.2	74.8	..	..
Nov	5 284	5 368	1 316.5	60.6	114.3	1 491.4	64.1	51.2	623.8	41.9	9.1	104.1	..	..
Dec	4 980	5 177	1 125.8	60.6	113.9	1 300.3	73.5	32.4	591.5	38.9	9.1	111.7	983.8	170.2
2006 Jan[6]	3 564	3 157	693.6	11.5	46.9	751.9	24.8	19.7	359.4	13.0	3.3	32.2	..	..
Feb	3 105	3 747	843.4	14.0	41.2	898.6	31.6	16.6	437.8	16.0	3.7	42.9	..	..
Mar	5 218	5 492	1 095.5	18.9	41.4	1 155.8	38.1	31.9	527.8	21.6	5.0	59.4	1 019.7	140.3
Apr	4 029	4 005	955.0	16.9	40.5	1 012.5	42.8	19.2	560.1	21.8	5.6	59.8	..	..
May	4 784	5 394	891.6	15.9	46.5	954.0	35.2	22.2	647.9	20.1	5.1	55.0	..	..
Jun	5 295	5 362	961.7	17.8	54.9	1 034.4	49.8	25.6	695.7	24.2	8.2	75.0	1 222.8	61.4
Jul	4 851	4 557	980.1	13.8	57.4	1 051.4	50.1	24.2	753.5	16.8	4.5	50.5	..	..
Aug	4 570	4 900	999.3	17.0	49.7	1 066.0	49.6	25.2	849.8	21.0	7.0	64.4	..	..
Sep	4 204	4 331	969.1	21.6	53.8	1 044.5	39.8	27.7	690.1	20.9	4.5	61.9	909.4	110.4
Oct	4 667[†]	4 740[†]	974.9	34.8	65.7	1 075.5	47.6	38.1	682.5	31.6	5.7	72.8	..	..
Nov	4 055	4 346	1 242.6	67.5	106.9	1 417.0	53.7	42.8	695.3	37.9	4.9	100.5	..	..

1 Data are available only quarterly.

2 Percentage alcohol by volume.

3 Made wine with alcoholic strength 1.2% to 5.5%, includes alcoholic lemonade of appropriate strength.

4 From 28 April 2002 duty on spirit-based ready-to-drink (RTDs) products is charged at the same rate as spirits per litre of alcohol. Until September 2002, RTDs were recorded under the imported and spirits. Customs and Excise have now been able to estimate the amount of RTDs under the spirits and remove them from the spirits clearances. Spirit-based RTDs were previously dutied at the made wine rate.

5 Includes imported spirits.

6 Due to the effect of the holiday period, these figures are subject to greater uncertainty than usual.

7 HMRC revised the beer production and clearances data back to April 2005 with the latest available information.

Source: HM Revenue and Customs: 020 7147 0593

7 Production, output and costs

7.1 Output of the production industries

Average 2003 = 100

	Total production industries	Mining and quarrying	Total manufact- uring industries	Food, drink and tobacco	Textiles, leather and clothing	Coke ref petrol and nuclear fuels	Chemicals and man-made fibres	Basic metals and metal products	Engineering and allied industries	Other manufact- uring	Electricity, gas and water
SIC 2003 Sub-section	Sect C+D+E	Sect C	Sect D	DA	DB_DC	DF	DG	DJ	DK_DM	DD_DN	Sect E
Weights	1000	118	792	118	26	13	87	81	237	229	90
	CKYW	CKYX	CKYY	CKZA	AGVO	CKZF	CKZG	CKZJ	AGXS	AGXQ	CKYZ
2001	102.3	105.0	102.5	98.0	110.1	106.9	99.3	101.4	107.1	100.4	98.0
2002	100.3	105.4	99.8	100.0	101.8	108.3	99.1	102.4	98.5	99.8	98.4
2003	100.0	100.0	100.0	100.0	100.0	100.0	100.0	100.0	100.0	100.0	100.0
2004	100.8	92.1	102.0	101.6	90.1	105.8	103.4	103.1	104.3	100.1	101.1
2005	98.9	84.1	100.9	102.3	87.9	109.7	104.5	103.2	103.2	96.6	100.8
2006	98.8	76.4	102.3	102.5	84.5	105.4	107.2	104.8	107.1	96.1	98.1
Seasonally adjusted											
2002 Q2	100.5	109.6	99.4	100.2	104.1	105.0	99.0	101.1	98.0	99.3	97.6
Q3	100.2	101.0	100.3	100.7	101.0	103.5	100.0	102.6	99.3	100.2	99.2
Q4	100.2	105.7	99.4	99.3	96.5	108.4	98.0	102.3	98.8	99.5	99.7
2003 Q1	99.9	105.0	99.3	100.1	99.4	104.4	98.3	100.0	98.9	99.3	98.1
Q2	99.4	99.8	99.4	99.4	99.5	100.7	99.4	99.5	99.4	99.1	98.9
Q3	100.0	98.9	100.0	100.2	101.6	97.9	99.8	99.7	99.6	100.5	100.6
Q4	100.8	96.3	101.3	100.4	99.5	97.0	102.5	100.7	102.1	101.1	102.3
2004 Q1	100.9	94.3	101.7	100.4	93.1	108.5	104.7	101.0	102.7	101.1	102.2
Q2	101.3	94.8	102.4	102.6	90.2	105.2	103.8	103.6	104.4	100.4	100.7
Q3	100.3	90.9	101.6	101.3	89.3	103.4	102.1	104.1	104.6	98.9	101.0
Q4	100.6	88.6	102.4	102.0	88.0	106.1	102.9	103.7	105.6	99.9	100.6
2005 Q1	99.7	86.9	101.6	102.2	86.2	122.4	104.1	103.4	102.9	98.9	100.0
Q2	99.4	87.5	100.9	103.0	89.2	110.0	103.5	103.4	103.1	96.4	101.8
Q3	98.6	80.8	100.9	102.2	88.0	106.0	105.0	102.8	104.1	96.0	101.1
Q4	98.1	81.3	100.3	102.0	88.2	100.4	105.5	103.0	102.8	95.3	100.4
2006 Q1	98.7	80.2†	101.2†	102.1†	85.3†	105.0†	105.1†	103.6†	105.1†	95.9†	100.8†
Q2	98.8	77.2	102.1	102.1	84.5	106.4	107.1	104.6	106.9	96.1	97.9
Q3	99.0	74.4	102.8	102.1	84.2	105.1	109.6	105.3	108.0	96.4	97.7
Q4	98.9	74.0	102.9	103.7	84.0	105.3	107.0	105.8	108.5	96.1	96.0
2005 Jan	100.3	86.6	102.4	102.4	86.1	118.0	105.5	103.3	104.0	100.1	99.5
Feb	100.3	86.5	102.3	103.1	86.5	123.4	104.1	104.7	103.4	99.9	100.3
Mar	98.6	87.6	100.0	101.0	85.9	125.7	102.5	102.0	101.4	96.7	100.2
Apr	99.5	87.4	101.0	101.4	88.5	117.1	103.0	104.3	103.1	97.1	102.0
May	99.3	88.4	100.7	103.3	88.6	108.0	102.5	104.1	103.1	96.1	101.3
Jun	99.4	86.7	101.0	104.2	90.5	105.0	105.1	101.8	103.3	96.2	102.2
Jul	99.2	83.4	101.3	103.5	88.9	106.4	104.7	103.2	104.2	96.4	101.6
Aug	98.1	76.5	101.0	101.8	87.6	105.2	106.0	102.6	104.4	95.7	101.2
Sep	98.4	82.5	100.5	101.4	87.5	106.4	104.2	102.7	103.7	95.8	100.6
Oct	97.6	82.1	99.9	101.4	87.0	101.9	105.1	103.1	102.3	94.8	97.9
Nov	98.1	80.7	100.3	101.9	87.8	99.1	104.4	102.6	103.1	95.5	102.2
Dec	98.4	80.9	100.7	102.5	89.8	100.3	107.1	103.3	102.9	95.6	101.2
2006 Jan	98.6†	81.7†	100.9	102.0†	86.0†	105.8†	105.6†	103.3†	103.9†	95.9†	100.4†
Feb	98.3	79.9	100.9	102.1	85.2	106.7	104.8	103.0	104.9	95.5	99.6
Mar	99.1	78.8	101.8†	102.1	84.8	102.5	105.0	104.5	106.6	96.3	102.4
Apr	98.5	78.5	101.6	102.1	83.8	104.1	105.8	104.1	106.4	95.5	98.2
May	98.9	77.8	102.3	101.7	84.3	107.7	108.1	103.8	106.9	96.7	97.5
Jun	98.9	75.3	102.5	102.6	85.3	107.3	107.3	105.9	107.5	96.0	98.0
Jul	98.9	74.8	102.6	101.9	83.8	104.6	109.2	105.2	107.7	96.2	98.1
Aug	98.9	72.8	102.9	101.7	84.6	106.1	109.4	105.7	107.8	96.9	97.6
Sep	99.3	75.7	103.0	102.7	84.4	104.7	110.3	104.8	108.5	96.0	97.4
Oct	98.7	75.0	102.7	103.5	83.3	100.7	107.0	105.6	108.3	96.1	94.2
Nov	99.0	74.8	102.9	103.2	84.1	106.2	106.7	106.3	108.6	96.1	96.8
Dec	98.9	72.1	103.1	104.4	84.5	109.1	107.4	105.6	108.5	96.0	97.1

47

7.1 Output of the production industries
continued

					Summary - Not seasonally adjusted						
						Manufacturing industries					
	Total production industries	Mining and quarrying	Total manufact-uring industries	Food, drink and tobacco	Textiles, leather and clothing	Coke ref petrol and nuclear fuels	Chemicals and man-made fibres	Basic metal and metal products	Engineering and allied industries	Other manufact-uring	Electricity, gas and water
SIC 2003 Sub-section	Sect C+D+E	Sect C	Sect D	DA	DB_DC	DF	DG	DJ	DK_DM	DD_DN	Sect E
Weights	1000	118	792	118	26	13	87	81	237	229	90
	AGVZ	AGVT	AGVV	AGUV	AGWR	AGUX	AGUZ	AGVF	AGXT	AGXR	AGVX
2001	102.3	105.0	102.5	98.0	110.1	106.9	99.3	101.4	107.1	100.4	98.0
2002	100.3	105.4	99.8	100.0	101.8	108.3	99.1	102.4	98.5	99.8	98.4
2003	100.0	100.0	100.0	100.0	100.0	100.0	100.0	100.0	100.0	100.0	100.0
2004	100.8	92.1	102.0	101.6	90.1	105.8	103.4	103.1	104.3	100.1	101.1
2005	99.0	84.2	101.0	102.3	88.0	109.6	104.5	103.1	103.4	96.6	101.1
2006	99.1	77.5	102.4	102.0	84.7	105.1	107.8	104.9	107.6	96.1	98.2
Not seasonally adjusted [3]											
2002 Q2	99.3	106.0	99.3	98.3	103.6	103.0	100.2	101.8	97.8	99.2	90.3
Q3	97.3	94.2	99.4	100.2	100.9	105.1	100.3	102.1	96.0	100.8	84.5
Q4	103.2	111.7	101.7	107.7	98.3	109.1	97.0	101.0	101.6	101.2	107.1
2003 Q1	100.8	109.6	98.2	94.1	98.6	104.1	98.9	101.7	98.9	97.8	113.3
Q2	98.0	97.2	99.0	97.5	98.7	98.3	100.0	100.2	99.2	98.9	91.1
Q3	97.6	91.7	99.8	100.3	101.4	98.9	100.5	100.0	97.4	101.7	85.6
Q4	103.5	101.5	103.0	108.1	101.3	98.7	100.5	98.1	104.5	101.7	110.0
2004 Q1	102.8	98.7	101.7	96.1	92.1	108.3	106.8	103.4	103.8	100.4	118.2
Q2	99.7	93.2	101.5	99.7	91.3	102.8	103.7	103.6	103.3	99.9	92.6
Q3	97.6	84.0	101.0	101.3	89.0	104.8	101.9	104.1	102.0	99.7	85.3
Q4	103.0	92.7	103.9	109.2	88.2	107.3	101.2	101.3	108.2	100.2	108.4
2005 Q1	100.3	91.0	99.9	97.2	83.8	122.5	103.5	104.7	101.6	96.9	116.0
Q2	98.8	86.9	101.2	100.4	89.6	107.8	105.0	104.2	104.2	96.9	93.7
Q3	96.7	74.4	101.1	102.7	89.4	107.7	106.3	102.8	102.4	97.4	86.4
Q4	100.2	84.3	101.7	108.8	89.1	100.2	103.4	100.7	105.4	95.3	108.3
2006 Q1	100.9	85.1†	101.4†	97.1†	84.5	103.7†	107.7†	106.8†	106.8†	95.6†	116.4
Q2	98.2†	77.9	102.1	100.4	84.5†	104.7	108.8	105.0	107.3	96.0	89.9
Q3	96.4	69.3	101.9	101.3	85.1	106.4	110.0	104.2	104.7	97.0	83.5†
Q4	100.9	77.6	104.2	109.4	84.7	105.7	104.8	103.8	111.6	95.7	103.1
2005 Jan	94.9	92.3	92.4	89.9	76.0	105.1	99.5	97.3	92.5	90.4	119.4
Feb	97.8	85.1	97.7	93.3	81.6	129.4	100.3	103.6	98.8	95.7	115.9
Mar	108.1	95.5	109.5	108.5	93.9	133.0	110.6	113.3	113.5	104.4	112.7
Apr	97.8	90.8	98.4	97.5	85.6	102.9	102.7	102.7	99.7	95.5	101.8
May	97.8	87.9	99.8	99.6	89.7	103.5	103.8	104.4	101.9	95.6	92.3
Jun	100.9	81.8	105.3	103.9	93.5	117.1	108.6	105.6	111.1	99.5	87.1
Jul	94.4	78.9	97.8	99.3	86.7	112.1	101.8	100.9	98.5	94.1	85.4
Aug	93.2	67.8	97.8	101.8	87.9	104.5	106.2	100.9	95.2	94.9	85.8
Sep	102.4	76.6	107.8	106.8	93.7	106.6	111.0	106.5	113.6	103.3	87.9
Oct	99.4	83.6	102.3	108.0	90.9	95.6	105.4	104.7	101.5	99.7	95.4
Nov	104.1	82.4	106.4	107.9	96.0	100.3	107.2	107.8	109.7	103.0	111.8
Dec	97.1	86.9	96.3	110.6	80.3	104.8	97.6	89.7	105.0	83.1	117.7
2006 Jan	95.7†	88.5	94.0†	90.5†	78.7	101.2†	104.6†	101.2	96.1†	88.5†	120.0
Feb	96.5	79.5	97.0	92.1	81.1	107.2	101.8	103.0	101.1	92.4	114.7
Mar	110.4	87.2	113.3	108.8	93.8	102.5	116.7	116.3†	123.3	105.9	114.6
Apr	93.6	82.2	94.9	95.2	78.1†	95.5	100.2	99.9	97.6	90.0	97.6
May	99.7	79.1†	104.0	102.4	87.8	102.9	113.4	106.5	107.5	98.7	88.8
Jun	101.2	72.3	107.5	103.5	87.5	115.6	112.8	108.5	116.8	99.3	83.2†
Jul	94.3	71.4	99.0	98.5	81.8	108.3	106.6	102.4	102.3	93.1	83.0
Aug	94.4	65.2	100.1	102.5	85.2	106.9	110.0	104.3	99.3	95.8	82.5
Sep	100.5	71.4	106.5	103.0	88.2	103.9	113.4	105.9	112.5	102.0	85.1
Oct	102.3	78.2	107.1	109.9	89.4	99.7	111.2	111.2	109.4	102.8	90.9
Nov	105.4	78.2	109.4	109.4	92.1	100.8	108.3	111.6	117.5	103.0	105.6
Dec	95.2	76.5	96.0	108.8	72.7	116.8	94.9	88.5	108.0	81.4	112.8

7.1 Output of the production industries
continued

	Mining and quarrying				Textiles, leather and clothing		Coke ref petrol and nuclear fuels	Chemicals and man-made fibres
	Oil and gas	Coal	Other mining and quarrying	Food, drink and tobacco	Textiles and textile products	Leather and leather products		
SIC 2003 Sub-section	C_1	C_11	CB	DA	DB	DC	DF	DG
Weights	107	3	8	118	24	3	13	87
	CKZO	CKZP	CKZQ	CKZA	CKZB	CKZC	CKZF	CKZG
2001	107.3	112.6	80.4	98.0	107.2	140.0	106.9	99.3
2002	105.9	105.9	98.7	100.0	99.7	122.5	108.3	99.1
2003	100.0	100.0	100.0	100.0	100.0	100.0	100.0	100.0
2004	91.6	85.8	101.4	101.6	91.8	74.6	105.8	103.4
2005	82.5	67.9	109.9	102.3	90.0	68.2	109.7	104.5
2006	74.1	64.1	109.8	102.5	86.1	69.5	105.4	107.2
Seasonally adjusted [1]								
2002 Q2	110.8	105.5	97.4	100.2	102.0	125.8	105.0	99.0
Q3	101.1	101.3	97.9	100.7	99.0	120.5	103.5	100.0
Q4	106.4	98.9	98.0	99.3	94.9	112.1	108.4	98.0
2003 Q1	105.1	106.6	101.7	100.1	97.8	115.3	104.4	98.3
Q2	99.5	104.8	101.6	99.4	99.2	102.5	100.7	99.4
Q3	99.1	89.9	100.2	100.2	102.6	91.5	97.9	99.8
Q4	96.3	98.7	96.5	100.4	100.4	90.7	97.0	102.5
2004 Q1	94.4	85.4	95.8	100.4	94.8	76.8	108.5	104.7
Q2	94.5	84.5	101.1	102.6	91.3	79.8	105.2	103.8
Q3	90.2	89.4	100.7	101.3	91.0	72.9	103.4	102.1
Q4	87.2	83.7	108.0	102.0	90.1	68.8	106.1	102.9
2005 Q1	85.6	66.3	110.5	102.2	88.1	68.2	122.4	104.1
Q2	86.3	63.9	109.9	103.0	91.2	70.1	110.0	103.5
Q3	78.9	70.5	108.0	102.2	90.1	68.3	106.0	105.0
Q4	79.2	70.9	111.1	102.0	90.5	66.0	100.4	105.5
2006 Q1	78.1[†]	77.2[†]	107.6[†]	102.1[†]	86.9[†]	70.9[†]	105.0[†]	105.1[†]
Q2	74.6	70.9	112.6	102.1	86.5	65.6	106.4	107.1
Q3	72.2	53.6	109.8	102.1	85.7	70.4	105.1	109.6
Q4	71.7	54.8	109.3	103.7	85.3	71.1	105.3	107.0
2005 Jan	85.5	61.1	109.0	102.4	88.0	68.3	118.0	105.5
Feb	85.0	70.0	110.1	103.1	88.5	68.3	123.4	104.1
Mar	86.2	67.9	112.3	101.0	87.8	68.0	125.7	102.5
Apr	86.1	63.8	110.7	101.4	90.3	71.6	117.1	103.0
May	87.4	63.5	109.5	103.3	90.6	69.6	108.0	102.5
Jun	85.4	64.4	109.4	104.2	92.8	69.2	105.0	105.1
Jul	82.1	57.7	107.7	103.5	91.0	69.8	106.4	104.7
Aug	74.2	71.0	108.1	101.8	89.8	67.5	105.2	106.0
Sep	80.5	82.6	108.2	101.4	89.6	67.7	106.4	104.2
Oct	80.3	71.6	109.2	101.4	89.5	63.1	101.9	105.1
Nov	78.5	70.4	112.2	101.9	90.2	65.2	99.1	104.4
Dec	78.8	70.8	111.8	102.5	91.9	69.9	100.3	107.1
2006 Jan	79.6[†]	86.3[†]	107.5[†]	102.0[†]	87.7[†]	70.6[†]	105.8[†]	105.6[†]
Feb	78.0	75.5	106.7	102.1	86.7	71.0	106.7	104.8
Mar	76.7	69.6	108.6	102.1	86.2	71.2	102.5	105.0
Apr	76.1	73.3	110.9	102.1	85.9	64.4	104.1	105.8
May	75.1	72.7	113.6	101.7	86.2	65.9	107.7	108.1
Jun	72.5	66.7	113.2	102.6	87.3	66.3	107.3	107.3
Jul	72.5	56.6	110.2	101.9	85.8	64.5	104.6	109.2
Aug	70.4	48.4	110.2	101.7	85.9	72.1	106.1	109.4
Sep	73.6	55.7	108.9	102.7	85.5	74.5	104.7	110.3
Oct	72.9	56.4	108.7	103.5	84.7	70.7	100.7	107.0
Nov	72.6	56.7	109.0	103.2	85.4	71.7	106.2	106.7
Dec	69.6	51.3	110.2	104.4	85.9	70.9	109.1	107.4

7.1 Output of the production industries

continued

Average 2003 = 100

| | | Detailed analysis (continued) | | | | | | | | |
| | Engineering and allied industries | | | | Other manufacturing | | | | | |
	Basic metal and metal products	Machinery and equipment	Electrical and optical equipment	Transport equipment	Wood and wood products	Pulp, paper, printing and publishing	Rubber and plastic products	Non-metallic mineral products	Other manufacturing NES	Electricity, gas and water
SIC 2003 Sub-section	DJ	DK	DL	DM	DD	DE	DH	DI	DN	Sect E
Weights	*81*	*66*	*85*	*86*	*15*	*108*	*41*	*30*	*35*	*90*
	CKZJ	CKZK	CKZL	CKZM	CKZD	CKZE	CKZH	CKZI	CKZN	CKYZ
2001	101.4	104.2	118.7	97.9	96.6	101.3	103.2	96.0	99.9	98.0
2002	102.4	98.3	102.6	94.8	99.2	101.4	99.2	94.5	100.5	98.4
2003	100.0	100.0	100.0	100.0	100.0	100.0	100.0	100.0	100.0	100.0
2004	103.1	105.8	101.8	105.8	101.8	99.1	98.5	105.8	99.3	101.1
2005	103.2	109.1	97.2	104.7	97.3	93.8	95.0	105.0	100.0	100.8
2006	104.8	114.0	97.6	111.3	94.7	91.4	97.6	107.3	99.7	98.1
Seasonally adjusted [1]										
2002 Q2	101.1	98.8	102.7	92.9	98.1	100.9	99.3	93.3	100.1	97.6
Q3	102.6	99.1	102.4	96.5	100.7	101.8	99.9	95.2	99.6	99.2
Q4	102.3	96.1	102.9	96.8	101.1	101.1	97.9	94.6	99.9	99.7
2003 Q1	100.0	96.6	100.5	99.0	97.6	100.0	97.8	98.3	100.2	98.1
Q2	99.5	100.1	99.5	98.8	97.1	99.6	98.2	98.4	99.8	98.9
Q3	99.7	100.8	99.8	98.6	101.2	99.9	101.5	101.0	100.7	100.6
Q4	100.7	102.5	100.2	103.5	104.1	100.5	102.5	102.3	99.2	102.3
2004 Q1	101.0	100.4	100.0	107.1	100.1	100.8	101.1	104.5	99.2	102.2
Q2	103.6	107.8	102.6	103.7	104.3	99.1	99.0	106.3	99.2	100.7
Q3	104.1	107.2	102.5	104.7	101.8	98.0	96.3	105.8	97.8	101.0
Q4	103.7	107.7	101.9	107.7	101.2	98.5	97.6	106.6	101.0	100.6
2005 Q1	103.4	109.5	97.3	103.4	100.7	96.1	96.5	108.3	101.3	100.0
Q2	103.4	108.6	97.1	105.0	95.8	93.9	94.6	103.8	100.4	101.8
Q3	102.8	108.6	98.3	106.4	97.8	93.4	94.6	102.7	99.0	101.1
Q4	103.0	109.5	96.1	104.1	95.1	91.8	94.2	105.0	99.1	100.4
2006 Q1	103.6[†]	110.9[†]	96.8[†]	108.8[†]	93.4[†]	92.4[†]	96.5[†]	104.8[†]	99.2[†]	100.8[†]
Q2	104.6	113.8	98.2	110.3	95.7	91.0	98.0	106.7	100.4	97.9
Q3	105.3	116.1	97.5	112.3	94.0	91.4	98.1	109.4	99.7	97.7
Q4	105.8	115.2	97.7	113.9	95.9	90.9	97.8	108.5	99.4	96.0
2005 Jan	103.3	109.7	98.1	105.5	103.8	97.8	96.2	110.5	101.2	99.5
Feb	104.7	110.7	98.0	103.0	102.3	96.8	98.5	109.3	101.7	100.3
Mar	102.0	108.0	95.9	101.7	96.0	93.7	94.8	105.2	101.1	100.2
Apr	104.3	108.8	97.2	104.5	96.6	94.0	95.3	104.5	102.5	102.0
May	104.1	108.2	96.8	105.4	95.6	93.8	93.8	103.7	99.5	101.3
Jun	101.8	108.8	97.2	105.0	95.1	93.9	94.9	103.2	99.1	102.2
Jul	103.2	109.0	97.8	106.9	96.1	94.6	94.5	102.4	99.4	101.6
Aug	102.6	108.6	99.0	106.6	99.5	92.8	94.7	102.8	98.4	101.2
Sep	102.7	108.3	98.0	105.7	97.9	92.8	94.6	102.8	99.2	100.6
Oct	103.1	108.1	96.6	103.5	95.3	91.9	93.4	104.1	97.2	97.9
Nov	102.6	109.7	96.8	104.2	95.6	92.1	94.1	104.8	100.0	102.2
Dec	103.3	110.7	94.9	104.7	94.4	91.5	95.0	106.2	100.2	101.2
2006 Jan	103.3[†]	110.7[†]	95.8[†]	106.8[†]	92.8[†]	92.4[†]	95.7[†]	106.3[†]	99.0[†]	100.4[†]
Feb	103.0	110.2	96.6	108.8	92.7	91.9	96.1	104.2	99.2	99.6
Mar	104.5	111.9	98.1	110.8	94.8	92.9	97.6	103.9	99.5	102.4
Apr	104.1	113.1	97.2	110.4	93.9	91.6	97.1	105.2	98.1	98.2
May	103.8	113.6	98.7	109.8	95.5	91.5	99.1	107.1	101.5	97.5
Jun	105.9	114.8	98.7	110.5	97.6	90.0	97.8	107.7	101.6	98.0
Jul	105.2	115.4	97.6	111.9	93.5	91.3	97.8	108.5	100.0	98.1
Aug	105.7	115.2	98.2	111.7	95.0	91.6	98.6	110.5	100.6	97.6
Sep	104.8	117.5	96.6	113.4	93.5	91.2	97.9	109.2	98.5	97.4
Oct	105.6	115.6	97.2	113.7	95.1	91.1	98.3	108.1	98.8	94.2
Nov	106.3	114.9	98.7	113.5	96.4	90.8	97.4	108.0	100.4	96.8
Dec	105.6	115.1	97.3	114.6	96.1	90.7	97.6	109.3	99.1	97.1

7.1 Output of the production industries

continued

	Consumer durables	Consumer non-durables	Capital goods industries	Intermediate goods and energy		
				Total	Energy	Intermediate goods
SIC 2003 Weights[2]	36	272	213	478	212	266
	UFIU	UFJS	UFIL	JMOH	UFJB	UFJL
2001	101.2	99.4	106.8	102.3	103.4	101.4
2002	101.7	99.9	98.2	101.5	102.9	100.4
2003	100.0	100.0	100.0	100.0	100.0	100.0
2004	104.6	100.0	103.7	99.7	96.4	102.3
2005	102.4	99.3	103.5	96.4	91.8	100.1
2006	104.1	99.3	107.3	94.4	86.1	101.1
Seasonally adjusted [1]						
2002 Q2	100.7	99.9	97.5	102.1	104.8	100.0
Q3	100.4	100.5	98.7	100.8	100.4	101.1
Q4	101.6	98.7	98.4	101.7	103.7	100.1
2003 Q1	99.7	99.0	98.7	101.0	102.2	99.9
Q2	99.3	99.2	99.1	99.6	99.4	99.6
Q3	99.9	100.6	99.8	99.7	99.5	99.8
Q4	101.2	101.2	102.4	99.8	98.9	100.6
2004 Q1	102.6	100.4	102.2	100.4	98.5	102.0
Q2	104.9	100.4	103.4	100.6	97.6	103.0
Q3	106.3	99.1	104.0	98.9	95.6	101.5
Q4	104.7	99.8	105.1	98.7	94.0	102.5
2005 Q1	105.0	99.5	102.9	98.0	93.7	101.4
Q2	101.9	99.4	103.4	97.5	94.1	100.2
Q3	101.0	99.3	104.7	95.3	89.9	99.6
Q4	101.6	99.2	103.1	94.9	89.4	99.3
2006 Q1	102.7†	99.3†	105.1†	95.2†	89.3†	99.8†
Q2	105.3	99.0	106.8	94.6	86.4	101.2
Q3	103.7	99.4	108.1	94.4	84.8	102.1
Q4	104.5	99.6	109.1	93.4	83.8	101.1
2005 Jan	104.5	100.6	103.9	98.1	93.1	102.1
Feb	106.7	100.2	103.4	98.5	93.7	102.3
Mar	103.8	97.8	101.5	97.3	94.3	99.8
Apr	104.5	98.4	103.0	98.1	94.5	101.0
May	100.9	99.2	103.3	97.5	94.2	100.1
Jun	100.4	100.5	103.9	96.8	93.4	99.5
Jul	100.6	100.3	104.9	96.0	91.5	99.6
Aug	100.7	98.9	105.0	94.4	87.5	100.0
Sep	101.6	98.6	104.2	95.4	90.6	99.3
Oct	100.8	98.4	102.7	94.6	89.0	99.1
Nov	101.4	98.9	103.6	95.0	89.7	99.3
Dec	102.5	100.1	103.1	95.1	89.5	99.6
2006 Jan	100.7†	99.3†	104.1†	95.5†	90.1†	99.9†
Feb	102.0	99.1	105.1	94.6	88.9	99.2
Mar	105.4	99.3	106.2	95.4	89.0	100.5
Apr	105.7	99.0	106.2	94.3	87.2	100.1
May	105.2	99.3	106.8	94.8	86.5	101.4
Jun	105.1	98.8	107.4	94.7	85.4	102.1
Jul	103.5	99.1	107.9	94.4	85.1	101.9
Aug	103.7	99.1	107.6	94.5	83.9	103.1
Sep	104.1	100.0	108.7	94.3	85.4	101.5
Oct	103.0	99.4	109.1	93.3	83.4	101.2
Nov	105.3	99.3	109.1	93.9	84.7	101.2
Dec	105.1	100.2	109.1	93.1	83.4	100.8

Note: The figures contain, where appropriate, an adjustment for stock changes.

1 Unadjusted data may be obtained from the Office for National Statistics, IOP Branch, Government Buildings, Cardiff Road, Newport, NP10 8XG.

2 These sum to the total of 1 000 for the production industries.
3 Includes adjustments to standardise the length of months.

Source: Office for National Statistics: 01633 812319

7.2 Productivity jobs and output per filled job[1]

2003 = 100

	Whole economy[2]	Total production industries	Total mining quarrying electricity gas & water supply	Total manufacturing industries	Food, drink and tobacco	Textiles, footwear, clothing and leather	Pulp, paper & paper products, printing and publishing	Chemicals and man-made fibres	Other non-metalic mineral products	Basic metals and fabricated metal products	Engineering and related industries	Other manufacturing
SIC 1992 Sub-section		Sect C+D+E	Sect C+E	Sect D	DA	DB_DC	DE	DG	DI	DJ	DK_DM	DD+DF+DH+DN
Productivity jobs												
	LNNM	LNOJ	LOIW	LNOK	LNOL	LOIS	LOIM	LOIN	LZYL	LZYP	LOIT	LOIZ
2000	97.8	115.1	105.7	115.5	109.2	155.5	105.1	106.7	112.9	115.1	121.1	110.1
2001	98.4	110.3	107.4	110.5	106.2	131.8	101.7	103.0	110.3	109.4	117.0	105.6
2002	99.1	105.4	104.1	105.5	103.6	117.1	100.6	103.7	103.3	105.1	108.4	103.2
2003	100.0	100.0	100.0	100.0	100.0	100.0	100.0	100.0	100.0	100.0	100.0	100.0
2004	100.8	95.6	94.3	95.6	99.1	89.5	95.1	95.8	99.5	95.9	94.6	96.2
2005	101.7	92.4	95.7	92.2	96.6	82.8	92.5	92.0	95.4	91.3	92.9	90.5
Seasonally adjusted												
2002 Q4	99.6	103.8	102.9	103.8	102.4	110.4	100.0	104.5	102.4	102.6	105.7	103.2
2003 Q1	99.7	102.5	102.8	102.4	101.5	106.3	100.5	102.9	102.0	101.7	103.3	102.5
Q2	99.9	100.7	101.4	100.7	100.4	102.6	100.2	99.9	100.2	100.7	100.7	101.0
Q3	100.1	99.2	99.3	99.2	99.2	97.8	100.3	98.7	99.1	99.1	99.0	99.2
Q4	100.2	97.6	96.5	97.7	99.0	93.3	99.1	98.5	98.7	98.4	97.0	97.3
2004 Q1	100.6	96.7	94.3	96.8	99.8	92.7	96.9	96.9	99.1	96.8	95.6	97.5
Q2	100.7	96.1	93.4	96.2	99.7	90.3	96.3	96.5	99.8	96.3	95.0	97.0
Q3	100.7	95.2	94.4	95.3	98.8	87.9	94.5	95.5	99.5	95.6	94.1	96.1
Q4	101.1	94.4	95.2	94.3	98.3	87.0	92.8	94.4	99.6	94.9	93.8	94.1
2005 Q1	101.4	93.5	95.0	93.6	97.5	87.3	92.3	93.3	98.4	93.0	93.8	91.9
Q2	101.6	92.6	94.8	92.5	96.7	83.3	93.3	92.7	96.3	90.7	93.1	90.6
Q3	101.8	92.0	96.4	91.7	96.4	81.1	92.3	91.7	94.6	90.2	92.7	90.0
Q4	101.8	91.5	96.7	91.2	96.0	79.5	92.1	90.2	92.3	91.3	92.0	89.5
2006 Q1	102.0	91.0	94.9	90.8	95.0	78.1	93.2	89.3	89.7	92.6	91.2	88.3
Q2	102.3	90.6	96.4	90.3	94.1	78.2	93.1	88.6	89.7	92.5	90.6	87.2
Q3	102.5	90.4	97.8	89.9	94.2	78.6	92.5	88.0	89.8	92.6	90.3	86.5
Output per filled job												
	LNNN	LNNW	LOJA	LNNX	LNNY	LNOG	LNOA	LNOB	LZYM	LZYQ	LNOH	LOJD
2000	95.8	90.2	99.2	89.8	88.6	79.3	96.5	87.7	84.6	90.0	91.4	94.7
2001	97.2	92.7	95.1	92.7	92.3	83.4	99.6	96.3	87.0	92.6	91.4	96.3
2002	98.2	95.2	98.5	94.6	96.6	86.7	100.8	95.5	91.4	97.5	90.8	97.7
2003	100.0	100.0	100.0	100.0	100.0	100.0	100.0	100.0	100.0	100.0	100.0	100.0
2004	102.5	105.4	101.6	106.7	102.5	100.5	104.2	107.8	106.3	107.5	110.2	104.1
2005	103.6	107.0	95.1	109.4	105.9	106.0	101.4	113.6	110.0	113.0	111.0	109.3
Seasonally adjusted												
2002 Q4	98.5	96.5	100.2	95.7	97.0	87.2	101.1	93.7	92.3	99.7	93.4	97.2
2003 Q1	99.0	97.5	99.3	96.9	98.6	93.2	99.5	95.4	96.3	98.3	95.7	97.0
Q2	99.4	98.6	98.1	98.7	99.0	96.8	99.4	99.5	98.2	98.8	98.6	97.9
Q3	100.3	100.8	100.3	100.8	101.0	103.6	99.6	101.1	101.9	100.6	100.6	101.4
Q4	101.3	103.2	102.4	103.5	101.4	106.4	101.4	104.1	103.7	102.3	105.1	103.7
2004 Q1	101.6	104.3	103.5	105.1	100.5	100.2	104.0	108.0	105.4	104.3	107.3	103.8
Q2	102.5	105.4	104.0	106.4	102.9	99.6	102.9	107.5	106.5	107.5	109.9	103.7
Q3	102.7	105.4	100.7	106.6	102.6	101.3	103.7	106.9	106.3	108.8	111.1	102.4
Q4	103.1	106.6	98.2	108.5	103.8	100.9	106.1	108.9	107.0	109.3	112.5	106.6
2005 Q1	103.0	106.5	97.2	108.5	104.9	98.5	104.1	111.4	110.1	111.1	109.6	111.1
Q2	103.3	107.3	98.5	109.1	106.5	106.9	100.6	111.6	107.8	113.9	110.7	108.9
Q3	103.6	107.1	92.5	110.0	106.1	108.2	101.1	114.4	108.5	114.0	112.2	108.8
Q4	104.4	107.1	92.2	109.9	106.2	110.6	99.7	116.9	113.8	112.8	111.6	108.2
2006 Q1	105.0	108.5	93.9	111.4	107.0	109.1	99.3	117.9	116.2	111.8	115.1	111.6
Q2	105.5	109.0	89.4	113.0	108.0	107.9	98.0	121.0	118.3	112.8	117.6	114.6
Q3	105.9	109.5	86.2	114.1	108.2	106.4	99.2	124.1	119.8	113.6	119.1	114.8

Note: The full productivity and unit wage costs data sets with associated articles can be found on the National Statistics website at: www.statistics.gov.uk/productivity.

1 Output per filled job is the ratio of the output index numbers published in Table 7.1 and productivity jobs. A monthly series for total manufacturing industries is presented in Table 7.3.

2 Whole economy output per job is based on Gross Value Added at Basic Prices.

Source: Office for National Statistics

7.3 Key Productivity Measures

2003=100

	Whole economy				Manufacturing industry	
	Implied GDP deflator[1]	Labour costs per unit of output	Wages and salaries per unit of output	Output per worker[2]	Wages and salaries per unit of output	Output per filled job
	YBGB	LNNL	LNNK	A4YM	LNNQ	LNNX
1998	88.9	86.4	88.1	92.1	101.2	81.2
1999	90.9	88.9	90.6	93.6	100.9	85.3
2000	92.1	91.5	92.9	96.1	99.5	89.8
2001	94.1	94.8	96.5	97.3	100.6	92.7
2002	97.0	97.0	98.2	98.3	102.1	94.6
2003	100.0	100.0	100.0	100.0	100.0	100.0
2004	102.6	102.1	101.3	102.2	97.2	106.7
2005	104.8	105.5	103.9	103.3	98.2	109.4
2000 Q3	92.3	92.2	93.7	96.3	99.5	90.1
Q4	92.5	93.0	94.5	96.8	99.1	92.0
2001 Q1	93.4	94.4	96.3	97.3	99.3	92.9
Q2	94.0	94.6	96.4	97.1	101.2	91.9
Q3	93.9	95.0	96.6	97.5	100.3	93.4
Q4	95.0	95.4	96.8	97.5	101.5	92.7
2002 Q1	96.1	96.1	97.5	98.0	101.7	93.4
Q2	96.9	97.0	98.4	97.9	102.9	93.5
Q3	97.3	97.2	98.4	98.6	101.5	95.7
Q4	97.6	97.8	98.5	98.6	102.3	95.7
2003 Q1	98.8	98.4	99.1	99.2	101.8	96.9
Q2	99.8	99.7	100.0	99.2	100.6	98.7
Q3	100.5	100.9	100.6	100.3	99.6	100.8
Q4	100.9	101.0	100.2	101.3	98.0	103.5
2004 Q1	101.1	101.2	100.3	101.4	97.4	105.1
Q2	102.4	101.6	101.0	102.3	97.3	106.4
Q3	103.0	102.0	101.4	102.5	97.4	106.6
Q4	103.9	103.5	102.5	102.7	96.7	108.5
2005 Q1	104.1	104.0	103.2	102.7	97.6	108.5
Q2	104.5	105.0	103.6	103.1	97.5	109.1
Q3	104.7	106.1	104.1	103.2	98.2	110.0
Q4	106.0	107.0	104.7	104.2	99.6	109.9
2006 Q1	105.8	108.2	105.3	104.5	99.8	111.4
Q2	106.7	107.7	105.5	105.1	99.7	113.0
Q3	107.9	108.3	106.1	105.6	99.6	114.1
2004 Apr	..	..	..	..	96.9	106.3
May	..	..	..	..	97.6	106.6
Jun	..	..	..	..	97.5	106.4
Jul	..	..	..	..	98.2	105.7
Aug	..	..	..	..	97.5	106.4
Sep	..	..	..	..	96.4	107.9
Oct	..	..	..	..	97.2	107.4
Nov	..	..	..	..	96.2	108.9
Dec	..	..	..	..	96.5	109.1
2005 Jan	..	..	..	..	96.7	108.9
Feb	..	..	..	..	96.8	109.3
Mar	..	..	..	..	99.2	107.3
Apr	..	..	..	..	97.9	108.6
May	..	..	..	..	97.4	108.9
Jun	..	..	..	..	97.3	109.6
Jul	..	..	..	..	97.4	110.4
Aug	..	..	..	..	98.2	110.1
Sep	..	..	..	..	99.0†	109.7
Oct	..	..	..	..	99.7	109.3†
Nov	..	..	..	..	99.7	109.8†
Dec	..	..	..	..	99.5	110.5
2006 Jan	..	..	..	..	99.7	110.9
Feb	..	..	..	..	100.2	111.0
Mar	..	..	..	..	99.5	112.1
Apr	..	..	..	..	100.5	112.1
May	..	..	..	..	99.2	113.1
Jun	..	..	..	..	99.4	113.7
Jul	..	..	..	..	99.1	113.7
Aug	..	..	..	..	99.7	114.2
Sep	..	..	..	..	99.9	114.5
Oct	..	..	..	..	100.3	114.3
Nov	..	..	..	..	99.7	115.0

Note: The full productivity and unit wage costs data sets with associated articles can be found on the National Statistics website at: www.statistics.gov.uk/productivity.

1 Based on the sum of expenditure components of GDP at current and constant market prices.

2 Whole Economy output per worker is the ratio of Gross Value Added (GVA) at Basic Prices and Labour Force Survey (LFS) total employment.

Source: Office for National Statistics

8.1 Inland energy consumption: primary fuel input basis

Million tonnes of oil equivalent

	Not seasonally adjusted[1]							Seasonally adjusted and temperature corrected (annual rate)[2,3]						
				Primary electricity							Primary electricity			
	Coal[4]	Petro-leum[5]	Natural gas[6]	Nuclear	Natural flow hydro[7]	Net imports	Total	Coal	Petro-leum	Natural gas	Nuclear	Natural flow hydro	Net imports	Total
	BHBB	BHBC	BHBD	BHBE	BHBF	BHBM	BHBA	BHBH	BHBI	BHBJ	BHBK	BHBL	BHBN	BHBG
2001	42.5	75.9	96.6	20.8	0.4	0.9	237.0	42.9	76.4	96.7	20.8	0.4	0.9	238.1
2002	39.3	73.5	95.4	20.1	0.5	0.7	229.5	40.1	74.9	98.7	20.0	0.5	0.7	235.0
2003	42.4	73.0	95.8	20.0	0.4	0.2	231.9	43.5	74.0	97.7	20.0	0.4	0.2	235.8
2004	41.2	75.3	97.7	18.2	0.6	0.6	233.6	41.7	76.4	100.0	18.1	0.6	0.6	237.5
2005	42.4	77.0	95.1	18.4	0.7	0.7	234.3	42.7	78.0	95.5	18.4	0.7	0.7	236.1
2002 Jun	2.5	6.0	5.7	1.8	–	0.1	16.2	36.3	74.0	87.0	20.5	0.9	1.0	219.7
Jul	3.0	6.8	5.6	1.5	–	–	16.9	45.2	82.9	85.9	20.1	0.7	0.1	234.8
Aug	2.1	6.1	5.3	1.6	–	–	15.1	34.4	69.5	88.3	19.6	0.5	0.2	212.4
Sep	2.9	6.3	5.9	1.7	–	–	16.9	35.6	76.6	92.8	20.0	0.4	0.4	225.8
Oct	3.5	5.6	8.3	1.3	–	0.1	29.4	42.4	69.6	99.0	17.4	0.3	0.6	229.4
Nov	4.0	5.9	9.0	1.4	–	0.1	20.5	46.3	72.8	102.5	17.5	0.5	1.2	240.7
Dec	4.7	5.8	10.7	2.1	–	0.1	23.4	42.0	70.5	107.9	21.9	0.3	1.4	244.0
2003 Jan	3.7	6.3	11.6	1.8	–	0.1	23.5	40.2	74.3	110.3	20.4	0.4	0.6	246.1
Feb	4.0	6.0	10.4	1.8	–	–	22.2	45.4	74.1	107.8	21.5	0.3	0.1	248.9
Mar	4.2	6.0	9.5	2.0	–	–	21.8	43.6	69.5	106.6	21.1	0.3	0.1	241.2
Apr	3.4	6.2	7.5	1.6	–	–	18.5	45.0	82.2	94.3	20.2	0.3	0.1	242.1
May	3.0	6.0	6.7	1.6	–	–	17.1	40.3	74.4	97.2	20.2	0.7	0.3	233.1
Jun	3.3	6.3	5.2	1.9	–	–	16.8	50.1	78.9	86.3	21.5	0.5	–0.1	237.2
Jul	2.8	6.1	5.2	1.3	–	–	15.4	44.0	74.4	82.7	18.1	0.5	–	219.7
Aug	2.6	6.1	5.0	1.4	–	–	15.1	42.7	69.7	82.8	17.7	0.5	0.3	213.7
Sep	3.2	6.4	5.9	2.0	–	–	17.4	39.4	77.1	91.3	23.5	0.4	–0.6	231.1
Oct	3.9	6.2	8.5	1.4	–	–	20.1	46.7	75.9	98.2	18.5	0.3	–	239.7
Nov	3.4	5.8	9.3	1.5	–	–	20.0	39.2	70.5	104.4	17.6	0.3	0.3	232.4
Dec	4.7	6.4	10.9	1.9	0.1	0.1	24.1	41.5	77.2	110.6	19.7	0.4	1.0	250.5
2004 Jan	3.8	6.2	11.3	1.6	0.1	0.1	23.1	41.3	73.3	110.6	18.5	0.6	0.7	245.1
Feb	3.8	5.8	10.5	1.6	0.1	–	21.8	44.5	73.4	114.1	19.7	0.5	0.6	252.5
Mar	4.5	6.0	10.5	2.1	–	–	23.1	44.4	68.3	110.7	22.4	0.4	–	246.2
Apr	3.2	6.2	8.2	1.5	–	–	19.2	42.7	81.0	102.1	18.2	0.5	0.5	245.0
May	2.7	7.0	6.8	1.3	–	–	17.8	37.4	86.3	100.0	16.7	0.6	0.4	241.5
Jun	2.9	5.7	5.6	1.5	–	0.1	15.8	41.6	70.8	89.4	16.8	0.6	0.8	219.9
Jul	2.6	7.3	5.7	1.5	–	0.1	17.2	38.8	88.9	86.3	19.7	0.8	0.8	235.1
Aug	2.6	5.9	5.3	1.4	–	0.1	15.3	42.7	67.3	84.6	17.3	0.9	0.7	213.4
Sep	3.4	6.2	5.6	1.4	0.1	–	16.7	41.7	75.3	89.3	16.8	0.8	0.6	224.5
Oct	3.7	7.2	8.2	1.4	0.1	0.1	20.6	44.9	89.3	100.4	18.1	0.5	1.2	254.5
Nov	3.8	6.0	9.5	1.4	–	0.1	20.8	43.7	72.5	106.0	16.8	0.5	0.7	240.3
Dec	4.5	7.0	10.6	1.6	0.1	0.1	23.9	40.0	84.6	108.7	17.0	0.5	0.7	251.5
2005 Jan	4.1	6.7	11.0	1.9	0.1	0.1	23.8	45.3	81.8	111.0	21.5	0.6	0.6	260.8
Feb	4.2	5.9	10.3	1.6	–	–	22.0	48.7	66.4	108.3	19.0	0.5	0.3	243.0
Mar	4.4	6.3	9.8	1.6	0.1	–	22.3	45.0	82.0	106.9	17.4	0.6	0.6	252.5
Apr	3.3	6.7	8.4	1.4	–	–	19.9	42.7	83.1	98.9	17.6	0.6	0.6	243.5
May	2.8	6.2	7.1	1.5	–	0.1	17.8	37.9	76.7	96.0	19.1	0.8	1.0	231.5
Jun	2.9	6.5	5.5	1.6	–	0.1	16.5	42.1	80.0	84.6	17.7	0.8	0.6	225.9
Jul	2.6	6.2	5.4	1.6	–	–	15.8	39.1	70.7	80.3	21.2	0.7	0.6	212.6
Aug	2.6	6.3	5.1	1.7	–	0.1	15.7	40.2	75.3	78.2	21.2	0.7	1.0	216.7
Sep	2.9	6.7	5.8	1.3	–	–	16.8	36.2	84.9	89.7	15.9	0.8	0.4	227.8
Oct	3.3	6.1	7.1	1.3	0.1	0.1	17.9	41.0	76.4	96.0	16.6	0.8	0.9	231.7
Nov	4.4	6.9	9.4	1.4	0.1	0.1	22.3	51.3	82.0	98.1	17.3	0.7	1.0	250.5
Dec	5.1	6.5	10.3	1.6	0.1	0.1	23.6	44.4	77.1	98.3	16.8	0.5	1.0	238.0
2006 Jan	5.0	6.2	10.7	1.7	0.1	0.1	23.7	53.2	72.6	97.7	19.8	0.7	0.8	244.7[†]
Feb	4.6	6.1	9.9	1.5	0.1	–	22.2	52.2	75.0	95.9	18.5	0.5	0.2	242.4[†]
Mar	5.0	7.3	10.2	1.8	0.1	0.1	24.4	49.3	80.8	95.6	18.7	0.6	0.9	245.8
Apr	3.3	6.2[†]	7.9	1.5	0.1	0.1	19.1[†]	43.4	79.0[†]	90.8	19.4	0.7	1.2	234.6
May	3.4[†]	6.9[†]	6.2	1.5	0.1	0.1	18.2[†]	46.7	84.7[†]	89.0	19.5	0.9	1.2	241.1
Jun	3.2[†]	5.8	5.0	1.4	–	–	15.5	47.0[†]	72.7	79.8	15.7	0.8	0.5	216.6
Jul	3.3	6.3	4.7[†]	1.4	–	–	15.7	52.1	78.1	73.9[†]	19.3	0.9	0.5	224.8
Aug	3.0	6.5	4.9	1.5	–	0.1	15.9	48.2	73.3	75.8	19.1	0.9	0.8	218.2
Sep	3.2	6.0	5.3	1.3	0.1	0.1	15.8	40.2	75.3	86.5	15.1	0.9	0.7	218.7
Oct	3.5	6.1	7.0	1.0	0.1	0.1	17.7	44.8	78.5	93.4	13.1	0.8	0.8	231.4
Nov	4.4	7.1	8.7	1.1	0.1	0.1	21.4	51.5	86.0	96.7	12.8	0.7	0.8	248.5

1 Not seasonally adjusted.
2 Coal, petroleum and natural gas are temperature corrected.
3 For details of temperature correction see DTI energy statistics website at www.dti.gov.uk/energy/inform/dukes/dukes2004/01longterm.pdf.
4 Includes solid renewable sources (wood, straw, waste, geothermal and active solar heat) net foreign trade and stock changes in other solid fuels.

5 Excludes non-energy use. A statistical month adjustment has been removed
6 Includes gas used during production, colliery methane, landfill gas and sewage gas. Excludes gas flared or re-injected and non-energy use of gas.
7 Includes generation at wind stations. Excludes generation from pumped storage stations.

Source: Department of Trade and Industry: 020 7215 2698

8.2 Supply and use of fuels[1]

Thousand tonnes of oil equivalent

		2004	2005	2004 Q4	2005 Q1	2005 Q2	2005 Q3	2005 Q4	2006 Q1	2006 Q2	2006 Q3
Supply											
Indigenous production	BHCE	238 034	215 447	60 607	59 891	55 269	45 865	54 421	56 748	49 004	42 314
Imports	DMNT	125 420	134 702	32 436	32 714	33 211	33 835	34 941	38 066	35 020	35 409
Exports	BHCH	−113 954	−100 519	−26 887	−25 609	−27 464	−22 967	−24 479	−23 589	−25 057	−23 845
Marine bunkers	DMNU	−2 220	−2 180	−568	−495	−566	−600	−519	−450	−583	−486
Stock change[2]	BHCI	−1 151	−637	1 194	4 721	−3 425	−4 771	2 838	2 917	−2 459	−2 974
Primary supply	LURA	246 130	246 813	66 782	71 223	57 027	51 361	67 202	73 692	55 925	50 418
Statistical difference[3]	BHCO	267	−71	−403	145	268	−546	63	155	104	−36
Primary demand	LURB	245 863	246 884	67 185	71 078	56 759	51 908	67 139	73 537	55 822	50 454
Transfers[4]	LURC	−140	−114	−100	135	−22	−33	−194	619	648	376
Transformation	LURD	−53 346	−54 371	−14 509	−14 868	−12 781	−12 345	−14 377	−15 706	−12 969	−12 549
Electricity generation	LURE	−50 187	−51 107	−13 512	−14 150	−11 779	−11 564	−13 614	−14 552	−11 757	−11 622
Heat generation[5]	SKYM	−864	−867	−256	−261	−195	−169	−241	−262	−183	−158
Petroleum refineries	YAPL	217	84	−77	135	−179	−7	134	−182	−349	−83
Coke manufacture	YAPM	−18	−38	−23	4	10	−24	−28	−12	−2	−25
Blast furnaces	YAPN	−2 502	−2 455	−643	−598	−643	−586	−629	−698†	−680	−660
Patent fuel manufacture	YAPO	8	11	1	1	5	4	1	−1	3	–
Energy industry use	YAPP	16 496	16 523	4 302	4 344	4 135	3 841	4 204	4 465	3 828	3 576
Losses	YAPQ	3 546	3 765	989	1 122	844	799	999	1 206	828	861
Final consumption	YAPR	172 335	172 111	47 291	50 878	38 976	34 891	47 366	52 775	38 844	33 846
Iron and steel	YAPS	1 918	1 762	465	450	457	413	443	476	461	426
Other industries	YAPT	31 061	31 333	7 903	9 427	7 266	6 410	8 229	10 084	7 154	6 261
Transport	YAPU	58 166	59 225	14 484	14 008	14 889	15 326	15 002	14 360	15 037	15 460
Domestic	YAPV	48 587	49 979	15 755	17 662	8 847	5 539	14 931	18 170	8 736	5 228
Public administration	YAPW	7 138	7 162	2 049	2 355	1 554	1 225	2 028	2 359	1 526	1 086
Commercial	YAPX	10 033	9 866	2 816	2 883	2 261	2 068	2 653	3 070	2 169	2 160
Agriculture	YAPY	918	990	225	268	221	224	277	221	236	201
Miscellaneous	YAPZ	2 084	2 211	611	726	517	347	620	796	412	258
Non energy use	BHCN	12 429	12 583	2 982	3 099	2 963	3 339	3 183	3 240	3 114	2 765

8.2 Supply and use of fuels[1]
continued

Thousand tonnes of oil equivalent

		2004	2005	2004 Q4	2005 Q1	2005 Q2	2005 Q3	2005 Q4	2006 Q1	2006 Q2	2006 Q3
Final consumption by user											
Iron and steel industry											
Coal and other manufactured fuels[6]	YAQA	582	593	139	130	149	149	165	151	155	142
Petroleum products	BHTF	35	15	8	6	2	2	5	9	–	14
Natural gas[7]	YAQB	835	723	201	206	198	153	166	207	197	162
Electricity	BHTE	465	–	117	109	107	108	108	109	109	108
Total[8]	YAPS	1 918	1 762	465	450	457	413	443	476	461	426
Other industries											
Coal and other manufactured fuels[6]	YAQC	1 373	1 427	377	340	330	364	393	403	367	398
Petroleum products	BHTM	6 717	7 051	1 567	2 076	1 618	1 503	1 853	2 403	1 621	1 579
Natural gas[7]	YAQD	12 378	12 067	3 357	4 207	2 612	1 992	3 255	4 567	2 492	1 728
Renewables and waste[9]	YAQE	265	151	68	49	35	28	38	49	35	29
Electricity	BHTL	9 496	9 785	2 326	2 542	2 458	2 309	2 476	2 449	2 426	2 314
Total[8]	YAPT	31 061	31 333	7 903	9 427	7 266	6 410	8 229	10 084	7 154	6 261
Transport											
Petroleum products	BHTQ	57 440	58 485	14 300	13 824	14 705	15 140	14 815	14 176	14 847	15 282
Electricity	BHTP	726	740	184	184	184	186	187	185	189	179
Total[8]	YAPU	58 166	59 225	14 484	14 008	14 889	15 326	15 002	14 360	15 037	15 460
Domestic											
Coal and other manufactured fuels[6]	YAQF	999	698	288	207	158	160	172	213	176	167
Petroleum products	BHTW	3 265	3 093	1 036	1 081	1 036	160	1 090	1 090	719	484
Natural gas[7]	YAQG	34 085	32 836	11 396	13 350	5 847	2 825	10 813	13 651	5 585	2 502
Renewables and waste[9]	YAQH	252	256	84	96	49	27	84	96	49	27
Electricity	BHTV	9 933	10 044	2 933	2 906	2 224	2 053	2 861	3 100	2 199	2 043
Total[8]	YAPV	48 587	49 979	15 755	17 662	8 847	5 539	14 931	18 170	8 736	5 228
Other final users[10]											
Coal and other manufactured fuels[6]	YAQI	16	23	2	6	5	5	7	14	11	7
Petroleum products	BHNC	1 438	1 754	333	460	474	496	325	394	353	472
Natural gas[7]	YAQJ	9 636	9 170	2 908	3 225	1 919	1 305	2 722	3 437	1 829	1 158
Renewables and waste[9]	YAQK	198	193	66	68	38	21	66	68	38	21
Electricity	BHNB	8 523	8 683	2 272	2 312	2 049	2 000	2 322	2 372	2 041	2 009
Total[8]	BHND	20 174	20 229	5 702	6 232	4 554	3 865	5 578	6 445	4 343	3 705
Total final users	BHNE	159 906	159 527	44 309	47 779	36 013	31 552	44 183	49 535	35 730	31 080

1 Layout comparable with annual balances published in Table 1.1 of DUKES 2005.
2 Stock fall (+), stock rise (-).
3 Primary supply minus primary demand.
4 Annual transfers should ideally be zero. For manufactured fuels differences occur in the rescreening of coke to breeze. For oil and petroleum products differences arise due to small variations in the calorific values used.
5 Generation of heat for sale under the provision of a contract.

6 Includes all manufactured solid fuels, benzole, tars, coke oven gas and blast furnace gas.
7 Includes colliery methane.
8 Includes heat sold.
9 Includes geothermal and solar heat. Latest quarter is estimated from the previous year and adjusted according to average annual rate of change over the last three years.
10 Includes public administration, commercial, agriculture and miscellaneous use.

Source: Department of Trade and Industry: 020 7215 2698

8.3 Coal supply

Thousand tonnes

	Production					
	Deep-mined	Opencast	Total[1]	Net imports	Imports[2]	Exports
	BHDC	BHDD	BHDB	BHDE	BHDF	BHDG
2001	17 347	14 166	31 930	34 992	35 542	550
2002	16 391	13 148	29 989	28 149	28 686	537
2003	15 633	12 126	28 279	31 349	31 891	543
2004	12 542	11 993	25 096	35 531	36 153	622
2005	9 563	10 445	20 498	43 433	43 968	536
2005 Jun	780	1 034	1 875	3 823	3 916	92
Jul	509	695	1 247	3 390	3 422	32
Aug	493	813	1 350	3 733	3 776	44
Sep	1 130	1 013	2 193	3 722	3 764	41
Oct	961	813	1 822	4 002	4 039	37
Nov	967	736	1 737	4 486	4 533	47
Dec	1 271	862	2 183	3 316	3 346	29
2006 Jan	1 029	526	1 588	4 355	4 390	35
Feb	995	833	1 871	4 169	4 206	37
Mar	1 048	1 012	2 108	4 325	4 361	37
Apr	761	679	1 477	3 944	3 998	54
May	840	757	1 640	3 920	3 943	23
Jun	940	860	1 848	4 224	4 267	42
Jul	615	497	1 154	4 560	4 599	39
Aug	326	566	929	4 431	4 450	18
Sep	692	813	1 554	3 946[†]	3 973[†]	27[†]
Oct	735	717	1 501	3 270[†]	3 355[†]	85[†]
Nov	754	704	1 511	4 047	4 091	44

1 Includes an estimate for slurry.
2 To December 1992, as recorded in the Overseas Trade Statistics of the United Kingdom (OTS). From January 1993, import figures include an additional estimate for unrecorded trade. From March 1994, Import figures are being estimated on the basis of information available for extra-EC trade until monthly statistics for intra-EC trade become available from the Office for National Statistics.

Source: Department of Trade and Industry: 020 7215 2698

8.4 Inland use and stocks of coal
Stocks: end of period

Thousand tonnes

		Inland use								
		Fuel producers (consumption)				Final users[1]				
			Secondary							
	Primary: collieries	Electricity generators[2]	Heat generation[3]	Coke ovens	Other conversion industries[4]	Industry[5]	Domestic[5,6]	Other[5,7]	Total inland consumption	Stocks[8]
	BHEB	BHEC	SKYY	BHED	BHEE	BHEF	BHEG	BHEI	BHEA	BHEJ
2001	10	50 931	750	7 896	496	1 826	1 874	70	63 853	1 583
2002	9	47 741	717	6 533	436	1 809	1 285	22	58 552	2 482
2003	6	52 463	622	6 612	396	1 856	1 042	24	63 021	1 624
2004	8	50 444	478	6 382	327	1 846	941	22	60 447	1 192
2005	6	52 084	453	6 603	266	1 791	614	34	61 850	1 101
2005 Jun	–	3 234	43	620	27	154	54	2	4 134	948
Jul	–	2 860	31	531	19	139	49	1	3 629	762
Aug	1	2 867	31	541	22	124	43	2	3 632	825
Sep	–	3 156	39	677	27	193	55	3	4 150	942
Oct	–	3 925	35	534	19	177	39	3	4 733	973
Nov	–	5 681	35	534	17	135	44	5	6 451	1 065
Dec	1	6 672	44	664	30	187	67	3	7 668	1 101
2006 Jan	–	6 609	39	538	27	153	58	7	7 430	1 060
Feb	1	6 020	39	513	23	170	61	6	6 832	1 080
Mar	–	6 499	46	674	28	155	67	6	7 475	1 040
Apr	–	4 075	35	550	23	137	39	6	4 865	1 046
May	–	4 123	35	537	24	144	60	4	4 928	992
Jun	–	3 537	43	675	26	142	57	5	4 487	1 000
Jul	–	3 899	31	568	20	162	46	4	4 731	879
Aug	–	3 469	31	542	22	146	53	5	4 269	813
Sep	–	3 562	39	676	31	164[†]	52[†]	2[†]	4 526[†]	969
Oct	1	4 083	38	553	16	198[†]	54[†]	5[†]	4 950[†]	1 091
Nov	–	5 829	38	540	20	187	70	2	6 687	920

1 Disposals by collieries and opencast sites.
2 Coal-fired power stations belonging to major electricity generating companies.
3 Generation of heat for sale under the provision of a contract.
4 Low temperature carbonisation and patent fuel plants.
5 Includes estimates of imports.
6 Including miners' coal.
7 Includes public administration and commerce.
8 Excluding distributed stocks held in merchant's yards, etc, mainly for the domestic market, and stocks held by the industrial sector.

Source: Department of Trade and Industry: 020 7215 2698

8.5 Natural gas production and supply

Gigawatt hours

	Upstream gas industry						Downstream gas industry					Percentage of net gas available for consumption in the UK	
		Less			Plus			Less					
	Gross gas production[1]	Producers own use[2]	Exports[3]	Stock change and other net losses	Imports	Gas available at terminals[4]	Gas input into transmission system[5]	Operators own use	Stock changes	Metering differences	Gas output from transmission system[6]	Indigenous	Imported
	BAWX	DMUE	BAWY	DMUF	BAWZ	BAXA	DMUG	DMUH	DMUI	DMUJ	BAXD	BAXB	BAXC
2001	1 230 533	78 457	138 330	–	30 464	1 044 208	1 044 900	6 549	661	1 798	1 035 892	97.1	2.9
2002	1 204 713	79 364	150 731	–	60 493	1 036 738	1 035 236	7 017	7 356	1 821	1 019 042	94.0	5.8
2003	1 196 117	76 839	177 039	–	86 298	1 028 538	1 029 922	7 828	–3 492	–1 280	1 026 866	91.6	8.4
2004	1 115 744	76 899	114 111	–	133 035	1 057 769	1 059 307	6 560	6 235	137	1 046 375	87.4	12.6
2005	1 017 813	73 652	96 181	–	173 328	1 021 308	1 023 472	6 555	–1 321	2 880	1 015 358	83.0	17.0
2005 Dec	99 282	6 731	6 236	–	24 445	110 760	111 046	712	–910	427	110 817	77.9	22.1
2006 Jan	98 248	6 604	5 582	–	24 590	110 651	110 730	687	–6 321	448†	115 916†	77.8	22.2
Feb	88 628†	5 973†	4 656	–	25 996†	103 996	103 960†	673	–4 470	255	107 502	75.0	25.0
Mar	95 362	6 595	5 868	–	25 342	108 241	108 216	643	–2 955	350	110 178	76.6	23.4
Apr	85 752	6 156	9 847	–	16 296	86 045	86 024	476	1 217	270	84 061	81.1	18.9
May	76 077	5 881	15 087	–	12 242	67 350	67 344	389	1 431	379	65 145	81.8	18.2
Jun	65 003	5 292	10 660	–	6 585	55 636	55 631	247	3 508	436	51 440	88.2	11.8
Jul	58 993	5 283	13 588	–	13 649	53 771	53 742	479	5 740	475	47 048	74.6	25.4
Aug	59 167	4 786	11 466	–	13 822	56 738†	56 697	353	5 759	365	50 220	75.6	24.4
Sep	67 810	5 165	16 623	–	12 353	58 375	58 344	481	3 107	502	54 254	78.8	21.2
Oct	74 644	5 404	13 777†	–	21 876	77 340	77 369	379	2 243	354	74 393	71.7	28.3
Nov	76 829	5 749	7 212	–	30 550	94 418	94 390	481	–75	342	93 642	67.6	32.4

1 Includes waste and own use but excludes gas flared.
2 Gas used for drilling, production and pumping operations.
3 Includes exports direct from the UKCS as well as others carried out by the downstream gas industry from the national transmission system.
4 Gas available at terminals for consumption in the UK as recorded by the terminal operators.
5 Gas input into inland transmission systems. It includes public gas supply, direct supply by North Sea producers, third party supplies, and stock changes. Figures differ from gas available for consumption in the UK mainly because of additional stock changes at local distribution zones. The figures also differ from total consumption (expressed in oil equivalent in Table 8.1) because they exclude producers' and operators' own use and losses.
6 Including public gas supply, direct supplies by North Sea producers, third party supplies and stock changes.

Source: Department of Trade and Industry: 020 7215 2698

8.6 Fuel used by and electricity production and availability from the electricity supply industry[1]

	Million tonnes of oil equivalent							Terawatt hours					
	Fuel used							Electricity supplied by type of plant					
	Coal[2]	Gas[2]	Nuclear electricity	Hydro-electricity	Total[3]	Electricity generated	Own use[4]	Conventional Thermal[5]	Combined Cycle Gas Turbine	Nuclear	Other[6]	Total	Total electricity available[7]
	FTAJ	WSFA	FTAL	FTAM	FTAN	BHJF	BHJJ	FTAB	BAYK	FTAC	FTAD	BHJK	BHJL
2001	30.57	23.80	20.77	0.28	76.49	353.06	19.28	132.57	115.89	82.99	3.02	333.78	352.99
2002	28.62	25.04	20.10	0.34	75.07	354.00	19.21	128.80	121.89	81.09	3.81	334.78	350.75
2003	31.57	24.48	20.04	0.22	77.34	362.60	20.29	140.20	118.55	81.91	2.71	343.37	356.39
2004	30.37	26.18	18.16	0.34	76.18	358.41	19.09	133.61	128.98	73.68	4.42	340.69	357.77
2005	31.65	25.42	18.37	0.34	77.43	362.10	19.69	136.00	128.18	75.17	5.65	345.00	361.26
2005 Dec	4.08	1.67	1.59	0.03	7.61	34.59	2.11	17.58	8.15	6.51	0.56	32.49	34.41
2006 Jan	4.05	1.74	1.73	0.03	7.81	35.60	2.18	17.19	8.75	7.09	0.44	33.42	35.22
Feb	3.68	1.67	1.53	0.02	7.20	32.80	1.97	15.87	8.47	6.25	0.29	30.83	32.01
Mar	3.96	1.60	1.75	0.02	7.65	35.02	2.21	17.15	8.22	7.18	0.30	32.81	34.68
Apr	2.46	1.90	1.55	0.03	6.02	28.31	1.57	10.86	9.19	6.34	0.39	26.74	28.86
May	2.49	1.79	1.52	0.02	5.92	27.58	1.62	10.74	8.71	6.23	0.30	25.96	28.00
Jun	2.10	2.01	1.38	0.01	5.59	26.39	1.53	9.32	9.74	5.63	0.20	24.86	26.25
Jul	2.36	1.97	1.43	0.01	5.87	27.49	1.71	10.31	9.52	5.84	0.12	25.78	27.18
Aug	2.09	1.95	1.52	0.01	5.63	26.57	1.58	9.03	9.65	6.21	0.13	24.99	26.64
Sep	2.12	2.26	1.27	0.02	5.75	27.17	1.53	9.10	11.03	5.19	0.31	25.64	27.18
Oct	2.47†	2.59†	0.99	0.03	6.25	29.11†	1.58	10.72†	12.36	4.04	0.39	27.53†	29.23†
Nov	3.56	2.15	1.05	0.04	6.97	31.76	1.77	15.16	9.99	4.32	0.57	30.00	31.80

1 Fuel used and electricity generated by major power producers and electricity available through the grid in England and Wales and from distribution companies in Scotland and Northern Ireland. For further information regarding major power producers please refer to Section 8 of the *Annual Supplement* in the January edition of *Monthly Digest*.
2 Including quantities used in the production of steam for sale.
3 Including oil, windpower and refuse-derived fuel.
4 Used in works and for pumping at pumped storage stations.
5 Includes gas turbines, oil engines and plants producing electricity from renewable sources other than hydro.
6 Includes wastes and renewable sources.
7 Electricity supplied from major power producers plus purchased from other UK producers plus net imports from overseas.

Source: Department of Trade and Industry: 020 7215 2698

8.7 Sales by the gas and public electricity supply systems

	Gas: Gigawatt hours							Electricity: Terawatt hours			
	Electricity generators[1]	Heat generation[2]	Iron and steel industry	Other industries	Domestic	Other[3]	Total	Industrial[4]	Domestic	Other[5]	Total
	BBKF	WSFM	BBKG	BBKH	BBKI	BBKJ	BBKK	FTAE	FTAG	FTAH	FTAI
2001	312 518	23 585	8 504	171 235	379 427	113 110	1 008 379	102.23	115.35	103.49	321.05
2002	329 442	22 010	8 791	156 285	376 372	100 833	993 733	101.59	114.52	103.68	319.80
2003	323 926	19 830	10 327	155 814	386 488	106 737	1 003 118	103.33	115.76	105.23	324.33
2004	340 516	19 886	9 716	143 894	396 411	112 065	1 022 488	103.24	115.53	104.94	323.71
2005	333 245	20 671	8 412	140 270	381 879	106 653	991 131	105.88	116.81	106.38	329.07
2002 Q3	83 238	3 865	1 934	28 663	32 744	14 280	164 724	23.87	23.28	24.55	71.70
Q4	80 858	6 416	2 121	43 313	128 269	28 654	289 631	25.57	32.57	26.72	84.87
2003 Q1	79 391	6 582	2 678	50 535	157 739	35 445	332 370	26.72	34.48	27.00	88.20
Q2	76 333	4 055	2 597	36 710	62 659	22 532	204 885	25.88	25.00	25.31	76.19
Q3	82 069	3 300	2 339	27 391	34 288	12 661	162 047	25.31	22.81	24.74	72.87
Q4	86 133	5 893	2 713	41 178	131 802	36 099	303 816	25.42	33.47	28.18	87.07
2004 Q1	83 746	6 258	2 573	50 138	159 663	39 538	341 915	27.86	35.18	27.62	90.66
Q2	81 929	4 287	2 559	31 557	66 867	23 158	210 358	24.52	23.81	24.16	72.49
Q3	85 684	3 659	2 248	23 175	37 341	15 555	167 662	25.73	22.42	25.35	73.51
Q4	89 157	5 682	2 336	39 024	132 540	33 814	302 553	25.13	34.11	27.82	87.06
2005 Q1	79 412	6 408	2 392	48 909	155 265	37 505	329 890	27.41	33.80	28.28	89.49
Q2	86 111	4 513	2 307	30 364	68 004	22 312	213 612	26.67	25.87	25.22	77.77
Q3	89 410	3 854	1 785	23 154	32 854	15 174	166 232	24.94	23.87	24.57	73.39
Q4	78 312	5 896	1 928	37 843	125 756	31 662	281 397	26.86	33.27	28.30	88.43
2006 Q1	66 507	6 408	2 406	53 099	158 755	39 970	327 145	26.56	36.05	29.01	91.62
Q2	73 953	4 513	2 288	28 960	64 950	21 274	195 939	26.47	25.57	25.21	77.25
Q3	78 978	3 854	1 879	20 077	29 103	13 463	147 353	25.43	23.76	24.62	74.54

1 Power stations belonging to major generating companies, industrial establishments and transport undertakings generating 1 gigawatt or more a year.
2 Generation of heat for sale under the provision of a contract.
3 Public administration, commerce and agriculture.
4 Manufacturing industry, construction, energy and water supply industries.
5 Commercial premises, transport, and other service sector customers. Agriculture, public lighting and combined domestic/commercial premises.

Source: Department of Trade and Industry: 020 7215 2698

8.8 Indigenous production, refinery receipts, arrivals and shipments of oil[1]

	Million tonnes			Thousand tonnes									
	Indigenous production			Refinery receipts		Foreign trade[2] [7]							
						Net imports/ exports[6] [8]	Crude oil and NGLs		Process oils		Petroleum products		
	Crude oil	NGLs	Total[3]	Total receipts[4] [8]	Indigenous[5]		Imports	Exports	Imports	Exports	Imports	Exports	Bunkers[7]
	BHMB	BHML	BHMA	G8GZ	BHMC	G8H2	BHMF	BHMG	BHMM	BHMH	BHMI	BHMJ	BHMK
2001	108.4	8.3	116.7	82 955	29 403	−35 233	48 992	84 441	4 559	2 489	17 234	19 088	2 274
2002	107.4	8.5	115.9	85 512	28 544	−38 720	52 042	85 028	4 926	2 116	14 900	23 444	1 913
2003	97.8	8.2	106.1	85 006	30 829	−27 571	48 589	72 526	5 588	2 372	16 472	23 323	1 764
2004	87.5	7.9	95.4	90 021	27 505	−13 712	55 858	63 412	6 659	1 091	18 545	30 270	2 085
2005	77.2	7.5	84.7	86 096	27 210	−2 424	52 211	52 106	6 675	1 992	22 511	29 722	2 055
2005 Sep	5.8	0.6	6.4	7 082	2 010	1 148	4 441	3 261	631	185	2 027	2 504	183
Oct	6.5	0.6	7.1	7 535	2 395	−197	4 485	4 112	655	202	1 662	2 685	178
Nov	5.9	0.6	6.5	6 295	1 969	29	3 814	3 471	512	242	2 020	2 604	157
Dec	6.5	0.6	7.0	6 960	2 582	−871	3 891	4 545	487	230	2 117	2 590	155
2006 Jan	6.8	0.6	7.4	7 132	2 685	149	3 875	4 341	572	182	2 262	2 038	153
Feb	5.9	0.6	6.5	6 035	1 896	192	3 735	3 650	404	180	2 097	2 214	121
Mar	6.4	0.6	7.0	6 695	1 522	97	4 672	5 270	501	173	2 482	2 115	149
Apr	6.1	0.6	6.7	6 702	1 816	393	4 256	4 395	630	373	2 232	1 956	170
May	6.0	0.6	6.5	7 051	1 287	1 353	5 047	3 882	717	193	2 086	2 422	196
Jun	5.3	0.6	5.9	6 406	1 867	−302	3 934	3 994	605	279	1 953	2 520	183
Jul	5.8	0.5	6.3	7 409	2 012	461	4 814	4 067	582	183	1 813	2 499	147
Aug	4.8	0.5	5.2	6 810	1 685	1 317[†]	4 563	2 854[†]	563	230	1 839	2 564[†]	158
Sep	5.3[†]	0.6	5.8[†]	6 746	2 329	566	3 914	3 812	503	186	2 363	2 216	154
Oct	5.9[†]	0.6	6.5[†]	6 878[†]	2 291	797	3 813[†]	3 554	774	321	2 572	2 487	182
Nov	5.8	0.6	6.4	6 397	1 806	1 248	3 786	3 176	805	146	2 404	2 425	135

1 The term indigenous is used in this table for convenience to include oil from the UK Continental Shelf as well as the small amounts produced on the mainland.
2 Foreign trade as recorded by the petroleum industry and may differ from figures published in the *Overseas Trade Statistics*.
3 Crude oil *plus* condensates and petroleum gases derived at onshore treatment plants.
4 Crude oil , natural gas liquids (NGLs) and process oils (i.e. partly refined oils).
5 Crude oil *plus* NGLs.
6 Net imports (+) or net exports (-) of oil and oil products.
7 From January 2000 arrivals of petroleum products and marine bunkers contain estimated additions to allow for (temporarily) missing imports data.
8 There have been some modest changes made to this table following the review announced by DTI in the June 2006 edition of *Energy Trends*. The review was aimed at improving data coherency and coverage between the different quarterly and monthly DTI tables. The 'net foreign imports' column has been replaced by 'total net imports/exports' and now covers oil 'total receipts' column for refinery receipts introduced instead.

Source: Department of Trade and Industry: 020 7215 2698

8.9 Deliveries of petroleum products for inland consumption

Thousand tonnes

	Butane and propane[1]	Other Petroleum Gases[2][6]	Naphtha (LDF)[6]	Motor Spirit[6]	Kerosene		Gas/diesel oil		Fuel oil[3]	Bitumen	Lubricating oils	Total[4]
					Aviation turbine fuel	Burning Oil[6]	Derv fuel	Other[6]				
	BHOB	G8GX	G8GY	BHOD	BHOE	BHOG	BHOI	BHOJ	BHOK	BHOL	BHOM	BHOA
2001	2 097	2 077	1 592	20 940	10 614	4 236	16 059	6 959	2 579	1 935	846	71 354
2002	2 553	2 181	1 592	20 808	10 519	3 578	16 926	6 099	1 723	2 002	829	70 557
2003	3 019	2 114	2 332	19 919	10 764	3 567	17 712	6 326	1 540	1 959	868	71 698
2004	3 115	1 918	2 029	19 484	11 862	3 948	18 514	6 023	2 064	1 991	914	73 867
2005	3 554	2 021	1 916	18 731	12 497	3 869	19 436	6 797	1 965	1 906	750	75 375
2005 May	289	172	218	1 479	1 006	245	1 671	549	169	161	60	6 162
Jun	252	148	109	1 606	1 132	187	1 649	588	132	177	63	6 188
Jul	238	158	226	1 561	1 200	144	1 639	614	140	169	62	6 323
Aug	312	154	175	1 501	1 244	206	1 564	606	201	183	64	6 343
Sep	288	157	191	1 584	1 136	243	1 639	568	180	171	70	6 405
Oct	320	171	206	1 497	1 054	253	1 629	549	131	162	56	6 204
Nov	329	167	114	1 590	831	471	1 774	585	159	171	60	6 364
Dec	276	173	140	1 694	1 094	505	1 579	589	183	129	51	6 622
2006 Jan	331	185	250	1 556	837	457	1 687	559	177	108	68	6 327
Feb	266	163	183	1 430	783	423	1 581	497	258	130	53	5 848
Mar	424	177	274	1 611	1 021	544	1 828	604	257	178	48	7 133
Apr	296	178	114	1 541	865	354	1 634	407	145	135	35	5 859
May	352	165	272	1 547	1 092	411	1 725	382	213	156	39	6 514
Jun	360	142	172	1 462	1 329	157	1 711	490	122	169	43	6 345
Jul	318	138	128	1 594	1 112	134	1 716	548	150	109	41	6 189
Aug	238	147	132	1 538	1 272	225	1 692	570	143[†]	138	42	6 264[†]
Sep	225	143	164	1 493	1 130	237	1 713	638	129	144	35	6 244
Oct	174	171	202	1 552	1 118	337	1 785	495	246	146	42	6 435
Nov	293	153	162	1 565	1 011	404	1 808	529	162	131	43	6 411

1 Including amounts for petro-chemicals.
2 Ethane and other petroleum gases (OPG)
3 Excluding Orimulsion and refinery fuel.
4 Including other petroleum gases, aviation spirit, wide-cut gasoline, industrial and white spirits, petroleum wax, non-domestic standard burning oil and miscellaneous products, but excluding refinery fuel.
5 Includes gas oil, marine diesel oil and middle distillate feedstock.

6 There have been some modest changes made to this table following the review announced by DTI in the June 2006 edition of *Energy Trends*. The review was aimed to improve data coherency and coverage between the different quarterly and monthly DTI tables. Other petroleum gases are now shown. Middle distillate feedstock is now included with other gas/diesel oil rather than previously being combined with naphtha. Unleaded motor spirit and premier burning oil have been discontinued.

Source: Department of Trade and Industry: 020 7215 2698

9 Chemicals

9.1 Fertilisers

	Deliveries to UK agriculture[1]					
	Straight		Compounds[2]			
	Nitrogen total weight	Nitrogen[2] six monthly	Nitrogen six monthly	$P_2 O_5$ (phosphate) six monthly	$K_2 O$ (potash) six monthly	Compounds[3] total weight
	BIAI	DMYC	DMYD	DMYE	DMYF	DMYG
2002 Jan	131.2	..	..	..	..	201.3
Feb	120.9	..	..	..	..	255.2
Mar	159.2	..	..	..	..	402.3
Apr	176.8	..	..	..	..	420.2
May	81.8	..	..	..	..	204.2
Jun	49.8	229.7	294.0	144.9	185.5	108.3
Jul	147.3	..	..	..	..	112.8
Aug	234.2	..	..	..	..	123.8
Sep	152.1	..	..	..	..	156.7
Oct	140.6	..	..	..	..	128.2
Nov	161.1	..	..	..	..	100.3
Dec	140.1	96.2	310.8	72.1	82.9	117.8
2003 Jan	180.1	..	..	..	..	190.0
Feb	175.1	..	..	..	..	280.6
Mar	213.2	..	..	..	..	416.8
Apr	152.3	..	..	..	..	339.3
May	89.5	..	..	..	..	182.5
Jun	59.2	256.5	280.7	136.7	172.1	117.2
Jul	160.0	..	..	..	..	93.3
Aug	188.5	..	..	..	..	144.6
Sep	175.5	..	..	..	..	160.6
Oct	195.9	..	..	..	..	175.3
Nov	181.7	..	..	..	..	145.2
Dec	157.3	350.2	147.1	96.6	106.5	136.8
2004 Jan	186.2	..	..	..	..	173.0
Feb	149.1	..	..	..	..	219.9
Mar	156.6	..	..	..	..	293.4
Apr	148.0	..	..	..	..	258.2
May	69.1	..	..	..	..	174.1
Jun	55.9	..	..	..	..	102.4
Jul	333.2	..	..	..	..	95.8
Aug	147.4	..	..	..	..	111.1
Sep	136.1	..	..	..	..	137.8
Oct	150.8	..	..	..	..	136.9
Nov	176.3	..	..	..	..	100.4
Dec	132.5	..	..	..	..	154.8
2005 Jan	159.0	..	..	..	..	149.1
Feb	147.3	..	..	..	..	155.4
Mar	193.9	..	..	..	..	251.4
Apr	169.7	..	..	..	..	253.1
May	73.4	..	..	..	..	182.4
Jun	53.9	..	..	..	..	140.4
Jul	288.7	..	..	..	..	87.8
Aug	210.2	..	..	..	..	127.6
Sep	179.7	..	..	..	..	143.1
Oct	188.6	..	..	..	..	126.8
Nov	169.1	..	..	..	..	116.8
Dec	119.8	..	..	..	..	108.9

1 Deliveries by F.M.A. members only for years ended 30 June.
2 Nutrient content.
3 Total weight of compound fertilisers.

Sources: HMRC;
Agricultural Industries Confederation

9.2 Sulphur and sulphuric acid
Production and consumption: monthly averages or calendar months; stocks: end of period

Thousand tonnes

| | Sulphur and other materials used for sulphuric acid manufacture | | | | Sulphuric acid (as 100 per cent acid) | |
| | Consumption | | Stocks | | | |
	Sulphur	Zinc concentrates[1]	Sulphur	Zinc concentrates[1]	Production	Consumption
	BIBA	BIBC	BIBD	BIBH	BIBF	BIBG
1998	30.9	14.6	75.6	34.0	95.4	95.6
1999	26.4	17.1	100.1	26.8	87.0	90.0
2000	27.0	13.2	113.5	21.2	88.2	88.9
2001	23.1	15.8	113.6	19.8	78.8	78.0
2002	14.2	15.7	106.7	24.7	52.7	55.6
1999 Mar	27.7	19.8	93.8	25.8	92.5	99.7
Apr	26.6	17.9	97.3	24.1	90.8	84.2
May	24.4	23.2	90.1	32.8	84.9	97.5
Jun	30.1	14.2	100.9	24.3	96.0	90.2
Jul	23.0	11.9	101.4	28.9	76.4	82.8
Aug	30.2	20.3	108.6	24.0	93.4	93.6
Sep	26.3	12.0	104.1	23.1	81.8	85.6
Oct	24.3	12.4	113.9	26.5	80.8	72.0
Nov	29.5	18.7	107.5	23.5	94.9	96.6
Dec	25.7	11.2	108.8	26.8	86.3	85.6
2000 Jan	27.6	12.0	115.0	26.3	90.3	77.6
Feb	26.1	16.0	106.7	23.8	85.1	91.5
Mar	30.2	17.0	113.4	20.7	97.7	103.0
Apr	25.2	11.3	106.7	21.2	82.7	78.0
May	27.4	16.7	114.6	19.0	89.6	100.2
Jun	27.8	13.8	107.6	19.8	92.1	94.3
Jul	25.2	13.2	113.7	21.5	84.7	87.4
Aug	28.8	13.5	120.5	19.0	90.9	93.2
Sep	29.3	11.7	114.8	19.5	93.3	92.2
Oct	28.1	11.6	115.9	19.5	91.6	84.7
Nov	26.0	10.4	119.1	20.3	83.9	85.1
Dec	22.4	10.6	114.1	24.2	75.9	79.1
2001 Jan	26.4	16.9	111.9	22.7	90.5	87.1
Feb	24.2	17.0	104.6	14.5	76.2	73.9
Mar	26.1	9.5	112.2	18.3	85.3	80.1
Apr	26.4	10.7	115.5	22.2	88.7	84.6
May	23.3	13.5	120.1	19.8	75.8	71.9
Jun	29.2	18.2	116.8	17.8	101.6	103.0
Jul	19.3	19.4	118.6	16.5	71.4	82.7
Aug	20.4	13.5	114.7	21.8	74.2	57.4
Sep	26.2	14.6	113.2	21.7	86.2	111.6
Oct	17.1	16.6	111.7	23.6	62.7	48.8
Nov	23.3	22.7	112.3	19.3	80.3	75.4
Dec	14.8	16.6	111.5	19.6	52.7	59.6
2002 Jan	12.0	17.6	107.7	24.7	50.7	55.1
Feb	13.0	14.6	106.9	24.7	49.1	60.5
Mar	21.6	11.0	107.6	24.7	73.2	63.8
Apr	23.9	12.2	107.1	24.7	79.5	92.4
May	10.7	19.7	106.3	24.7	48.2	54.8
Jun	11.7	17.4	106.4	24.7	43.8	51.4
Jul	12.8	15.5	106.6	24.7	48.0	51.0
Aug	12.6	21.7	106.9	24.7	55.1	51.6
Sep	14.0	15.4	106.2	24.7	49.7	42.2
Oct	13.0	16.5	105.9	24.7	44.7	62.3
Nov	12.4	15.0	106.2	24.7	46.1	32.1
Dec	12.7	11.7	106.4	24.7	43.8	49.5
2003 Jan	12.8	11.8	107.4	22.0	42.1	42.4
Feb	13.1	..	107.3	..	32.8	43.4
Mar	13.5	..	106.9	..	35.9	38.8
Apr	13.3	..	107.3	..	34.4	31.8
May	13.7	..	107.5	..	36.1	41.2
Jun	14.1	..	106.7	..	37.0	36.0
Jul	11.6	..	107.5	..	29.1	25.9
Aug	13.3	..	107.1	..	36.4	42.5
Sep	13.5	..	107.2	..	35.2	42.7
Oct	16.7	..	107.5	..	33.8	37.8

1 From February 2003 these data are no longer available.

Source: National Sulphuric Acid Association

9.3 Basic chemicals, pesticides and other agro-chemical products[1]
Total UK manufacturers' sales by industry

£ Thousand

	Industrial gases	Dyes & pigments	Inorganic basic chemicals	Organic basic chemicals	Fertilisers & nitrogen compounds	Plastics	Synthetic rubber in primary forms	Pesticides & other agro-chemical products
Subclass (SIC 92)	24110	24120	24130	24140	24150	24160	24170	24200
	CKOM	CKON	CKOO	CKOP	CKOQ	CKOR	CKOS	CKOT
2002	508 320	1 021 589	1 108 953	5 448 893	699 232	3 426 657	..	483 274
2003	524 737	971 120	1 080 555	5 163 343	811 258	3 473 664	..	474 365
2004	528 194	935 693	1 089 852	5 824 800	786 130	3 739 668	..	470 069
2001 Q3	132 579	258 550	289 580	1 367 720	127 572	824 128	72 462	231 030
Q4	129 962	251 056	271 589	1 210 807	115 304	832 475	69 143	164 652
2002 Q1	126 078	271 288	276 965	1 444 131	240 384	804 568	75 699	142 921
Q2	129 957	267 774	291 208	1 345 502	187 600	941 368	72 897	136 247
Q3	127 322	252 441	275 448	1 469 071	135 047	863 006	80 677	99 130
Q4	124 962	230 087	265 332	1 190 189	136 202	817 716	..	104 976
2003 Q1	129 010	257 201	272 810	1 467 524	257 992	888 288	..	133 160
Q2	130 647	258 344	274 693	1 259 769	196 925	896 333	..	129 173
Q3	134 298	235 323	273 321	1 262 444	168 185	857 837	..	107 031
Q4	130 782	220 252	259 731	1 173 606	188 155	831 206	..	105 001
2004 Q1	135 245	241 672	276 144	1 581 316	253 383	935 744	..	135 681
Q2	133 256	251 657	271 560	1 557 527	190 625	970 095	..	132 577
Q3	128 752	229 504	265 429	1 380 591	168 019	928 094	..	96 243
Q4	130 941	212 860	276 719	1 305 366	174 103	905 735	..	105 568

1 As from the end of 2004, quarterly data will not be collected for these industries. Annual results for the years 2005 onwards will be published on the ONS website at http://www.statistics.gov.uk/ as and when available.

Source: Office for National Statistics: 01633 813395

9.4 Pharmaceutical products, soaps and other cleaning preparations and perfumes[1]
Total UK manufacturers' sales by industry

£ Thousand

	Pharmaceutical products			Perfumes and essential oils	
	Basic products	Preparations	Soap & detergents, cleaning & polishing preparations	Perfumes & toilet preparations	Essential oils
Subclass (SIC 92)	24410	24420	24510	24520	24630
	CKOU	CKOV	CKOW	CKOX	CKOY
2002	867 982	8 318 855	1 886 334	2 377 369	548 137
2003	748 613	9 230 833	1 718 431	2 313 845	..
2004	733 658	8 760 908	1 804 671	2 171 087	..
2001 Q3	157 878	1 996 266	493 958	686 084	137 590
Q4	192 833	2 244 546	498 112	692 288	141 500
2002 Q1	229 737	2 066 964	494 464	550 720	135 312
Q2	216 451	2 098 634	480 091	575 818	139 684
Q3	230 131	2 123 729	493 566	605 893	..
Q4	191 663	2 029 527	418 214	644 938	..
2003 Q1	193 191	2 160 693	409 896	563 410	..
Q2	203 869	2 285 852	423 457	552 000	..
Q3	171 582	2 326 716	469 822	581 783	..
Q4	179 971	2 457 573	415 255	616 652	..
2004 Q1	224 562	2 288 169	445 503	520 673	..
Q2	174 383	2 131 549	455 364	550 685	..
Q3	163 555	2 157 741	459 350	542 423	..
Q4	171 158	2 183 449	444 454	557 306	..

1 As from the end of 2004, quarterly data will not be collected for these industries. Annaul results for the years 2005 onwards will be published on the ONS website at http://www.statistics.gov.uk/ as and when available.

Source: Office for National Statistics: 01633 813395

9.5 Other chemical products[1]
Total UK manufacturers' sales by industry

£ Thousand

	Paints, varnishes, & similar coatings; printing ink, mastic & sealants	Explosives	Glues & gelatines	Photographic chemical materials	Prepared unrecorded media	Other chemical products	Man made fibres
Subclass (SIC 92)	24300	24610	24620	24640	24650	24660	24700
	CKOZ	CKPA	CKPB	CKPC	CKPD	CKUX	CKUY
2002	2 725 613	..	357 805	305 373	124 354	1 941 449	601 807
2003	2 792 607	105 579	371 321	316 808	127 414	1 768 073	618 236
2004	2 775 868	109 706	400 400	249 915	75 297	1 992 133	586 931
2001 Q3	639 252	30 601	93 998	83 654	..	496 398	146 061
Q4	600 333	30 679	87 721	72 319	26 762	511 098	139 144
2002 Q1	663 064	..	91 933	73 174	28 553	503 830	152 738
Q2	700 851	23 388	88 789	77 156	32 122	491 787	166 738
Q3	717 210	21 454	91 637	76 799	28 940	492 446	154 579
Q4	644 487	25 649	85 446	78 244	34 740	453 386	127 751
2003 Q1	699 164	24 431	95 886	81 095	32 666	436 132	151 992
Q2	734 341	26 225	91 215	80 446	34 179	427 189	164 116
Q3	717 652	28 047	95 086	79 738	32 200	452 312	157 736
Q4	641 449	26 877	89 134	75 529	28 369	452 440	144 393
2004 Q1	685 386	25 221	100 975	65 487	26 612	501 207	160 128
Q2	712 036	27 150	95 564	63 485	..	497 776	145 842
Q3	731 486	27 665	105 216	61 747	..	486 379	144 281
Q4	646 960	29 670	98 645	59 196	..	506 771	136 680

1 As from the end of 2004, quarterly data will not be collected for these industries. Annual results for the years 2005 onwards will be published on the ONS website at http://www.statistics.gov.uk/ as and when available.

Source: Office for National Statistics: 01633 813395

10 Metals, engineering and vehicles

10.1 Iron and steel
Weekly averages   Stocks: end of period

Thousand tonnes

	Consump-tion of imported iron ore[2]	Iron			Stocks[1]		Crude steel production	Finished steel products	
		Production in blast furnaces[3]	Consump-tion in steel-making	Total stocks[4]	Consump-tion in steel-making	Total stocks		Net home and export deliveries	At producers' works[5]
	BJAB	BJAC	BJAD	BJAE	BJAF	BJAG	BJAH	BJAI	BJAJ
2003	296.0	193.0	187.8	106.2	84.4	207.9	250.3	234.5	1 622.8
2004	306.4	195.8	192.5	29.5	97.1	242.4	264.7	256.6	1 666.1
2005	305.9	195.9	192.0	7.3	87.1	228.4	254.6	247.6	1 816.4
2005 Jan	319.6	200.9	197.4	22.1	80.7	251.2	258.4	216.9	1 771.0
Feb	316.9	190.4	181.6	15.0	89.2	251.9	254.2	258.4	1 672.7
Mar	271.6	187.0	177.9	12.9	90.7	234.5	252.1	267.7	1 608.8
Apr	335.8	213.2	209.5	9.8	110.5	221.0	290.5	256.5	1 733.2
May	342.1	211.2	208.0	8.2	106.3	242.0	286.9	270.2	1 786.1
Jun	295.9	192.2	188.7	6.2	76.1	210.1	242.9	247.3	1 766.4
Jul	294.8	190.1	188.6	5.3	75.1	223.0	241.6	235.3	1 791.6
Aug	303.2	186.3	184.2	5.0	77.1	237.5	238.7	219.9	1 804.4
Sep	287.0	180.5	177.8	5.2	90.1	271.0	239.9	290.2	1 602.9
Oct	268.4	192.3	187.7	8.9	85.3	262.2	246.5	238.4	1 605.2
Nov	310.0	203.4	201.5	5.5	77.0	257.4	255.9	253.7	1 583.9
Dec	337.0	207.4	204.6	7.3	70.4	225.8	257.0	222.4	1 816.4
2006 Jan	325.9	211.7	208.3	7.5	81.7	269.3	263.4	253.0	1 712.3
Feb	325.4	210.4	204.8	8.8	90.3	245.3	269.3	265.4	1 712.3
Mar	328.7	216.8	210.8	11.8	101.5	245.0	284.1	276.7	1 708.1
Apr	317.6	207.5	203.6	7.4	106.8	237.0	282.1	277.3	1 701.4
May	318.4	204.5	202.5	4.6	99.0	246.9	277.0	261.4	1 698.0
Jun	319.9	208.8	207.3	5.0	107.9	250.1	287.8	297.4	1 615.8
Jul	292.0	200.5	197.6	9.1	98.7	292.4	269.4	251.6	1 641.1
Aug	307.8	207.2	201.5	17.0	84.5	251.1	261.2	215.9	1 951.3

1 Excludes iron foundries and refined iron works.
2 Including manganese ore.
3 Includes blast furnace ferro-alloys.

4 Includes blast furnace ferro-alloys, but excludes iron foundries and refined iron works.
5 Stocks of ingots, semi-finished and finished steel.

Source: UK Iron and Steel Statistics Bureau

10.2 Supplies and deliveries of steel
Weekly averages

Thousand tonnes (crude steel equivalent)

	Supply from home sources							
	Crude steel production		Producers' stock changes[1]	Re-usable material[2]	Total	Imports[3]	Exports[3]	Net home disposals
	Total	of which: alloy						
	BJBA	BJBB	BJBC	BJBD	BJBE	BJBF	BJBG	BJBH
2001	260.4	20.4	−2.6	–	263.0	173.6	145.1	291.5
2002	224.4	19.4	−1.1	–	225.5	185.8	141.4	269.9
2003	250.3	18.4	–	–	250.3	175.5	153.2	272.7
2004	264.7	18.4	0.7	–	264.0	187.4	163.7	287.8
2005	254.5	16.6	3.6	–	251.0	159.8	182.3	228.5
2006	277.5	14.9	−85.9	–	363.5	178.1	190.6	351.0
2003 Q3	253.8	16.1	11.3	–	242.5	161.8	154.7	249.6
Q4	254.2	17.6	−11.1	–	265.4	167.8	161.6	271.5
2004 Q1	260.1	19.2	−12.8	–	272.8	174.2	157.5	289.6
Q2	283.2	19.7	−5.5	–	288.7	182.9	171.2	300.4
Q3	261.9	18.2	17.1	–	244.8	197.1	160.8	281.2
Q4	253.7	16.6	4.0	–	249.8	195.5	165.2	280.0
2005 Q1	254.7	20.6	−5.4	–	260.1	191.3	180.7	270.7
Q2	271.1	20.8	14.9	–	256.2	175.4	194.0	237.6
Q3	238.9	12.2	−15.5	–	254.4	131.3	168.9	216.8
Q4	253.4	12.7	20.2	–	233.2	141.6	185.5	189.3
2006 Q1	273.2	13.9	−10.2	–	283.4	176.2	187.4	272.2
Q2	281.9	15.9	−19.4	–	301.3	180.0	193.8	287.6

1 Increases in stock are shown as + and decreases in stock (ie deliveries from stock) as -.
2 Currently mainly old rails for re-rolling.
3 Derived from HM Customs statistics.

Source: UK Iron and Steel Statistics Bureau

10.3 Aluminium
Monthly averages or calendar months; stocks: end of period

Thousand tonnes

	Production		Despatches to customers				
	Primary[1]	Secondary[2]	Primary[1]	Secondary	Rolled products	Extrusions and tubes	Castings
	BJDH	BJDI	BJDJ	BJDK	BJDN	BJDO	BJDM
2000	25.4	19.8	25.1	19.8	34.9	15.4	11.2
2001	28.4	20.7	28.1	20.7	32.1	14.0	10.8
2002	28.7	17.3	29.0	17.3	26.0	13.2	10.6
2003	28.6	17.2	28.3	17.2	22.9	12.4	10.6
2004	30.0	15.5	30.6	15.5	22.2	12.3	9.2
2005	30.7	11.2	30.7	11.2	22.5	11.0	8.6
2002 Oct	28.3	17.3	30.7	17.3	26.0	14.7	10.6
Nov	27.8	17.3	27.2	17.3	26.0	13.5	10.6
Dec	31.1	17.3	25.1	17.3	26.0	8.2	10.6
2003 Jan	28.8	17.2	34.9	17.2	22.9	13.2	10.6
Feb	29.7	17.2	29.8	17.2	22.9	12.3	10.6
Mar	28.7	17.2	29.0	17.2	22.9	13.1	10.6
Apr	28.0	17.2	29.7	17.2	22.9	12.4	10.6
May	31.3	17.2	32.3	17.2	22.9	12.3	10.6
Jun	28.1	17.2	28.1	17.2	22.9	12.5	10.6
Jul	28.7	17.2	26.3	17.2	22.9	13.7	10.6
Aug	29.4	17.2	27.1	17.2	22.9	10.5	10.6
Sep	25.9	17.2	27.0	17.2	22.9	13.4	10.6
Oct	26.8	17.2	28.1	17.2	22.9	14.4	10.6
Nov	29.1	17.2	27.4	17.2	22.9	13.0	10.6
Dec	28.5	17.2	20.2	17.2	22.9	8.3	10.6
2004 Jan	29.5	15.5	34.7	15.5	22.2	12.7	11.6
Feb	27.6	15.5	29.4	15.5	22.2	12.6	11.6
Mar	29.6	15.5	31.2	15.5	22.2	14.7	11.6
Apr	28.8	15.5	30.9	15.5	22.2	12.8	11.6
May	30.1	15.5	27.5	15.5	22.2	12.1	11.6
Jun	29.6	15.5	35.6	15.5	22.2	13.1	11.6
Jul	30.7	15.5	27.8	15.5	22.2	13.7	11.6
Aug	30.9	15.5	32.2	15.5	22.2	10.9	11.6
Sep	30.2	15.5	32.1	15.5	22.2	13.2	11.6
Oct	31.1	15.5	28.9	15.5	22.2	12.4	11.6
Nov	30.2	15.5	32.2	15.5	22.2	12.2	11.6
Dec	31.4	15.5	24.5	15.5	22.2	7.2	11.6
2005 Jan	31.3	11.2	30.2	11.2	22.5	11.2	8.6
Feb	28.0	11.2	26.9	11.2	22.5	11.1	8.6
Mar	31.5	11.2	29.4	11.2	22.5	12.0	8.6
Apr	30.4	11.2	30.0	11.2	22.5	12.0	8.6
May	31.1	11.2	31.3	11.2	22.5	11.6	8.6
Jun	30.4	11.2	30.3	11.2	22.5	11.8	8.6
Jul	31.5	11.2	29.8	11.2	22.5	10.8	8.6
Aug	31.2	11.2	34.2	11.2	22.5	10.8	8.6
Sep	30.5	11.2	30.3	11.2	22.5	11.9	8.6
Oct	31.1	11.2	32.1	11.2	22.5	11.3	8.6
Nov	30.4	11.2	31.8	11.2	22.5	11.1	8.6
Dec	30.8	11.2	31.8	11.2	22.5	6.9	8.6
2006 Jan	30.9	..	31.5	..	..	10.8	..
Feb	27.9	..	28.3	..	..	10.8	..
Mar	31.1	..	31.7	..	..	12.6	..
Apr	30.0	..	28.4	..	..	10.0	..
May	30.1	..	33.3	..	..	11.8	..
Jun	29.7	..	31.9	..	..	11.6	..
Jul	31.2	..	29.0	..	..	11.0	..
Aug	31.2	..	30.7	..	..	10.3	..
Sep	28.7	..	31.6	..	..	11.2	..

1 Including the pure content of primary alloys.
2 Including the primary content used in the production of secondary metal.

Source: Aluminium Federation: 0121 456 1103

10.4 Total engineering
Total turnover of UK - based manufacturers[1,2,3]
Standard Industrial Classification 2003

£ millions

Division Description		2004	2005	2005 Q2	2005 Q3	2005 Q4	2006 Q1	2006 Q2	2006 Q3
Division 29 : Manufacture of machinery and equipment not elsewhere classified									
2911 Manufacture of engines and turbines except aircraft, vehicle & cycle engines	MXVO	1 632	1 847	486	448	467	443	483	511
2912 Manufacture of pumps and compressors	MXXO	2 957	2 872	720	720	745	767	796	822
2913 Manufacture of taps and valves	MXZH	1 206	1 231	329	319	290	301	302	308
2914 Manufacture of bearings, gears, gearing and driving elements	MYCT	988	947	244	224	238	257	259	259
2922 Manufacture of lifting and handling equipment	MYLS	3 142	3 542	902	844	932	992	962	987[†]
2923 Manufacture of non-domestic cooling and ventilation equipment	MYPT	3 340	3 393	833	873	873	853	966	986
2924 Manufacture of other general purpose machinery not elsewhere classified	MYRM	2 946	3 027	823	745	702	752	769	731[†]
2941 Manufacture of metalworking machine tools	MYWY	603	730[†]	182	183	208[†]	198	214	211
2949 Manufacture of other machine tools	MYYP	789	699	168	185	182	175	198	204
2952 Manufacture of machinery for mining, quarrying and construction	MZCE	2 933	3 098	744	801	824	798	847[†]	794
2953 Manufacture of machinery for food, beverage and tobacco processing	MZFS	950	948	242	232	225	240	241	251
2954 Manufacture of machinery for textile, apparel and leather production	MZJP	172	123	31	30	29	28	32	31[†]
2956 Manufacture of other special purpose machinery not elsewhere classified	MZQF	2 218	2 262	533	576	593	543	546	542
2971 Manufacture of electric domestic appliances	MZTZ	2 460	2 642	630	637	737	654	635	687
Division 30 : Manufacture of electrical and optical equipment									
3001 Manufacture of office machinery	MZXQ	815	889	222	215	226	327	319	291
3002 Manufacture of computers and other information processing equipment	VBCE	5 014	4 234	981	1 077	1 080	886	777	731[†]
Division 31 : Manufacture of electrical machinery and apparatus not elsewhere classified									
3110 Manufacture of electric motors, generators and transformers	VBEB	2 331	2 348	616	591	571	642[†]	617	656
3120 Manufacture of electricity distribution and control apparatus	VBFU	3 641	3 459	867	891	860	929	957	930[†]
3130 Manufacture of insulated wire and cable	VBHW	1 123	959	251	237	245	301	326	351
3140 Manufacture of accumulators, primary cells and primary batteries	VBJW	461	440	105	111	119	87	85	86[†]
3150 Manufacture of lighting equipment and electric lamps	VBLP	1 442	1 331	322	336	311	332	322	358[†]
3161 Manufacture of other electrical equipment for engines and vehicles not otherwise classified	VBNI	1 006	1 004	259	239	248	242	243	223
3162 Manufacture of other electrical equipment not elsewhere classified	VBPK	2 639	2 694	674	685	659	702	665	728[†]
Division 32 : Manufacture of radio, television and communication equipment and apparatus									
3210 Manufacture of electronic valves and tubes and other electronic components	VBRI	4 215	3 917	939	975	1 006	996	953	966
3220 Manufacture of television and radio transmitters and apparatus for line telephony and line telegraphy	VBTF	4 170	3 610	865	942	956	1 053	1 056	879[†]
3230 Manufacture of television and radio receivers, sound or video recording or reproducing apparatus and associated goods	VBVJ	3 781	3 229	718	732	963	896	915	832
Division 33 : Manufacture of medical, precision and optical instruments, watches and clocks									
3310 Manufacture of medical and surgical equipment and orthopaedic appliances	VBXH	3 543	3 749	957	956	879	974	900	839
3320 Manufacture of instruments and applicances for measuring, checking, testing, navigating and other purposes, except industrial process control equipment	VBZF	7 046	6 853	1 804	1 668	1 732	1 580	1 619	1 661[†]
3340 Manufacture of optical instruments and photographic equipment	VCCV	1 070	1 060	275	274	253	277	277	261

1 The figures shown represent the output of UK - based manufacturers classified to Subsections DK and DL of the Standard Industrial Classification 2003. The figures shown are derived from the monthly production inquiry (MPI) and include estimates for non-responders and for establishments which are not sampled.

2 Orders on hand figures are given for the end of the period to which they relate.
3 The data on this table are not seasonally adjusted.

Source: Office for National Statistics: 01633 813351

67

10.5 Mechanical, instrument and electrical engineering industries[1]
Seasonally adjusted volume index numbers of turnover: Standard Industrial Classification 2003

2000 average monthly sales = 100

	Total engineering			Machinery and equipment			Electrical and optical equipment		
	Total	Home	Export	Total	Home	Export	Total	Home	Export
	JIQJ	JIQD	JIQG	JINY	JINS	JINV	JIPR	JIPL	JIPO
2002	84.5	91.8	74.8	95.4	97.2	92.3	79.5	89.0	68.5
2003	81.6	90.2	70.3	97.2	97.6	96.4	74.5	86.2	60.9
2004	82.1	89.3	72.6	102.5	100.2	106.7	72.7	83.3	60.5
2005	80.8	89.0	70.0	105.6	100.9	113.8	69.5	82.4	54.4
2005 Feb	81.6	91.1	69.0	106.3	104.8	109.1	70.3	83.7	54.8
Mar	79.6	87.6	69.0	104.8	102.1	109.4	68.0	79.6	54.6
Apr	81.8	89.8	71.3	105.1	101.0	112.5	71.1	83.7	56.6
May	80.2	88.7	68.9	103.3	100.3	108.6	69.6	82.3	54.7
Jun	80.1	89.1	68.1	105.5	102.9	110.1	68.4	81.6	53.2
Jul	80.5	88.7	69.7	103.7	101.0	108.6	69.9	81.9	55.9
Aug	81.0	89.1	70.3	103.4	97.0	114.7	70.8	84.8	54.5
Sep	82.6	89.0	74.1	111.2	100.7	129.7	69.5	82.6	54.3
Oct	80.0	88.8	68.3	104.6	100.0	112.8	68.7	82.7	52.5
Nov	80.5	87.8	71.0	107.0	99.5	120.4	68.4	81.4	53.3
Dec	80.6	87.2	72.0	108.0	99.3	123.4	68.1	80.5	53.7
2006 Jan	80.4	86.1	72.8	105.0	97.9	117.6	69.1	79.7	56.8
Feb	80.8	87.7	71.8	104.6	99.1	114.2	70.0	81.5	56.7
Mar	82.4	88.2	74.7	108.1	100.4	121.6	70.7	81.6	58.0
Apr	81.2	88.7	71.4	107.2	101.6	117.2	69.3	81.6	55.1
May	83.1	90.9	72.8	110.9	105.3	121.0	70.3	83.0	55.6
Jun	83.5	91.6	72.7	112.1	106.5	122.0	70.4	83.5	55.2
Jul	81.9	90.4	70.7	110.6	104.2	122.0	68.8	82.9	52.4
Aug	82.9	91.2	71.8	111.5	105.5	122.1	69.7	83.4	54.0
Sep	82.1	88.7	73.4	115.1	107.9	128.0	67.0	78.2	54.0

1 Footnotes as 1 and 2 on Table 10.4.

Source: Office for National Statistics: 01633 813351

Note: From December 2006 (reference period October 2006) constant price seasonally adjusted data will no longer be available.

10.6 Mechanical, instrument and electrical engineering industries[1]
Seasonally adjusted volume index numbers of orders on hand: Standard Industrial Classification 2003

2000 average monthly sales = 100

	Total engineering			Machinery and equipment			Electrical and optical equipment		
	Total	Home	Export	Total	Home	Export	Total	Home	Export
	JIQI	JIQC	JIQF	JINX	JINR	JINU	JIPQ	JIPK	JIPN
2002	92.6	104.5	72.4	100.5	100.4	101.0	87.5	107.6	58.4
2003	92.6	108.4	65.8	111.4	120.5	91.1	80.5	99.3	53.3
2004	88.9	102.5	65.8	106.5	111.9	94.5	77.6	95.5	51.6
2005	92.9	103.7	74.6	103.9	101.4	109.3	85.9	105.5	57.5
2005 Feb	89.2	102.6	66.4	105.5	111.3	92.6	78.7	96.0	53.5
Mar	89.8	101.3	70.3	108.5	110.8	103.6	77.7	94.2	53.9
Apr	89.2	102.9	65.9	107.1	111.2	98.1	77.6	96.6	50.1
May	89.7	101.6	69.5	108.5	110.4	104.5	77.6	95.1	52.3
Jun	89.8	100.6	71.4	108.6	110.5	104.3	77.7	93.3	55.2
Jul	89.7	99.6	72.9	107.3	108.5	104.7	78.3	92.9	57.3
Aug	92.0	103.0	73.3	108.0	109.0	105.9	81.7	98.5	57.2
Sep	91.6	102.9	72.4	105.2	106.3	102.8	82.9	100.4	57.4
Oct	92.4	104.0	72.6	105.6	107.3	101.9	83.9	101.6	58.2
Nov	92.3	103.5	73.3	105.9	106.7	104.2	83.6	101.2	58.1
Dec	92.9	103.7	74.6	103.9	101.4	109.3	85.9	105.5	57.5
2006 Jan	91.7	101.0	75.9	101.9	97.3	112.2	85.1	103.8	58.1
Feb	94.3	105.1	75.9	104.1	99.7	113.7	88.0	109.2	57.3
Mar	92.3	102.5	75.1	102.6	97.8	112.9	85.8	106.0	56.5
Apr	93.2	103.1	76.3	102.6	96.9	115.4	87.1	107.7	57.1
May	92.9	102.3	77.1	104.0	98.7	115.6	85.9	105.0	58.1
Jun	94.0	104.6	76.0	104.2	100.4	112.7	87.4	107.7	57.9
Jul	93.1	103.7	75.1	104.3	100.3	112.9	85.9	106.3	56.5
Aug	94.3	105.3	75.5	104.3	100.7	112.2	87.8	108.8	57.4
Sep	95.3	105.6	77.8	105.2	100.3	116.0	89.0	109.6	59.1

1 Footnotes as 1 and 2 on Table 10.4.

Source: Office for National Statistics: 01633 813351

Note: From December 2006 (reference period October 2006) constant price seasonally adjusted data will no longer be available.

10.7 Mechanical, instrument and electrical engineering industries[1,2]

Seasonally adjusted volume index numbers of new orders: Standard Industrial Classification 2003

2000 average monthly sales = 100

	Total engineering			Machinery and equipment			Electrical and optical equipment		
	Total	Home	Export	Total	Home	Export	Total	Home	Export
	JIQH	JIQB	JIQE	JINW	JINQ	JINT	JIPP	JIPJ	JIPM
2002	80.8	87.9	71.2	95.6	98.7	90.2	74.2	82.4	64.4
2003	78.9	87.9	66.8	99.7	105.2	90.2	69.6	79.0	58.4
2004	78.3	83.9	70.8	99.2	96.3	104.2	69.0	77.5	58.8
2005	79.3	85.8	70.5	102.9	96.0	114.8	68.7	80.5	54.6
2005 Feb	76.8	80.2	72.2	111.7	110.3	114.2	61.2	64.7	57.1
Mar	79.1	79.2	79.0	116.8	99.2	147.1	62.3	68.8	54.5
Apr	77.0	92.4	56.3	97.3	102.6	88.2	67.9	87.1	44.8
May	79.4	80.4	77.9	108.1	95.9	129.0	66.6	72.4	59.5
Jun	77.8	81.9	72.3	104.2	103.2	105.8	66.0	70.9	60.2
Jul	77.4	81.1	72.5	96.6	91.0	106.3	68.9	76.0	60.3
Aug	86.4	98.8	69.8	105.0	98.9	115.6	78.1	98.7	53.3
Sep	78.6	85.3	69.6	97.1	87.4	113.7	70.4	84.3	53.6
Oct	80.1	89.7	67.2	104.9	104.4	105.7	69.0	82.1	53.3
Nov	77.7	82.5	71.3	106.9	96.4	125.1	64.7	75.3	51.9
Dec	80.1	84.5	74.1	97.3	73.4	138.5	72.4	90.3	51.0
2006 Jan	73.5	72.3	75.1	94.9	77.6	124.7	64.0	69.6	57.3
Feb	87.2	100.1	69.9	112.6	110.5	116.2	75.9	94.8	53.2
Mar	72.8	74.6	70.4	99.7	91.1	114.5	60.8	66.1	54.5
Apr	81.5	87.5	73.4	106.1	96.5	122.6	70.5	82.9	55.7
May	79.6	84.4	73.2	115.3	113.7	117.9	63.7	69.2	57.0
Jun	84.3	96.7	67.5	111.5	114.1	107.0	72.1	87.8	53.3
Jul	76.3	83.6	66.4	109.2	103.6	118.7	61.6	73.3	47.6
Aug	84.2	93.9	71.1	109.9	106.8	115.2	72.7	87.2	55.2
Sep	83.1	86.4	78.8	117.5	105.5	138.2	67.8	76.5	57.4

1 The figures shown represent the output of UK - based manufacturers classified to Subsections DK and DL of the Standard Industrial Classification 2003. The figures shown are derived from the monthly production inquiry (MPI) and include estimates for non-responders and for establishments which are not sampled.

2 Orders on hand figures are given for the end of the period to which they relate.

Note: From December 2006 (reference period October 2006) constant price seasonally adjusted data will no longer be available.

Source: Office for National Statistics: 01633 813351

10.8 Passenger cars

Number

	Total production					Production for export				
	1000cc and under	Over 1000cc and not over 1600cc	Over 1600cc and not over 2500cc	Over 2500cc	Total	1000cc and under	Over 1000cc and not over 1600cc	Over 1600cc and not over 2500cc	Over 2500cc	Total
	GKAB	GKAD	GKAF	GKAH	JCYM	GKAC	GKAE	GKAG	GKAI	JCYL
2000	96 043	676 438	723 294	145 677	1 641 452	56 556	375 528	509 591	121 315	1 062 990
2001	93 695	632 747	634 573	131 350	1 492 365	56 426	329 944	400 648	107 236	894 254
2002	79 545	711 553	720 067	118 579	1 629 744	35 866	442 975	470 285	98 158	1 047 284
2003	23 985	750 840	740 486	142 247	1 657 558	12 380	503 950	509 050	118 379	1 143 759
2004	15 471	796 174	690 759	144 346	1 646 750	10 316	560 505	492 564	116 371	1 179 756
2005	6 111	854 687	546 744	188 155	1 595 697	4 925	625 929	405 204	148 445	1 184 503
2006	–	792 187	446 143	203 755	1 442 085	–	622 205	324 880	159 008	1 106 093
2005 Oct	–	66 379	42 397	16 007	124 783	–	51 316	34 742	13 366	99 424
Nov	–	82 428	43 406	23 828	149 662	–	64 942	34 971	19 474	119 387
Dec	–	52 536	30 092	12 655	95 283	–	43 779	24 739	9 358	77 876
2006 Jan	–	62 453[†]	40 296[†]	16 373[†]	119 122	–	45 143[†]	28 836[†]	12 488[†]	86 467
Feb	–	74 048	38 871	18 274	131 193[†]	–	53 830	27 236	14 145	95 211[†]
Mar	–	88 834	46 000	24 154	158 988	–	67 774	32 449	19 499	119 722
Apr	–	65 103	37 156	16 313	118 572	–	52 814	28 855	13 504	95 173
May	–	74 864	39 886	17 566	132 316	–	60 704	30 835	13 901	105 440
Jun	–	78 055	42 280	18 991	139 326	–	61 508	30 749	14 566	106 823
Jul	–	66 528	37 241	14 060	117 829	–	51 223	26 995	10 692	88 910
Aug	–	35 223	23 067	14 669	72 959	–	25 963	15 333	10 810	52 106
Sep	–	67 981	37 354	16 977	122 312	–	53 839	25 365	13 083	92 287
Oct	–	66 536	35 526	14 081	116 143	–	56 995	27 157	11 620	95 772
Nov	–	68 181	38 726	21 656	128 563	–	56 623	28 557	17 085	102 265
Dec	–	44 381	29 740	10 641	84 762	–	35 789	22 513	7 615	65 917

Source: Office for National Statistics: 01633 812394

10.9 Commercial motor vehicles[1]

Number

	Total production						Production for export					
	Light Commercial vehicles	Gross Vehicle Weight Trucks		Motive units	Buses, coaches and mini-buses	Total	Light Commercial vehicles	Gross Vehicle Weight Trucks		Motive units	Buses, coaches and mini-buses	Total
		Under 7.5 tonnes	Over 7.5 tonnes					Under 7.5 tonnes	Over 7.5 tonnes			
	GKDH	GKDJ	GKDL	GKCV	GKDN	JCYG	GKDI	GKDK	GKDM	GKCW	GKDO	JCYF
2000	145 655	5 160	6 849	2 673	12 105	172 442	65 636	1 032	3 059	129	6 325	76 181
2001	169 705	5 000	7 359	2 539	8 270	192 873	87 208	1 307	3 315	151	4 238	96 224
2002	168 311	4 600	7 357	1 795	9 204	191 267	104 902	1 157	3 474	70	4 631	114 234
2003	166 359	4 151	7 779	2 095	8 487	188 871	94 887	806	3 494	130	3 709	102 917
2004	178 887	4 977	8 537	2 558	14 334	209 293	113 076	659	3 626	164	10 582	128 107
2005	171 866	5 533	9 756	2 755	16 843	206 753	112 647	763	4 258	190	12 415	130 273
2006	175 713	4 418	11 447	2 230	13 896	207 704	118 632	844	5 523	179	11 041	136 219
2005 Dec	11 171	339	741	135	1 172	13 558	8 910	70	313	18	990	10 301
2006 Jan	15 456	570	987	169	978	18 160	10 645	72	346	22	720	11 805
Feb	15 521	472	1 033	142	1 046	18 214	10 843	62	411	13	807	12 136
Mar	18 323	464	1 146	207	1 179	21 319	12 270	99	489	20	904	13 782
Apr	13 443	457	875	167	1 347	16 289	10 091	98	368	24	1 222	11 803
May	12 751	404	978	169	793	15 095	9 102	43	465	10	688	10 308
Jun	14 129	362	1 096	175	1 282	17 044	8 923	52	527	15	1 100	10 617
Jul	13 287	227	840	98	866	15 318	9 283	33	436	8	665	10 425
Aug	6 804	391	944	183	426	8 748	3 454	107	521	8	273	4 363
Sep	18 614	477	950	301	1 089	21 431	11 564	67	457	4	645	12 737
Oct	16 526	207	871	308	1 957	19 869	11 308	69	515	14	1 651	13 557
Nov	17 953	249	967	220	1 869	21 258	12 665	87	560	28	1 533	14 873
Dec	12 906	138	760	91	1 064	14 959	8 484	55	428	13	833	9 813

Source: Office for National Statistics: 01633 812394

11 Textiles and other manufactures

11.1 Index numbers of textile and clothing industries
Standard Industrial Classification 2003

2003=100, seasonally adjusted

	Textile industry (production)							
	Man-made fibres	All textiles[1]	Preparation and spinning of textile fibres	Textile weaving	Manufacture of knitted and crocheted fabrics	Finishing of textiles	Manufacture of other textiles	Manufacture of made-up textile articles except apparel
SIC 2003 classification	2 470	17	171	172	176	173	175	174
	AHXI	AIMS	AIOE	AIOF	AHGJ	AHGE	AHGQ	AHGF
2003	100.0	100.0	100.0	100.0	100.0	100.0	100.0	100.0
2004	99.2	93.0	89.5	89.1	94.1	96.4	86.6	101.4
2005	79.9	91.4	66.3	89.8	84.9	93.3	87.6	100.2
2006	79.4	84.8	67.0	80.5	80.6	99.6	87.5	93.8
2004 Q3	95.5	93.6	88.2	86.0	94.4	96.6	86.1	103.7
Q4	98.3	92.5	92.5	87.2	85.7	96.5	83.1	102.7
2005 Q1	95.0	90.9	70.6	89.9	91.8	83.9	87.9	100.4
Q2	80.7	93.2	72.8	89.9	86.3	91.7	91.3	99.9
Q3	68.9	90.9	57.8	92.7	79.1	98.7	84.7	99.6
Q4	74.9	90.5	64.1	86.6	82.5	98.9	86.6	100.9
2006 Q1	72.0†	86.3†	71.9†	83.6†	88.4†	105.4†	84.6†	93.2†
Q2	76.5	84.7	66.7	86.1	80.4	98.7	85.8	93.1
Q3	85.5	85.2	66.8	80.1	79.5	100.6	89.6	94.5
Q4	83.7	83.1	62.5	72.1	74.1	93.5	89.9	94.7

	Clothing industry (production)				
	Manufacture of wearing apparel, dressing and dyeing of fur[2]	Manufacture of other outerwear	Manufacture of workwear	Manufacture of underwear[3]	Manufacture of other wearing apparel and accessories nec[3]
SIC 2003 classification	18	1822	1821	1823	1824
	AIMT	AHGU	AHGT	AHGV	AHGW
2003	100.0	100.0	100.0	100.0	100.0
2004	89.5	95.4	87.5	105.3	71.6
2005	87.4	93.4	83.2	..	..
2006	88.5	90.9	89.0	..	..
2004 Q3	86.2	92.7	86.0	102.8	66.0
Q4	85.6	89.7	89.8	107.0	65.2
2005 Q1	82.8	84.9	90.9	..	..
Q2	87.5	92.3	81.6	..	..
Q3	88.6	95.7	78.8	..	..
Q4	90.7	100.6	81.7	..	..
2006 Q1	88.0†	98.6†	91.8†	..	..
Q2	89.9	92.3	91.0	..	..
Q3	86.7	86.1	84.8	..	..
Q4	89.6	86.8	88.5	..	..

1 In addition to the sectors listed, this includes throwing, texturing, etc of continuous filament yarn; spinning and weaving of flax, hemp and ramie; jute and polypropylene yarns and fabrics, and miscellaneous textiles (ie lace; rope, twine and net; narrow fabrics and other miscellaneous textiles).

2 In addition to the sectors listed, this includes hats, caps and millinery; gloves, other dress industries (ie swimwear and foundation garments; umbrellas and miscellaneous industries).

3 For confidentiality reasons, the data for these industries have been suppressed from 2005 Q1 onwards. This suppression is required because both industries now fail ONS disclosure rules and are therefore no longer available for publication.

Source: Office for National Statistics: 01633 812319

11.2 Household textiles, non-woven products, canvas and ropes[1]
Total UK manufacturers' sales by industry

£ Thousand

	Household textiles			Non-woven excluding apparel	Canvas goods, sacks etc	Cordage rope, twine & netting
	Soft furnishings	Household textiles	Carpets & rugs			
Subclass (SIC 92)	17401	17403	17510	17530	17402	17520
	CKPE	CKPF	CKPG	CKPH	CKPI	CKPJ
2002	528 697	782 536	840 517	166 592	120 351	..
2003	592 748	732 362	750 697	153 389	112 987	87 328
2004	575 923	653 816	689 945	149 188	100 897	75 530
2001 Q3	139 992	218 937	220 282	39 120	37 494	26 665
Q4	140 519	209 244	217 960	39 987	32 045	23 132
2002 Q1	122 140	199 646	212 015	40 620	32 507	21 817
Q2	130 164	192 838	204 124	43 259	32 835	..
Q3	139 134	196 336	206 386	..	29 327	22 776
Q4	137 259	193 716	217 992	..	25 683	18 954
2003 Q1	146 353	190 581	202 480	39 756	32 422	21 392
Q2	151 610	191 622	180 666	38 773	30 294	25 864
Q3	144 138	177 868	172 041	37 299	26 868	22 499
Q4	150 647	172 292	195 510	37 561	23 404	17 574
2004 Q1	139 764	171 148	172 601	36 966	28 536	18 766
Q2	146 324	164 879	172 927	37 355	25 795	24 002
Q3	138 436	159 291	166 521	38 523	25 890	17 244
Q4	151 399	158 498	177 896	36 344	20 676	15 518

1 As from the end of 2004, quarterly data will not be collected for these industries. Annual results for the years 2005 onwards will be published on the ONS website at http://www.statistics.gov.uk/ as and when available.

Source: Office for National Statistics: 01633 813395

11.3 Knitted and crocheted products, lace and narrow fabrics[1]
Total UK manufacturers' sales by industry

£ Thousand

	Knitted and crocheted			Lace	Narrow fabrics
	Fabrics	Hosiery	Pullovers, cardigans & similar articles		
Subclass (SIC 92)	17600	17710	17720	17541	17542
	CKPK	CKPL	CKPM	CKPN	CKPO
2002	244 325	..	351 497	23 442	186 972
2003	202 657	..	312 099	17 875	160 099
2004	197 031	..	219 492	15 708	144 529
2001 Q3	..	..	127 850	6 770	47 658
Q4	..	86 485	115 043	6 386	48 669
2002 Q1	65 797	..	82 900	7 243	48 223
Q2	..	..	66 684	6 204	47 339
Q3	57 695	..	102 224	4 781	47 210
Q4	..	..	99 689	5 214	44 200
2003 Q1	52 227	..	63 284	4 797	40 789
Q2	52 505	..	57 490	4 769	39 976
Q3	..	..	98 523	3 967	39 150
Q4	..	..	92 802	4 341	40 184
2004 Q1	49 422	..	53 084	4 582	37 952
Q2	51 654	..	49 323	3 978	36 709
Q3	..	..	65 855	3 462	35 535
Q4	..	..	51 230	3 686	34 333

1 As from the end of 2004, quarterly data will not be collected for these industries. Annual results for the years 2005 onwards will be published on the ONS website at http://www.statistics.gov.uk/ as and when available.

Source: Office for National Statistics: 01633 813395

11.4 Wearing apparel, dressing and dying of fur, leather clothes[1]
Total UK manufacturers' sales by industry

£ Thousand

| | Workwear | Outerwear | | Underwear | | Hats | Other & accessories | Dressing & dyeing of fur & articles of fur | Leather clothes |
		Men's	Women's	Men's	Women's				
Subclass (SIC 92)	18210	18221	18222	18231	18232	18241	18249	18300	18100
	CKPP	CKPQ	CKPR	CKPS	CKPT	CKPU	CKPV	CKPW	CKPX
2002	270 898	296 498	880 357	223 664	552 991	45 156	425 260	6 297	..
2003	287 771	291 549	844 730	195 604	461 974	37 122	362 097	3 937	8 689
2004	262 704	248 638	791 793	171 319	392 406	34 712	314 667	4 467	6 678
2001 Q2	60 134	83 455	194 177	48 963	154 188	..	116 518	1 598	2 714
Q3	60 489	84 999	202 813	59 780	143 195	13 771	108 079	2 139	2 799
Q4	57 186	77 656	181 318	68 637	158 177	12 059	116 107	2 119	4 891
2002 Q1	66 362	64 310	215 242	47 365	148 329	11 851	102 330	1 907	1 997
Q2	64 937	71 555	221 890	48 950	139 487	11 212	103 520	1 980	..
Q3	66 025	80 550	225 896	60 781	126 194	11 318	110 235	1 225	2 528
Q4	73 573	80 083	217 330	66 568	138 981	10 776	109 174	1 185	3 471
2003 Q1	72 092	71 176	209 821	51 870	110 878	10 373	92 390	855	2 307
Q2	72 748	61 875	195 087	40 850	118 041	9 003	90 151	821	1 812
Q3	73 757	77 148	233 484	55 365	110 387	8 657	93 798	1 086	1 899
Q4	69 174	81 349	206 338	47 519	122 668	9 089	85 757	1 174	2 671
2004 Q1	65 211	66 356	207 891	43 589	102 862	10 902	67 798	1 008	1 435
Q2	66 093	73 748	185 842	41 275	98 446	7 465	73 161	840	1 287
Q3	62 388	62 890	210 358	45 456	98 009	8 360	86 834	1 595	1 874
Q4	69 012	45 644	187 702	40 999	93 089	7 985	86 874	1 024	2 082

1 As from the end of 2004, quarterly data will not be collected for these industries. Annual results for the years 2005 onwards will be published on the ONS website at http://www.statistics.gov.uk/ as and when available.

Source: Office for National Statistics: 01633 813395

11.5 Miscellaneous products - goods not classified elsewhere[1]
Total UK manufacturers' sales by industry

£ Thousand

	Pumps	Compressors	Taps & valves
Subclass (SIC 92)	29121	29122	29130
	CKPY	CKPZ	CKQA
2002	1 021 177	1 112 086	1 147 070
2003	1 133 685	1 142 365	1 104 915
2004	1 156 732	1 176 542	1 164 351
2001 Q2	244 232	310 975	292 890
Q3	240 094	293 727	296 275
Q4	262 583	276 820	285 859
2002 Q1	231 218	271 009	291 337
Q2	249 411	277 824	288 487
Q3	252 339	283 137	293 403
Q4	288 209	280 116	273 844
2003 Q1	258 145	299 143	267 460
Q2	292 316	289 568	283 105
Q3	280 253	273 389	286 829
Q4	302 970	280 265	267 522
2004 Q1	287 588	292 119	290 613
Q2	282 748	296 709	287 634
Q3	280 844	299 516	295 911
Q4	305 552	288 198	290 193

1 As from the end of 2004, quarterly data will not be collected for these industries. Annual results for the years 2005 onwards will be published on the ONS website at http://www.statistics.gov.uk/ as and when available.

Source: Office for National Statistics: 01633 813395

12 Construction

12.1 Volume of construction output by all agencies[1] by type of work at constant 2000 prices (seasonally adjusted)

Great Britain

£ millions

	New work							Repair and maintenance						All work (seasonally adjusted volume index numbers)
	New housing for			Other new work for				Housing		Other work for		Total repair and main-tenance	Total all work	
				Public sector	Private sector		Total new work							
	Public sector	Private sector	Infrastr-ucture	Public sector	Industri-al	Commerci-al		Public	Private	Public sector	Private sector			
	BLAC	BLAD	BAXF	BLAE	BLAF	BLAG	BLAB	BLBK	BLBL	BLAJ	BLAK	BLAH	FGAY	SFZX
2002	1 483	8 449	7 438	6 018	2 863	12 692	38 944	5 898	11 719	6 174	11 355	35 146	74 090	106.0
2003	1 637	9 568	6 734	7 274	3 064	12 095	40 372	6 334	12 264	6 919	11 963	37 480	77 852	112.0
2004	1 972	10 791	5 851	8 062	3 371	12 757	42 804	6 845	12 418	6 643	11 534	37 441	80 245	115.0
2005[3]	1 833[†]	11 239[†]	5 328[†]	7 341	3 582	12 888	42 211[†]	6 730	12 044	7 003	11 562	37 338	79 549[†]	114.0
2003 Q4	426	2 603	1 589	1 997	885	3 051	10 551	1 611	3 175	1 682	2 909	9 377	19 927	114.0
2004 Q1	507	2 615	1 479	2 003	818	3 085	10 507	1 889	3 165	1 843	2 988	9 884	20 391	117.0
Q2	547	2 672	1 542	2 045	818	3 245	10 870	1 627	3 093	1 529	2 779	9 028	19 898	114.0
Q3	493	2 768	1 536	2 018	841	3 211	10 867	1 626	3 099	1 615	2 850	9 191	20 058	115.0
Q4	425	2 737	1 293	1 997	894	3 216	10 560	1 704	3 061	1 656	2 917	9 338	19 898	114.0
2005 Q1	485	2 669	1 297	1 867	819	3 053	10 189[†]	1 930	3 015	1 905	2 902	9 752	19 941	114.0
Q2	482	2 875	1 322	1 844	874	3 221	10 618	1 778	3 069	1 712	2 872	9 432	20 050	115.0
Q3	431	2 873	1 386	1 796	909	3 241	10 636	1 530	2 941	1 711	2 958	9 140	19 776	114.0
Q4[3]	436[†]	2 822[†]	1 323[†]	1 833	980	3 373	10 767	1 492	3 019	1 674	2 830	9 015	19 782[†]	114.0
2006 Q1[3]	562	2 762	1 266	1 777[†]	964	3 383	10 715	1 714	2 971	1 753	2 790	9 227	19 942	114.0
Q2[2]	604	2 877	1 261	1 733	959	3 553	10 986	1 589[†]	3 020[†]	1 621[†]	2 836[†]	9 066[†]	20 051	115.0
Q3[2]	583	2 951	1 312	1 757	991	3 608	11 203	1 630	2 805	1 665	2 892	8 992	20 195	116.0

1 Classified to construction in the *Standard Industrial Classification 1992*. Estimates of unrecorded output by small firms and self-employed workers, and output by the public sector's direct labour department are included.
2 Provisional.
3 Revised

Source: Department for Trade and Industry: 020 7215 1953

12.2 Value of new orders obtained by contractors for new work[1] at current prices

Great Britain

£ millions

| | New housing[2] | | | Other new work | | | | | New work total |
	Public and housing association	Private	Total	Infrastructure	Other public	Private industrial	Private commercial	Total	
	BLBC	BLBD	FGAU	BAWT	BAWU	BAWV	BAWW	BLBE	FHAA
2002	1 129	8 088	9 217	5 555	5 910	2 247	10 482	24 194	33 411
2003	1 340	9 471	10 812	4 894	6 142	2 383	9 721	23 139	33 951
2004	1 697	12 153	13 850	3 772	6 847	2 593	12 026	25 238	39 089
2005	1 951	13 171	15 122	5 532	6 694	3 421	13 163	28 811	43 932
2003 Q4	289	2 373	2 661	780	1 463	610	2 208	5 061	7 722
2004 Q1	549	3 168	3 717	964	1 643	553	3 382	6 543	10 260
Q2	444	2 893	3 338	1 164	1 834	589	2 827	6 414	9 752
Q3	335	3 234	3 569	816	1 572	717	3 099	6 203	9 773
Q4	368	2 858	3 226	828	1 797	735	2 719	6 078	9 305
2005 Q1	552	3 203	3 756	1 483	1 606	679	3 283	7 052	10 807
Q2	448	3 605	4 053	1 463	1 693	856	3 248	7 259	11 312
Q3	390	3 626	4 016	1 488	1 867	842	3 114	7 311	11 328
Q4	560	2 737	3 297	1 098	1 528	1 044	3 518	7 188	10 485
2006 Q1	833	3 333	4 166	1 025	1 625	961	4 410	8 021	12 187
Q2	586	3 704	4 290	1 279	1 375	804	5 133	8 590	12 880
Q3[4]	696[†]	3 317[†]	4 014[†]	1 089[†]	1 672[†]	955[†]	4 386[†]	8 102[†]	12 116[†]
2006 Apr	302	978	1 280	490	463	231	980	2 164	3 444
May	130	1 168	1 299	497[†]	364	336	2 466	3 663	4 962
Jun	154[†]	1 558[†]	1 711[†]	292[†]	547[†]	236[†]	1 687[†]	2 763[†]	4 474[†]
Jul	205	1 112	1 316	272	669	394	1 534	2 869	4 186
Aug[4]	203	1 054	1 257	524	580	285	1 696	3 084	4 342
Sep[3]	289	1 151	1 440	293	424	275	1 157	2 149	3 589
Oct[3]	244	1 173	1 416	436	596	329	1 170	2 532	3 949
Nov[3]	129	1 141	1 270	235	433	299	1 359	2 327	3 596

1 Including the value of speculative building when work starts on site.
2 Excluding orders for home improvement work.
3 Provisional.
4 Revised

Source: Department for Trade and Industry

12.3 Building materials and components
Great Britain

monthly averages or calendar months

	Building bricks (millions)		Concrete blocks (000 sq m)		Concrete roofing tiles (000 sq m of roof covered)		Slate[1] (tonnes)		Cement[2] (tonnes)		RMX[3] (000 cu m)	Sand and gravel (000 tonnes)
	Production	Deliveries	Production	Deliveries	Production	Deliveries	Production	Deliveries	Production	Deliveries	Deliveries	Deliveries
	BLDA	QXIH	BLDM	QXII	BLDN	QXIJ	BLDQ	QXIK	QXIM	QXIL	BLDP	BLDS
1996	254	244	6 322	6 365	2 054	2 004	9 147	8 930	1 018	974	1 741	6 339
1997	250	254	6 878	6 837	2 080	2 090	8 859	8 636	1 053	996	1 861	7 062
1998	250	248	7 055	7 041	2 082	2 132	8 742	8 546	1 034	988	1 915	7 148
1999	245	252	7 314	7 154	2 164	2 114	8 239	8 330	1 058	978	1 963	6 819
2000	239	241	7 518	7 377	2 230	2 087	7 155	7 495	1 038	988	1 920	7 322
2001	230	235	7 327	7 376	2 069	2 036	7 760	7 852	924	888	1 917	8 121
2002	229	235	7 623	7 612	2 085	2 033	7 913	7 972	924	897	1 883	7 126
2003	231	245	7 973	8 032	1 786	1 783	6 591	6 543	935	923	1 857	6 896
2004	239	236	8 021	7 905	1 728	1 617	..	..	950	923	1 905	6 779
2005	229	215	7 500	7 463	2 143	2 041	..	..	871	860	1 869	6 708
2004 Q2	241	253	8 238	8 392	1 755	1 600	6 903	6 987	1 008	958	1 977	7 323
Q3	236	246	8 179	8 509	1 640	1 711	..	..	990	954	2 013	7 217
Q4	236	215	7 577	7 240	1 685	1 725	..	..	929	884	1 851	6 338
2005 Q1	238	210	7 712	7 331	1 929	1 593	..	..	848	852	1 712	6 318
Q2	242	236	7 943	8 129	2 501	2 135	..	..	1 008	995	1 998	7 392
Q3	226	220	7 616	7 818	2 045	2 368	..	..	980	974	1 973	7 107
Q4	210	190	6 729	6 573	2 098	2 067	..	..	903	847	1 794	6 017
2006 Q1	206	189	7 052	6 940	2 264	2 262	..	..	882	912	1 815	5 930
Q2	219	213	7 478	7 520	2 001	1 903	..	..	994	965	1 998	7 179
Q3	213	216	7 538	7 870	2 210	2 124	..	..	997	972	2 016	6 905
2004 Dec	243	200	5 905	5 805	..	..	..	..	822	710	..	..
2005 Jan	214	194	7 160	6 936	..	..	..	..	687	749	..	..
Feb	235	212	7 786	7 508	..	..	..	..	892	863	..	..
Mar	264	224	8 189	7 548	..	..	..	..	965	943	..	..
Apr	237	224	7 984	8 082	..	..	..	..	1 014	991	..	..
May	230	225	7 511	8 060	..	..	..	..	1 018	972	..	..
Jun	260	260	8 333	8 244	..	..	..	..	992	1 023	..	..
Jul	213	219	7 669	7 875	..	..	..	..	1 028	946	..	..
Aug	207	208	7 128	7 720	..	..	..	..	954	951	..	..
Sep	259	233	8 050	7 859	..	..	..	..	958	1 026	..	..
Oct	228	217	7 525	7 442	..	..	..	..	1 013	911	..	..
Nov	219	185	7 207	6 904	..	..	..	..	925	939	..	..
Dec	182	169	5 454	5 374	..	..	..	..	770	690	..	..
2006 Jan	144	169	6 221	6 527	..	..	..	..	743	810	..	..
Feb	210	183	6 788	6 726	..	..	..	..	891	902	..	..
Mar	264	216	8 146	7 568	..	..	..	..	1 013	1 025	..	..
Apr	211	187	6 333	6 489	..	..	..	..	942	868	..	..
May	211	206	7 679	7 623	..	..	..	..	1 018	994	..	..
Jun	235	247	8 424	8 447	..	..	..	..	1 021	1 044	..	..
Jul	201	210	7 567	7 836	..	..	..	..	1 042	949	..	..
Aug	194	208	7 313	7 864	..	..	..	..	986	977	..	..
Sep	244	230	7 650	7 786	..	..	..	..	962	991	..	..
Oct	211	197	7 717[4]	7 499[4]	..	..	..	..	1 030	985	..	..
Nov	212	199	7 797[4]	7 382[4]	..	..	..	..	..	..	..	..

1 Excluding slate residue used as fill.
2 United Kingdom; Great Britain from January 2002.
3 United Kingdom; RMX stands for ready mixed concrete.
4 Provisional

Source: Department of Trade and Industry: 020 7215 1555

12.4 Permanent dwellings started and completed

Number

	Starts				Completions			
	Private enterprise	Registered social landlords[1]	Local Authorities	All dwellings	Private enterprise	Registered social landlords[1]	Local Authorities	All dwellings
United Kingdom								
	LMDB	LMDD	LMDE	LMDF	LMDG	LMDW	LMDY	LMDZ
2001/02	177 453	17 341	192	194 986	153 141	21 747	225	175 113
2002/03	178 791	16 430	185	195 406	163 880	19 682	301	183 863
2003/04	193 273	18 771	289	212 333	171 840	18 375	207	190 422
2004/05	204 884	20 665	239	225 788	183 933	22 716	131	206 780
2005/06	211 033	24 166	255	235 454	188 934	24 393	326	213 653
2003 Q4	43 256	3 098	74	46 428	50 503	4 294	59	54 856
2004 Q1	50 607	7 344	39	57 990	37 656	5 293	36	42 985
Q2	55 745	4 378	143	60 266	46 849	4 413	51	51 313
Q3	54 196	3 811	13	58 020	47 426	5 447	19	52 892
Q4	46 928	4 664	9	51 601	50 264	5 891	27	56 182
2005 Q1	48 015	7 812	74	55 901	39 394	6 965	34	46 393
Q2	56 570	4 563	74	61 207	48 402	5 437	138	53 977
Q3	51 519	4 614	6	56 139	43 903	5 811	12	49 726
Q4	48 003	5 537	37	53 577	53 544	5 863	54	59 461
2006 Q1	54 941	9 452	138	64 531	43 085	7 282	122	50 489
England								
	BLHC	BLHM	BAEP	BLHA	BLHK	BLHO	BAEX	BLHI
2001/02	138 536	11 110	118	149 764	115 533	14 171	63	129 767
2002/03	139 552	11 019	159	150 730	124 278	13 242	199	137 719
2003/04	148 614	12 381	275	161 270	129 797	13 662	191	143 650
2004/05	159 947	14 391	205	174 543	139 132	16 661	100	155 893
2005/06	167 432	17 226	248	184 906	144 937	18 162	299	163 398
2004 Q2	43 815	3 982	128	47 925	35 727	3 581	51	39 359
Q3	42 351	3 020	2	45 373	36 124	3 764	19	39 907
Q4	36 804	4 040	1	40 845	37 812	5 017	25	42 854
2005 Q1	36 977	3 349	74	40 400	29 469	4 299	5	33 773
Q2	44 995	4 145	74	49 214	36 712	4 096	114	40 922
Q3	40 755	4 020	5	44 780	33 545	4 372	9	37 926
Q4	38 379	4 325	31	42 735	42 011	4 768	54	46 833
2006 Q1	43 303	4 736	138	48 177	32 669	4 926	122	37 717
Q2	43 011	4 134	93	47 238	38 221	4 872	52	43 145
Q3	36 806	3 931	26	40 763	32 297	5 340	57	37 694
Wales								
	BLIC	BLIM	BAEQ	BLIA	BLIK	BLIO	BAEY	BLII
2001/02	8 375	715	6	9 096	7 494	711	68	8 273
2002/03	9 014	497	11	9 522	7 522	782	6	8 310
2003/04	9 480	566	14	10 060	7 863	417	16	8 296
2004/05	9 095	381	34	9 510	7 986	475	31	8 492
2005/06	8 613	359	1	8 973	7 883	347	27	8 257
2004 Q1	2 240	126	–	2 366	1 836	141	–	1 977
Q2	1 970	149	15	2 134	2 116	146	–	2 262
Q3	2 571	66	11	2 648	1 972	143	–	2 115
Q4	2 141	66	8	2 215	2 424	114	2	2 540
2005 Q1	2 413	100	–	2 513	1 474	72	29	1 575
Q2	2 215	104	–	2 319	1 975	81	24	2 080
Q3	2 328	62	1	2 391	1 877	79	3	1 959
Q4	1 683	124	–	1 807	2 085	80	–	2 165
2006 Q1	2 387	69	–	2 456	1 946	107	–	2 053
Q2	2 529	87	–	2 616	2 089	71	–	2 160

12.4

Permanent dwellings started and completed

Number

	Starts				Completions			
	Private enterprise	Registered social landlords[1]	Local Authorities	All dwellings	Private enterprise	Registered social landlords[1]	Local Authorities	All dwellings
Scotland								
	BLFC	BLFM	BAER	BLFA	BLFK	BLFO	BAEZ	BLFI
2001/02	18 483	4 744	43	23 270	18 037	5 479	65	23 581
2002/03	18 510	4 270	15	22 795	18 519	4 695	94	23 308
2003/04	22 352	4 718	–	27 070	19 926	3 727	–	23 653
2004/05	22 643	4 864	–	27 507	21 875	4 752	–	26 627
2005/06	21 033	5 352	6	26 391	19 486	5 102	–	24 588
2003 Q4	5 147	620	–	5 767	5 595	853	–	6 448
2004 Q1	5 749	3 112	–	8 861	4 436	747	–	5 183
Q2	6 358	186	–	6 544	5 449	619	–	6 068
Q3	6 058	557	–	6 615	5 800	1 411	–	7 211
Q4	4 882	524	–	5 406	5 689	706	–	6 395
2005 Q1	5 345	3 597	–	8 942	4 937	2 016	–	6 953
Q2	5 488	311	–	5 799	5 210	1 157	–	6 367
Q3	4 844	478	–	5 322	4 769	1 170	–	5 939
Q4	4 937	867	6	5 810	5 114	888	–	6 002
2006 Q1	5 764	3 696	–	9 460	4 393	1 887	–	6 280
Northern Ireland								
	BLGC	BLGM	BAES	BLGA	BLGK	BLGO	BAFA	BLGI
2001/02	12 065	772	25	12 862	12 072	1 386	29	13 487
2002/03	11 573	669	–	12 242	13 387	1 026	2	14 415
2003/04	12 671	1 140	–	13 811	13 951	560	–	14 511
2004/05	13 199	1 029	–	14 228	14 940	828	–	15 768
2005/06	13 955	1 229	–	15 184	16 628	782	–	17 410
2004 Q1	3 435	870	–	4 305	3 717	163	–	3 880
Q2	3 602	61	–	3 663	3 557	67	–	3 624
Q3	3 216	168	–	3 384	3 530	129	–	3 659
Q4	3 101	34	–	3 135	4 339	54	–	4 393
2005 Q1	3 280	766	–	4 046	3 514	578	–	4 092
Q2	3 872	3	–	3 875	4 505	103	–	4 608
Q3	3 592	54	–	3 646	3 712	190	–	3 902
Q4	3 004	221	–	3 225	4 334	127	–	4 461
2006 Q1	3 487	951	–	4 438	4 077	362	–	4 439
Q2	3 901	69	–	3 970	–	–	–	–

1 Includes non-registered social landlords.

Sources: Department for Communities and Local Government;
0117 372 8055;
National Assembly for Wales;
Scottish Development Department;
Department for Social Development (Northern Ireland)

13 Transport

13.1 Motor vehicles: new registrations in Great Britain

Thousands

	Private and light goods (PLG)	of which: PLG: Bodytype cars	of which: PLG: Others (mainly light goods)	Motorcycles	Other goods vehicles	Public transport vehicles	Agriculture machines	Other licensed vehicles	Vehicles exempt from tax	All vehicles	Of which bodytype cars
	BMAK	BMAA	BMAE	BMAL	BBJY	BBJZ	BBKA	BBKB	BBKC	BBKD	BBKE
2001	2 704.2	2 426.4	278.0	177.5	49.0	7.1	19.9	8.2	170.9	3 136.6	2 579.4
2002	2 815.6	2 528.9	286.9	162.3	44.9	7.7	23.1	8.5	167.2	3 229.5	2 682.1
2003	2 820.7	2 497.1	323.5	157.3	48.4	8.4	24.1	9.5	163.5	3 231.9	2 646.0
2004	2 784.9	2 437.5	347.2	133.7	48.0	8.1	25.0	8.5	177.2	3 185.3	2 599.0
2005	2 603.7	2 266.2	337.0	132.1	51.3	8.9	23.5	9.2	193.2	3 021.4	2 443.5
2004 Oct	187.5	157.5	30.0	9.6	4.5	0.6	1.6	0.7	15.6	220.1	171.8
Nov	190.8	161.5	29.3	9.0	4.9	0.6	1.4	0.8	15.3	222.7	175.6
Dec	163.9	138.7	25.1	6.5	3.4	0.6	1.0	0.5	11.9	187.8	149.5
2005 Jan	191.3	168.4	22.8	5.9	3.4	0.6	1.3	0.6	12.8	215.7	180.0
Feb	87.4	71.4	16.0	5.2	3.0	0.5	1.2	0.7	7.3	105.3	77.5
Mar	464.8	414.8	49.9	15.2	5.2	1.1	3.2	0.8	27.0	517.3	440.4
Apr	192.7	164.0	28.7	14.1	4.4	0.8	2.7	0.9	16.5	232.0	178.9
May	201.2	174.2	27.0	13.1	4.6	0.9	2.4	0.8	16.5	239.4	189.2
Jun	241.6	210.6	30.9	15.3	4.7	1.0	2.3	0.8	17.2	282.8	226.3
Jul	185.9	161.6	24.3	13.2	4.3	0.7	2.6	0.8	15.2	222.7	175.3
Aug	90.2	74.2	16.0	10.5	3.5	0.4	1.8	0.8	11.3	118.5	84.2
Sep	441.6	395.0	46.6	15.7	5.5	1.1	2.1	0.8	24.0	490.8	417.6
Oct	163.8	138.4	25.4	9.3	4.4	0.6	1.5	0.8	16.8	197.3	153.9
Nov	172.4	146.1	26.2	8.4	4.8	0.7	1.4	0.8	15.9	204.3	160.3
Dec	170.8	147.5	23.2	6.2	3.5	0.5	1.0	0.6	12.7	195.3	159.2
2006 Jan	164.6	142.7	22.0	6.4	3.4	0.5	1.2	0.6	12.5	189.2	154.0
Feb	81.4	66.8	14.6	4.9	3.0	0.5	1.3	0.6	9.2	100.9	74.8
Mar	462.4	409.8	52.6	16.8	5.5	1.4	3.2	0.8	24.7	514.8	432.9
Apr	173.5	149.0	24.5	12.1	10.5	1.0	2.5	1.0	15.0	215.6	163.0
May	202.0	173.7	28.3	13.8	1.5	0.6	2.6	0.8	16.6	237.8	189.0
Jun	232.7	201.7	31.0	14.3	2.4	0.5	2.4	0.8	17.5	270.7	217.9
Jul	177.8	153.2	24.6	13.5	2.8	0.4	2.6	0.8	17.2	215.3	169.2
Aug	82.9	66.2	16.6	10.6	2.9	0.3	1.8	0.8	12.8	112.1	77.8
Sep	439.0	392.8	46.2	14.7	7.5	0.8	2.1	0.9	23.8	488.8	415.4

Source: Department for Transport

13.2 Motor vehicles currently licensed as at 31 December[1]

Thousands

	Private and light goods: Private cars	Private and light goods: Other vehicles	Motor-cycles, scooters and mopeds	Public transport vehicles[1]	Goods vehicles[2]	Special concession group[3]	Other vehicles[4]	Crown and exempt vehicles	All vehicles
	BMBJ	BMBK	BMBB	BMBE	BMBD	BMBC	BMBF	BMBL	BMBI
1994	20 479	2 192	630	107	434	309	50	1 030	25 231
1995	20 505	2 217	594	74	421	274	44	1 169	25 369
1996	21 172	2 267	609	77	413	254	40	1 424	26 302
1997	21 681	2 317	626	79	414	249	38	1 522	26 974
1998	22 115	2 362	684	80	412	243	37	1 558	27 538
1999	22 785	2 427	760	84	415	241	36	1 573	28 368
2000	23 196	2 469	825	86	418	233	34	1 590	28 898
2001	23 899	2 544	882	89	422	233	33	1 602	29 747
2002	24 543	2 622	941	92	425	243[5]	32	1 855[5]	30 557
2003	24 985	2 730	1 005	96	426	258[5]	32	1 887[5]	31 207
2004	25 754	2 900	1 060	100	434	275[5]	32	1 929[5]	32 259
2005	26 208	3 019	1 075	103	433	283	81	1 978	32 897

1 Includes taxis for years up to 1994. Taxation group now restricted to only vehicles with 9 or more seats.
2 Includes agricultural vans and lorries and showman's goods vehicles licensed to draw trailers.
3 Includes combine harvesters, mowing machines, digging machines, mobile cranes and works trucks. Taxation group subject to revision from 1st July 1995, formerly termed the "agricultural and special machines" group.
4 Includes three-wheelers, pedestrian controlled vehicles and showman's haulage.
5 Vehicles in this taxation class are exempt from duty and form part of the "Crown and Exempt" taxation class with effect from January 2002.

Source: Department for Transport

13.3 Index numbers of road traffic and goods transport by road

Average 1995=100

	Index of vehicle kilometres travelled on roads in Great Britain								Index of tonne-kilo-metres of road goods transport[3]
	Motor traffic					Other goods vehicles			
	All motor traffic	Two-wheeled motor vehicles	Cars	Buses and coaches	Light vans[1]	Total	Articulated[2]	Pedal cycles	
	BLUV	BMCO	BMCJ	BMCP	BMCK	BMCL	BMCQ	BMCM	BMCN
1998	107	110	106	107	114	110	114	96	106
1999	109	120	107	108	116	111	115	98	104
2000[4]	109	122	107	105	117	111	115	100	105
2001	110	128	109	105	120	111	115	102	104
2002	113	135	112	106	123	112	116	106	104
2003	115	..	112	..	130	113	..	103	106
2004	..	..	..	..	..	..	..	..	106
2005	..	..	..	..	..	..	..	..	106
Seasonally adjusted									
2003 Q1	114	..	112	..	126	112	..	101	105
Q2	114	..	112	..	130	111	..	106	107
Q3	116	..	113	..	133	114	..	113	105
Q4	115	..	113	..	133	114	..	90	105
2004 Q1[5]	114	..	112	..	126	112	..	101	105
Q2[5]	114	..	112	..	130	111	..	106	106
Q3[5]	116	..	113	..	133	114	..	113	104
Q4[5]	115	..	113	..	133	114	..	90	109
2005 Q1	114	..	112	..	126	112	..	101	105
Q2	114	..	112	..	130	111	..	106	107
Q3	116	..	113	..	133	114	..	113	107
Q4	115	..	113	..	133	114	..	90	105

1 Not exceeding 3.5 tonnes gross vehicle weight. Includes all car based vans and those of the next larger capacity such as transit vans.
2 Goods vehicles up to 3.5 tonnes gross vehicle weight.
3 The figures for road goods transport are estimated from a continuing sample enquiry; excluding estimates of work done by vehicles under 3.5 tonnes gross vehicle weight. The quarterly figures relate to 13-week periods and not three calendar months.
4 Figures affected by September 2000 fuel protest.
5 Provisional.

Source: Department for Transport: 020 7944 3095

13.4 Road casualties in Great Britain

Number

	Total casualties		Severity			All severities			
	All ages	Under 16 years	Killed	Seriously injured	Slightly injured	Pedestrians	Pedal cyclists	Motor cyclists and their passengers[1]	Other drivers and their passengers
	BMDA	BMDB	BMDC	BMDD	BMDE	BMDF	BMDG	BMDH	BMDI
1999	320 310	42 051	3 423	39 122	277 765	42 888	22 840	26 192	228 390
2000	320 283	39 715	3 409	38 155	278 719	42 033	20 612	28 212	229 426
2001	313 309	38 269	3 450	37 110	272 749	40 577	19 114	28 810	224 808
2002	302 605	34 689	3 431	35 976	263 198	38 784	17 107	28 353	218 361
2003	290 607	31 988	3 508	33 707	253 392	36 405	17 033	28 411	208 758
2004	280 840	31 000	3 221	31 130	246 489	34 881	16 648	25 641	203 670
2005	271 017	28 126	3 201	28 954	293 858	33 281	16 561	24 824	196 351
2003 Q2	71 244	8 745	831	8 424	61 989	8 830	4 547	7 527	50 340
Q3	74 681	8 888	937	8 778	64 966	8 655	5 356	8 641	52 029
Q4	76 888	7 299	900	8 563	67 425	9 732	3 813	6 520	56 823
2004 Q1	65 814	6 853	703	7 096	58 015	8 833	3 168	4 794	49 019
Q2	69 430	8 616	825	8 184	60 421	8 531	4 752	7 287	48 860
Q3	70 743	8 400	801	7 968	61 974	7 969	5 031	7 518	50 225
Q4	74 853	7 131	892	7 882	66 079	9 548	3 697	6 042	55 566
2005 Q1	62 037	6 066	740	6 301	54 996	8 097	2 884	4 692	46 364
Q2	67 547	7 787	727	7 322	59 498	8 549	4 527	7 006	47 465
Q3	68 616	7 646	818	7 598	60 200	7 618	5 249	7 304	48 445
Q4	72 817	6 627	916	7 733	64 168	9 017	3 901	5 822	54 077
2006 Q1	59 430	5 640	710	7 680	52 460	7 500	2 990	4 060	44 880
Q2	62 240	7 000	740	7 010	54 490	7 410	4 250	6 200	44 380

1 Includes riders and passengers of mopeds, motor scooters and combinations.

Source: Department for Transport: 020 7890 3078

13.5 Local (stage) bus services: vehicle kilometres and passenger journeys
Great Britain

Millions

	London[1]	English metropolitan areas	English shire counties	England	Scotland	Wales	All Great Britain	All outside London	All outside London and English metropolitan areas
Vehicle kilometres[2]									
	BAJO	BAJP	BAJQ	BAJR	BAJS	BAJT	BAJU	BAJV	BAJW
1998/99	358	684	1 123	2 165	358	118	2 642	2 284	1 600
1999/00	366	661	1 160	2 186	363	123	2 673	2 307	1 646
2000/01	373	656	1 134	2 164	369	126	2 659	2 286	1 630
2001/02	379	647	1 102	2 128	368	126	2 622	2 243	1 596
2002/03	406	631	1 082	2 119	374	123	2 616	2 210	1 579
2003/04	474	596	1 063	2 133	369	113	2 615	2 141	1 545
2004/05	470	575	1 088	2 133	366	116	2 614	2 144	1 569
2005/06	465	547	1 086	2 098	357	115	2 570	2 105	1 558
Passenger journeys[2]									
	BAJX	BAJY	BAJZ	BAKA	BAKB	BAKC	BAKD	BAKE	BAKF
1998/99	1 266	1 195	1 242	3 702	413	116	4 231	2 965	1 770
1999/00	1 294	1 178	1 250	3 722	442	114	4 278	2 984	1 806
2000/01	1 347	1 166	1 247	3 761	443	116	4 319	2 972	1 806
2001/02	1 422	1 154	1 222	3 798	449	104	4 352	2 930	1 776
2002/03	1 527	1 145	1 210	3 882	452	110	4 444	2 917	1 772
2003/04	1 692	1 114	1 189	3 995	457	111	4 564	2 872	1 758
2004/05	1 782	1 083	1 167	4 032	465	113	4 609	2 827	1 744
2005/06	1 810	1 117	1 198	4 125	477	118	4 719	2 909	1 792

1 Passenger journey statistics for London may not be consistent with those published by Transport for London.
2 There have been revisions to kilometres and journeys based on new data from bus operators and local authorities.

Source: Department for Transport: 020 7944 3076

13.6 Local (stage) bus services: fare indices
Great Britain

1995=100

	London	English metropolitan areas	English shire counties	England	Scotland	Wales	All Great Britain	All outside London	All outside London and English metropolitan areas
1998/99[1]	113.7	118.7	116.7	116.5	121.8	116.3	117.1	118.2	117.8
1999/00[1]	117.2	124.6	122.0	121.5	125.3	122.2	122.0	123.4	122.8
2000/01[1]	117.2	129.9	128.6	125.9	129.9	127.5	126.4	129.2	128.9
2001/02[1]	115.5	137.4	135.1	130.3	131.8	133.5	130.6	135.3	134.4
2002/03[1]	114.8	142.7	141.7	134.2	134.5	139.5	134.5	140.8	139.9
2003/04[1]	116.9	148.0	148.5	139.1	136.8	145.5	139.1	146.3	145.4
2004/05[1]	126.8	154.2	155.7	146.2	140.4	152.4	145.7	152.5	151.4
2005/06[1]	139.7	167.0	165.9	159.2	144.6	160.2	156.3	162.4	160.3
	BAKG	BAKH	BAKI	BAKJ	BAKK	BAKL	BAKM	BAKN	BAKO
2002 Q3	114.7	142.2	141.2	133.9	133.3	138.9	134.0	140.1	139.3
Q4	114.7	143.6	142.2	134.7	135.9	140.3	135.0	141.5	140.6
2003 Q1	115.1	144.3	143.7	135.6	136.1	141.2	135.9	142.6	141.7
Q2	115.1	145.6	146.1	136.8	136.4	142.1	137.0	144.1	143.4
Q3	115.1	147.2	147.5	137.8	136.6	144.1	137.9	145.4	144.6
Q4	115.1	148.2	148.9	138.6	137.0	147.6	138.8	146.7	146.0
2004 Q1	122.2	151.3	151.4	143.0	137.3	148.3	142.5	148.9	147.6
Q2	122.2	152.2	153.1	143.9	139.7	149.5	143.6	150.4	149.4
Q3	122.2	152.8	154.2	144.5	139.7	151.9	144.2	151.3	150.4
Q4	122.2	154.6	157.0	146.0	140.8	153.7	145.7	153.4	152.2
2005 Q1	140.7	157.0	158.5	150.4	141.3	154.3	149.4	155.0	153.4
Q2	140.7	161.3	160.2	152.2	142.6	157.1	151.1	157.5	155.0
Q3[2]	131.7	163.9	162.2	152.8	143.3	158.0	151.8	159.4	156.6
Q4	131.7	166.7	166.5	153.5	144.0	159.5	152.6	162.5	159.8
2006 Q1	148.7	172.8	161.7	162.1	146.3	164.4	160.4	167.3	164.1
Q2	148.6	167.2	158.5	155.6	149.3	164.5	155.3	160.3	156.4
Q3	147.6	166.3	155.1	153.8	149.3	164.6	153.7	158.5	154.2
Q4	145.7	166.3	158.1	154.3	149.3	164.6	154.2	162.0	166.3

1 Owing to rounding financial year data may differ slightly from that published by DfT.
2 London bus fares reduced overall in Q3 owing to TfL's free travel scheme for children, introduced in September 2005.

Source: Department for Transport: 020 7944 3076

13.7 National Rail and London Underground

Millions

	National Rail: passenger kilometres			London Underground: passenger journeys[1,2]		
	Ordinary fares	Season tickets	Total	Full and reduced fares	Season tickets	Total
1999/00	28 030	10 443	38 472	477	450	927
2000/01	27 245	10 933	38 179	486	484	970
2001/02	28 149	10 992	39 141	491	462	953
2002/03	28 394	11 284	39 678	495	446	942
2003/04	29 003	11 908	40 911	491	457	948
2004/05	29 487	12 275	41 762	486	490	976
2005/06	30 405	12 805	43 211	460	510	970
	BMGB	BMGD	BMGA	BMGF	BMGG	BMGE
2002 Q4	7 072	2 934	10 006	128	111	239
2003 Q1	6 662	3 030	9 691	113	112	225
Q2	7 204	2 831	10 036	120	111	231
Q3	7 674	2 752	10 426	127	106	233
Q4	7 143	3 078	10 221	127	119	246
2004 Q1	6 983	3 247	10 229	117	121	238
Q2	7 250	2 927	10 177	120	119	239
Q3	7 747	2 818	10 565	124	117	241
Q4	7 463	3 296	10 759	127	125	252
2005 Q1	7 027	3 233	10 260	115	129	244
Q2	7 720	3 112	10 832	120	131	251
Q3	7 632	2 940	10 572	107	127	234
Q4	7 721	3 354	11 075	121	119	240
2006 Q1	7 332	3 399	10 731	113	133	246
Q2	8 008	3 087	11 095	119	126	245
Q3	8 515	2 932	11 447	125	125	250

1 The annual figures are greater than the the sum of the four quarters owing
to year end revision by Transport for London.
2 London Underground data partly estimated.

Sources: Office of Rail Regulation: 020 7282 2192;
Department for Transport: 020 7944 3076

13.8 National Rail: freight traffic

	National Rail[1]			
	Freight lifted: million tonnes			
	Coal and coke	Other traffic	Total	Net tonne kilometres: millions
	BMHB	BMHD	BMHA	BMHE
2001	46.9	48.5	95.4	19 200
2002	41.1	46.4	87.5	18 634
2003	42.4	46.9	89.3	18 748
2004	48.8	49.9	98.7	20 175
2005	53.5	49.3	102.8	21 689
2002 Q4	8.4	12.7	21.2	4 382
2003 Q1	9.5	13.6	23.1	4 752
Q2	8.7	13.2	21.9	4 618
Q3	8.6	13.8	22.4	4 737
Q4	8.9	13.0	21.9	4 641
2004 Q1	9.0	13.7	22.6	4 878
Q2[2]	10.6	14.4	25.0	4 994
Q3	10.7	14.3	25.0	5 153
Q4	11.7	14.3	26.1	5 149
2005 Q1	12.0	13.8	25.8	5 278
Q2[3]	11.2	14.5	25.7	5 504
Q3	10.6	13.9	24.6	5 388
Q4	13.0	15.0	27.9	5 519
2006 Q1	12.8	14.3	27.1	5 698
Q2	13.0	15.1	28.1	5 654
Q3	11.4	14.6	26.1	5 399

1 Freight train traffic only.
2 There is a break in the series between 2004 Q1 and 2004 Q2, due to a
change in the method of data collection.
3 There is a further break in the series between 2005 Q1 and 2005 Q2, since
the 2005 Q2 figures onwards include some of the tonnes lifted by and additional Freight Operating Company.

Note: The freight lifted series has been revised. Previously the coal figure for
one of the freight operating companies included iron ore. Iron ore has now been
correctly assigned to the 'other' category. This revision did not affect the total
figures, but caused a reduction in the coal and coke figures and an increase in
the 'other' category.

Source: Office of Rail Regulation: 020 7282 2192

81

13.9 UK airlines: aircraft kilometres flown, passengers and cargo uplifted[1]
Tonne-kilometres and seat kilometres used on scheduled services

Monthly averages or calendar months: thousands or tonnes

	All services			Domestic services			International services		
	Aircraft kilometres flown (000's)	Passengers uplifted (000's)	Cargo uplifted (tonnes)	Aircraft kilometres flown (000's)	Passengers uplifted (000's)	Cargo uplifted (tonnes)	Aircraft kilometres flown (000's)	Passengers uplifted (000's)	Cargo uplifted (tonnes)
	BMIA	BMIB	BMIC	BMID	BMIE	BMIF	BMIG	BMIH	BMII
1999	947 285	65 424.30	860 405	119 984	17 173.2	26 039	827 301	48 250.9	834 366
2000	1 012 008	70 066.50	897 864	120 270	17 987.9	24 361	891 738	50 120.8	873 512
2001	1 054 939	70 034.40	741 623	128 125	18 331.6	19 560	920 814	51 703.1	722 061
2002	1 047 400	72 708.90	769 519	125 758	19 992.7	16 800	921 643	52 717.6	752 713
2003	1 082 392	76 207.10	799 406	121 260	20 730.9	17 158	964 140	55 476.3	782 232
2004	1 204 698	86 048.90	905 622	138 790	22 539.7	14 928	1 065 908	63 508.6	890 720
2005	1 324 222	93 602.90	921 406	147 468	23 128.4	10 012	1 177 063	70 474.7	911 405
2002 Apr	87 026	5 879.10	61 546	10 655	1 590.2	1 326	76 371	4 289.0	60 218
May	91 178	6 148.00	65 794	11 187	1 737.2	1 507	79 991	4 410.8	64 287
Jun	88 145	6 439.30	64 609	10 374	1 723.5	1 395	77 771	4 715.5	63 214
Jul	92 652	7 296.20	65 039	11 335	1 998.6	1 502	81 318	5 297.6	63 536
Aug	92 358	6 974.10	61 077	11 069	1 904.6	1 384	81 289	5 069.5	59 692
Sep	89 175	6 612.00	62 876	10 687	1 769.2	1 375	78 488	4 842.8	61 501
Oct	90 789	6 490.50	68 992	10 585	1 772.9	1 712	80 204	4 717.6	67 279
Nov	85 508	5 564.00	70 269	10 458	1 675.9	1 449	75 050	3 888.1	68 820
Dec	84 826	5 488.00	63 633	9 901	1 589.0	1 330	74 925	3 900.6	62 303
2003 Jan	85 332	5 182.70	57 971	10 038	1 463.6	1 326	75 294	3 719.1	56 645
Feb	78 501	5 339.90	61 203	8 758	1 396.7	1 241	69 743	3 943.2	59 962
Mar	87 376	5 957.40	67 861	9 772	1 570.4	1 315	77 604	4 387.0	66 546
Apr	86 156	6 093.30	61 770	9 720	1 725.6	1 238	76 436	4 367.7	60 533
May	90 575	6 401.70	66 677	10 212	1 760.6	1 263	80 363	4 641.2	65 414
Jun	91 856	6 839.10	65 786	9 676	1 799.0	1 451	82 188	5 040.1	64 317
Jul	96 889	7 218.40	63 722	10 563	1 911.9	1 852	86 326	5 306.5	61 870
Aug	97 406	7 439.60	64 416	10 359	1 958.3	1 305	87 047	5 481.3	63 112
Sep	91 047	6 979.00	67 516	10 147	1 868.1	1 676	83 900	5 110.9	65 841
Oct	97 142	6 893.00	73 931	11 365	1 905.6	1 455	85 777	4 987.4	72 476
Nov	91 188	6 082.70	76 014	10 671	1 761.8	1 649	80 517	4 320.9	74 364
Dec	88 924	5 780.30	72 539	9 979	1 609.3	1 387	78 945	4 171.0	71 152
2004 Jan	92 746	5 637.40	66 296	10 547	1 507.3	1 360	82 199	4 130.2	64 937
Feb	88 947	6 028.90	72 876	10 421	1 678.8	1 328	78 526	4 350.2	71 548
Mar	96 622	6 784.00	81 275	11 550	1 853.3	1 524	85 072	4 930.1	79 751
Apr	97 529	7 063.00	71 670	11 554	1 898.0	1 402	85 975	5 165.0	70 267
May	103 598	7 270.90	77 514	11 336	1 822.3	995	92 262	5 448.6	76 518
Jun	103 317	7 893.40	73 777	11 883	1 971.0	1 209	91 434	5 922.5	72 562
Jul	106 835	8 326.10	75 298	11 741	2 098.0	1 208	95 094	6 228.0	74 090
Aug	105 970	8 234.70	71 329	11 984	2 073.6	1 341	93 986	6 161.1	69 989
Sep	111 635	7 848.40	74 822	12 730	1 997.5	1 253	98 905	5 850.9	73 570
Oct	103 048	7 602.00	79 973	11 868	1 976.1	1 559	91 180	5 625.8	78 415
Nov	96 562	6 665.70	80 749	12 000	1 875.1	858	84 562	4 790.6	79 891
Dec	97 889	6 694.40	80 043	11 176	1 788.7	891	86 713	4 905.6	79 182
2005 Jan	100 014	6 375.30	71 101	11 494	1 590.7	646	88 828	4 784.6	70 455
Feb	92 589	6 291.80	72 159	10 815	1 654.4	773	81 774	4 637.3	71 387
Mar	104 148	7 559.20	80 413	11 880	1 904.8	908	92 268	5 654.4	79 506
Apr	107 874	7 646.60	74 458	12 240	1 914.8	782	95 634	5 731.8	73 676
May	115 294	8 073.20	75 346	12 700	1 970.2	806	102 594	6 103.0	74 548
Jun	115 219	8 560.30	74 797	11 537	2 003.5	1 259	103 682	6 556.9	73 538
Jul	121 204	9 098.20	76 106	13 472	2 145.5	930	107 732	6 952.8	75 176
Aug	119 731	9 009.40	69 614	13 136	2 109.2	820	106 595	6 900.2	68 794
Sep	117 645	8 738.10	76 595	12 972	2 067.5	806	104 673	6 670.7	75 789
Oct[2]	118 905	8 367.90	83 404	12 983	2 045.7	781	105 922	6 322.2	82 624
Nov	105 519	6 982.00	83 838	12 641	1 920.0	806	92 879	5 062.0	83 032
Dec	106 080	6 900.90	83 575	11 598	1 802.1	695	94 482	5 098.8	82 880
2006 Jan	109 075	6 609.20	77 962	12 233	1 623.6	605	96 842	4 985.6	80 358
Feb	100 291	6 663.50	73 672	11 511	1 679.7	733	420 780	4 983.9	72 939
Mar	112 867	7 766.20	87 093	12 942	1 923.8	785	99 928	5 842.3	86 308
Apr	115 300	8 344.60	78 401	12 287	1 934.9	667	103 013	6 409.8	77 734
May	123 296	8 610.70	78 236	13 276	1 997.0	673	110 020	6 613.6	77 562
Jun	122 920	9 022.30	78 635	13 388	2 026.2	824	109 532	6 996.1	77 811
Jul	127 129	9 560.00	78 551	12 358	2 110.6	733	114 771	7 449.3	77 818
Aug	125 321	9 130.70	72 886	12 664	1 985.2	706	112 657	7 145.5	72 180
Sep	122 265	8 884.61	78 162	12 372	1 963.9	735	109 893	2 560.7	77 426
Oct	122 796	8 547.26	80 873	12 833	1 998.2	696	109 963	6 549.1	80 178

13.9 UK airlines: aircraft kilometres flown, passengers and cargo uplifted[1]
Tonne-kilometres and seat kilometres used on scheduled services

continued

Monthly averages or calendar months: thousands or tonnes

	All services (thousand tonne-kilometres)				Domestic services (thousand tonne-kilometres)				International services (thousand tonne-kilometres)			
	Mail	Freight	Passenger	Seat kilometres used (millions)	Mail	Freight	Passenger	Seat kilometres used (millions)	Mail	Freight	Passenger	Seat kilometres used (millions)
	BMIJ	BMIK	BMIL	BMIM	BMIN	BMIO	BMIP	BMIQ	BMIR	BMIS	BMIT	BMIU
1999	153 014	4 924 973	15 516 752	160 331.3	4 027	5 995	612 378	7 214.8	148 987	4 929 003	14 904 410	153 110.4
2000	179 239	5 160 794	16 495 712	170 323.5	3 647	5 712	636 636	7 843.8	175 522	5 155 082	15 857 076	162 800.6
2001	101 886	4 548 053	15 264 370	158 717.7	3 539	4 089	649 744	7 658.1	98 347	4 544 144	14 614 626	151 059.6
2002	56 551	4 940 528	15 042 639	156 582.0	2 797	3 605	703 521	8 330.2	53 754	4 936 923	14 339 173	148 252.0
2003	55 082	5 250 490	15 376 586	166 445.2	3 067	3 480	737 791	8 991.9	52 015	5 247 010	14 638 795	157 453.4
2004	80 859	5 698 327	16 480 406	182 728.0	2 619	2 621	780 832	9 530.6	78 240	5 695 706	15 699 574	173 196.7
2005	89 299	5 998 434	14 980 897	200 333.0	277	2 652	783 849	9 789.5	90 022	5 995 782	14 197 048	190 543.3
2002 Apr	4 342	389 765	1 197 624	12 535.3	238	267	55 891	663.0	4 104	389 498	1 141 733	11 872.4
May	4 100	420 537	1 195 162	12 518.8	250	329	60 964	723.4	3 850	420 208	1 134 198	11 795.5
Jun	3 564	414 912	1 295 769	13 526.3	202	346	61 301	725.9	3 362	414 566	1 234 468	12 800.5
Jul	3 590	418 248	1 372 205	14 320.0	239	324	66 976	791.5	3 351	417 924	1 305 229	13 528.5
Aug	3 578	398 948	1 402 147	14 579.9	227	290	67 382	797.1	3 351	398 658	1 334 765	13 782.8
Sep	3 271	406 311	1 322 762	13 784.6	223	284	63 078	744.5	3 048	406 027	1 259 684	13 040.1
Oct	3 558	444 868	1 292 158	13 530.0	257	413	62 963	745.4	3 301	444 455	1 229 195	12 784.6
Nov	4 930	459 477	1 168 848	12 176.8	241	305	59 379	705.3	4 689	459 172	1 109 469	11 471.5
Dec	6 728	403 994	1 215 863	12 534.3	234	267	56 925	672.2	6 494	403 727	1 158 993	11 862.0
2003 Jan	3 533	369 564	1 172 313	12 041.0	242	243	52 215	617.5	3 291	369 321	1 120 098	11 423.5
Feb	3 085	397 296	1 052 275	11 693.1	224	265	49 328	612.8	2 861	397 031	1 002 947	11 080.2
Mar	4 598	533 143	1 166 617	14 960.3	248	294	56 243	740.2	4 350	532 849	1 110 374	14 220.0
Apr	3 988	406 914	1 156 061	12 905.8	238	261	62 030	764.1	3 750	406 653	1 094 031	12 141.8
May	3 817	440 143	1 208 511	13 560.2	211	292	62 484	792.5	3 606	439 851	1 146 027	12 767.6
Jun	3 798	421 332	1 347 807	14 704.8	258	330	63 998	812.9	3 540	421 002	1 283 809	13 892.0
Jul	3 819	401 962	1 440 374	15 096.7	275	344	68 416	812.4	3 544	401 618	1 371 958	14 284.4
Aug	4 239	409 606	1 496 153	15 650.2	257	235	69 041	819.3	3 982	409 371	1 427 112	14 831.0
Sep	4 072	439 999	1 381 055	14 470.8	290	323	66 310	788.1	3 782	439 676	1 314 745	13 682.7
Oct	4 943	480 572	1 376 472	14 477.7	224	342	67 791	805.5	4 719	480 230	1 308 681	13 672.2
Nov	5 942	494 564	1 281 768	13 435.3	296	293	62 346	742.8	5 646	494 271	1 219 422	12 692.5
Dec	9 248	455 395	1 297 180	13 449.3	304	258	57 589	683.8	8 944	455 137	1 239 591	12 765.5
2004 Jan	5 761	414 252	1 267 150	13 135.4	253	239	53 713	637.9	5 508	414 013	1 213 437	12 497.5
Feb	5 471	460 557	1 217 118	12 803.9	269	255	59 429	708.8	5 202	460 302	1 157 689	12 095.2
Mar	5 777	517 345	1 401 143	14 742.4	303	284	65 862	784.2	5 474	517 061	1 335 281	13 958.2
Apr	6 703	458 300	1 424 524	15 126.0	257	280	67 361	805.6	6 446	458 020	1 357 163	14 320.3
May	7 198	498 549	1 433 682	15 158.5	229	173	64 696	769.1	6 969	498 376	1 368 986	14 389.4
Jun	6 319	453 879	1 467 827	16 431.9	249	213	67 354	835.6	6 070	453 666	1 400 473	15 596.4
Jul	5 769	475 814	1 561 267	17 455.4	243	171	72 371	892.5	5 526	475 643	1 488 896	16 562.7
Aug	5 375	452 747	1 534 320	17 127.3	218	144	71 348	874.3	5 157	452 603	1 462 972	16 253.0
Sep	6 133	454 903	1 436 946	16 150.1	236	245	67 842	841.6	5 897	454 658	1 369 104	15 308.0
Oct	6 272	500 028	1 353 461	15 680.3	242	262	67 378	838.0	6 030	499 766	1 286 083	14 842.3
Nov	7 895	513 126	1 170 601	14 220.2	46	182	63 090	789.4	7 849	512 944	1 107 511	13 430.7
Dec	12 186	498 827	1 212 367	14 696.6	74	173	60 388	753.6	12 112	498 654	1 151 979	13 943.0
2005 Jan	8 289	454 494	1 094 967	14 699.2	21	155	53 344	670.1	8 268	454 339	1 041 623	14 029.1
Feb	8 020	447 848	966 971	13 593.9	52	194	55 423	696.1	7 968	447 654	911 548	12 897.8
Mar	7 579	509 500	1 220 722	16 298.2	61	211	63 589	803.1	7 518	509 289	1 157 133	15 495.1
Apr	7 218	491 439	1 217 590	16 194.3	33	188	64 716	809.8	7 185	491 251	1 152 874	15 384.5
May	6 788	494 285	1 255 675	16 670.6	16	223	66 491	833.9	6 772	494 062	1 189 184	15 836.7
Jun	6 316	473 969	1 346 995	17 990.4	16	297	67 803	848.8	6 300	473 672	1 279 192	17 141.6
Jul	6 007	492 762	1 443 304	19 205.0	17	258	73 143	916.0	5 990	492 504	1 370 161	18 289.0
Aug	5 372	461 964	1 408 459	18 659.9	18	232	71 931	893.5	5 354	461 732	1 336 528	17 766.4
Sep	5 278	509 140	1 371 706	18 258.1	15	226	70 309	876.8	6 263	508 914	1 301 397	17 381.2
Oct[2]	6 827	561 179	1 330 099	17 735.2	7	240	69 946	869.6	6 820	560 939	1 260 153	16 865.6
Nov	8 307	551 188	1 146 070	15 396.4	10	237	65 479	809.4	8 297	550 951	1 080 591	14 587.0
Dec	13 298	550 666	1 178 339	15 631.8	11	191	61 675	762.4	13 287	550 475	1 116 664	14 869.3
2006 Jan	7 171	511 467	111 055	15 538.5	7	154	55 749	686.1	7 164	511 313	1 054 801	14 852.4
Feb	6 279	479 895	1 092 572	14 402.3	6	189	57 906	712.9	6 273	479 706	1 034 666	13 689.4
Mar	7 034	576 299	1 277 807	16 981.8	9	193	65 822	813.3	7 025	576 106	1 211 985	16 168.5
Apr	7 156	518 472	1 384 501	18 179.9	5	158	65 911	819.6	7 151	518 314	1 318 590	17 360.3
May	6 964	517 240	1 387 018	18 215.0	7	178	68 700	850.1	6 957	517 062	1 318 318	17 364.8
Jun	7 307	521 956	1 484 171	19 469.5	6	223	72 245	862.4	7 301	521 733	1 411 926	18 607.2
Jul	6 799	519 585	1 561 182	20 625.5	6	212	65 463	900.8	6 793	519 373	1 495 719	19 724.7
Aug	7 192	486 701	1 495 475	19 719.7	8	216	64 117	836.3	7 184	486 485	1 431 628	18 883.3
Sep	8 425	513 291	1 437 492	19 028.6	8	216	64 181	833.1[†]	8 417	513 075	1 373 311	18 195.5
Oct	9 063	518 126	1 390 842	18 405.3	8	193	66 096	853.3	9 055	517 933	1 324 746	17 551.9

1 The annual figures are the sum of the monthly figures provided by the CAA. All kilometre statistics are based on standard (Great Circle) distance. Including weight of freight and mail, excess baggage and diplomatic bags, but excluding passengers' and crews' permitted baggage.

2 Provisional figures.

Source: Civil Aviation Authority

13.10 Merchant vessels registered in the United Kingdom (500 gross tons and over)[1]

	Bulk, tanker and dry			Other			Total		
	Number	Grt million	Dwt million	Number	Grt million	Dwt million	Number	Grt million	Dwt million
	BMJG	BMJH	BMJI	BMJJ	BMJK	BMJL	BMJM	BMJN	BMJO
2000	167	4.8	8.6	304	4.7	3.4	471	9.5	12.0
2001	194	5.5	9.6	340	5.2	4.0	534	10.7	13.6
2002	229	6.0	10.3	381	6.5	5.3	610	12.5	15.6
2003	262	7.7	12.7	461	8.3	7.0	723	16.0	19.7
2004	293	8.5	14.0	461	8.4	7.5	754	16.9	21.6
2005	323	9.9	16.7	472	8.9	8.2	795	18.8	24.9
2006	331	10.4	17.3	483	9.4	8.7	814	19.8	26.1
End Quarter									
2002 Q1	196	5.3	9.4	353	5.5	4.2	549	10.8	13.6
Q2	203	5.6	9.8	360	5.7	4.3	563	11.2	14.1
Q3	225	5.9	10.2	368	6.1	4.8	593	12.0	15.0
Q4	229	6.0	10.3	381	6.5	5.3	610	12.5	15.6
2003 Q1	236	6.4	11.0	392	6.7	5.5	628	13.1	16.4
Q2	246	7.0	11.5	431	7.4	6.2	677	14.4	17.7
Q3	250	7.2	11.8	442	7.7	6.5	692	14.9	18.4
Q4	262	7.7	12.7	461	8.3	7.0	723	16.0	19.7
2004 Q1	268	8.1	13.4	475	8.9	7.7	743	17.0	21.1
Q2	279	8.3	13.8	470	9.0	7.8	749	17.3	21.6
Q3	295	8.5	14.1	465	8.7	7.7	760	17.2	21.8
Q4	293	8.5	14.0	461	8.4	7.5	754	16.9	21.6
2005 Q1	306	9.1	15.2	459	8.4	7.7	765	17.5	23.0
Q2	311	9.4	15.8	472	8.6	7.9	783	18.0	23.8
Q3	315	9.8	16.5	474	8.8	8.0	789	18.5	24.5
Q4	323	9.9	16.7	472	8.9	8.2	795	18.8	24.9
2006 Q1	326	9.8	16.4	466	8.7	7.9	792	18.5	24.3
Q2	335	10.3	17.2	467	9.0	8.3	802	19.2	25.5
Q3	334	10.4	17.3	478	9.4	8.7	812	19.8	26.0
Q4	331	10.4	17.3	483	9.4	8.7	814	19.8	26.1

1 Covers vessels registered within the United Kingdom, the Channel Isles and the Isle of Man.

Source: Department for Transport

13.11 UK passenger movement by sea and air[1]

Thousands

Inward

	Sea						Air				
	Irish Republic	Other EU	Rest of Europe and Mediterranean Sea area	Rest of world	Pleasure cruises[2]	Total	Irish Republic	Other EU	Rest of Europe and Mediterranean Sea area	Rest of world	Total[3]
	BMKC	BMKD	BMKE	BMKF	BMKG	BMKB	BMKI	BMKJ	BMKK	BMKL	BMKH
2003	1 893	11 210	121	13	348	13 584	5 093	39 724	9 262	22 838	76 919
2004	1 808	10 950	119	19	384	13 279	5 423	41 729	11 215	24 903	83 268
2005	1 665	10 039	104	..	..	11 808	5 898	43 522	13 640	25 891	88 951
2003 Q4	325	2 332	21	4	56	2 737	1 287	8 758	2 142	5 630	17 818
2004 Q1	263	1 664	20	–	5	1 953	1 225	7 459	2 025	5 739	16 447
Q2	517	3 116	36	10	154	3 831	1 368	11 060	2 806	6 368	21 602
Q3	713	3 981	44	6	165	4 909	1 485	14 072	3 749	6 907	26 213
Q4	315	2 189	19	3	60	2 586	1 345	9 138	2 635	5 889	19 006
2005 Q1	278	1 548	18	..	..	1 844	1 310	8 072	2 644	6 004	18 030
Q2	444	2 752	34	..	..	3 230	1 502	11 633	3 471	6 693	23 299
Q3	653	3 696	41	..	..	4 390	1 627	14 446	4 439	7 167	27 679
Q4	289	2 044	10	..	..	2 344	1 459	9 371	3 086	6 027	19 943
2006 Q1	224	1 568	11	..	..	1 803	1 412	8 109	3 082	6 128	18 731
Q2	..	..	..	..	..	..	1 627	12 071	3 940	7 017	24 655
Q3	..	..	..	..	..	..	1 674	14 733	4 893	7 341	28 642

Outward

	Sea						Air				
	Irish Republic	Other EU	Rest of Europe and Mediterranean Sea area	Rest of world	Pleasure cruises[2]	Total	Irish Republic	Other EU	Rest of Europe and Mediterranean Sea area	Rest of world	Total[3]
	BMKO	BMKP	BMKQ	BMKR	BMKS	BMKN	BMKU	BMKV	BMKW	BMKX	BMKT
2003	1 908	11 271	120	12	350	13 662	5 071	39 643	9 231	22 807	76 752
2004	1 848	10 955	119	22	383	13 327	5 403	41 584	11 155	24 880	83 021
2005	1 715	10 066	104	..	..	11 885	5 894	43 400	13 556	25 788	88 637
2003 Q4	344	2 378	21	2	61	2 806	1 296	8 488	2 119	6 040	17 943
2004 Q1	255	1 597	20	–	8	1 879	1 199	7 353	1 951	5 581	16 083
Q2	527	3 181	35	12	138	3 894	1 347	11 505	2 874	5 928	21 654
Q3	731	3 930	44	5	169	4 879	1 507	13 864	3 712	7 030	26 113
Q4	335	2 247	20	4	69	2 675	1 350	8 862	2 618	6 341	19 171
2005 Q1	290	1 539	19	..	..	1 848	1 300	8 102	2 589	5 924	17 915
Q2	448	2 740	33	..	..	3 221	1 472	11 918	3 490	6 091	22 971
Q3	675	3 672	41	..	..	4 388	1 656	14 288	4 405	7 276	27 624
Q4	302	2 116	11	..	..	2 428	1 466	9 092	3 072	6 497	20 127
2006 Q1	239	1 508	10	..	..	1 756	1 401	7 984	2 937	5 981	18 303
Q2	..	..	..	..	..	..	1 600	12 550	4 001	6 479	24 631
Q3	..	..	..	..	..	..	1 691	14 486	4 825	7 445	28 447

Note: Sea and Air passenger numbers are seasonal, which should be taken into account when comparing figures within a year.

1 Excluding movement by land across the frontier between the Irish Republic and Northern Ireland, passengers travelling between the Channel Islands and Great Britain, passengers carried in aircraft chartered by British government departments and as far as possible, passengers travelling by sea on day trips and HM and other Armed Forces travelling in the course of their duties.

2 Passengers on pleasure cruises beginning and/or ending at UK seaports (excluding QE2 passengers between Southampton and New York which are included in rest of world).

3 The figures do not include oil rigs.

Sources: Department for Transport;
Civil Aviation Authority

13.12 UK passenger movement by sea and air
Analysis of countries of landing and of embarkation

Thousands

		2004	2005	2005 Q1	2005 Q2	2005 Q3	2005 Q4	2006 Q1	2006 Q2	2006 Q3
European continent and Mediterranean Sea area										
By sea										
Belgium	BMLB	739	778	152	198	241	187	145	..	..
France[1]	BMLC	18 566	16 834	2 511	4 614	6 294	3 415	2 550	..	..
Netherlands	BMLD	2 001	1 848	337	503	588	419	323	..	..
Irish Republic	BMDJ	3 655	3 380	568	892	1 328	591	463	..	..
Germany	BMDK	118	98	19	30	37	13	..	..	..
Denmark	BMDL	97	91	15	25	34	18	14	..	..
Sweden	BMDM	75	79	10	24	29	15	6	..	..
Spain	BMDN	309	378	43	98	145	92	37	..	..
Norway	BMDO	231	200	37	65	79	19	20	..	..
Other Europe	BMLE	6	8	–	2	3	2	..	..	..
Total	A4N3	25 799	23 693	3 692	6 451	8 778	4 772	3 559	..	..
By air										
Austria	BMLH	1 747	1 797	554	416	483	344	546	409	454
Belgium	BMLI	1 863	1 711	401	443	439	428	401	439	389
Denmark	BMLJ	2 186	2 255	481	576	639	559	493	601	624
Eastern Europe[2]	BMLM	5 634	7 871	1 604	1 945	2 372	1 950	1 978	2 442	2 912
Finland	BMLK	812	799	163	172	175	289	173	204	210
France	BMLL	10 887	11 009	2 513	2 883	3 265	2 348	2 584	3 113	3 414
Germany	BMLN	10 279	10 912	2 498	2 779	2 971	2 664	2 442	3 060	3 187
Greece	BMLO	5 824	5 598	224	1 656	3 008	710	206	1 623	2 995
Irish Republic	BMLP	10 825	11 791	2 610	2 974	3 282	2 925	2 813	3 227	3 365
Italy	BMLQ	9 672	10 715	2 193	2 885	3 411	2 226	2 074	2 917	3 348
Malta	BMLR	1 096	1 110	186	295	395	234	183	286	349
Netherlands	BMLS	7 896	7 887	1 797	2 053	2 081	1 956	1 813	2 173	2 185
Norway	BMLT	1 605	1 753	400	454	456	443	435	483	514
Portugal	BMLU	3 796	4 088	615	1 183	1 464	826	623	1 267	1 548
Spain	BMLV	25 621	27 302	4 093	7 749	10 035	5 425	4 128	8 040	10 119
Sweden	BMLW	2 253	2 321	536	616	607	562	496	624	601
Switzerland	BMLX	4 119	4 502	1 432	1 024	1 068	978	1 558	1 172	1 146
Turkey	BMLY	2 788	3 551	203	1 004	1 746	598	206	938	1 692
Yugoslavia	BMLZ	150	150	29	38	50	33	32	45	53
Other countries[3]	BMMA	1 276	1 484	232	418	526	308	250	492	622
Total	BMLG	110 333	118 607	22 764	31 564	38 474	25 805	23 435	33 555	39 729
Mediterranean area										
By air										
Cyprus	BMMC	2 769	2 990	324	867	1 182	617	309	913	1 191
Near East[4]	BMMD	1 069	1 056	231	270	312	243	232	283	270
North Africa[5]	BMME	2 016	2 776	598	680	765	733	793	866	927
Total	BMMB	5 854	6 822	1 153	1 817	2 259	1 593	1 334	2 061	2 388
Rest of World										
By sea										
United States & Canada[17]	BMDQ	39	..	..	..	..	..	..	..	..
Rest of World[17]	RVCO	–	..	..	..	..	..	..	..	..
Pleasure cruises[18]	LUQZ	767	..	..	..	..	..	..	..	..
Total	A4N4	807	..	..	..	..	..	..	..	..
By air										
Australia and New Zealand	BMMP	1 055	1 397	369	336	352	340	350	323	350
Canada	BMMQ	3 307	3 610	590	974	1 356	690	606	1 026	1 317
Canary Islands	BMMR	7 719	7 261	1 832	1 670	1 893	1 866	1 794	1 752	1 901
Caribbean[6]	BMMS	1 880	1 824	528	423	406	467	527	442	433
Central Africa[7]	BMMT	66	67	15	18	18	16	15	18	19
Central America[8]	BMMU	1 152	1 271	227	341	425	278	259	370	429
East Africa[9]	BMMV	662	736	181	152	222	181	187	170	224
Far East[10]	BMMW	4 611	4 787	1 128	1 191	1 294	1 174	1 205	1 237	1 354
Indian Continent[11]	BMMX	2 109	2 691	694	590	642	765	954	789	834
Japan	BMMY	1 189	1 185	290	294	314	287	272	278	289
Middle East[12]	BMMZ	3 421	3 774	949	868	1 052	905	946	970	1 215
Southern Africa[13]	BMNA	1 768	1 733	487	373	416	457	480	381	425
South America[14]	BMNB	394	379	100	83	97	99	102	93	75
United States of America	BMNC	18 005	18 310	3 861	4 874	5 252	4 323	3 720	4 966	5 144
West Africa[15]	BMND	856	916	240	195	236	245	249	222	252
Other countries[16]	BMNE	865	852	208	199	229	216	208	206	222
Oil rigs	BMNF	565	627	134	158	172	163	154	176	188
Total	BMMO	49 625	51 420	11 835	12 738	14 376	12 471	12 027	13 419	14 671

Note: Sea and Air passenger numbers are seasonal, which should be taken into account when comparing figures within a year.

1 Includes Hovercraft passengers.
2 Including Albania, Bulgaria, Czech Republic, Hungary, Poland, Romania and Commonwealth of Independent States.
3 Including Faroes, Gibraltar, Iceland, Luxembourg, Croatia, Slovenia and Bosnia-Herzegovina.
4 Including Jordan, Lebanon, Israel and Syria.
5 Including Algeria, Egypt, Libya, Morocco and Tunisia.
6 Including Bahamas, Barbados, Bermuda, Cayman Islands, French Antilles, Jamaica, Leeward Islands, Netherlands Antilles, Puerto Rico, Trinidad and Tobago, Turks and Caicos Islands, US Virgin Islands and Windward Islands.
7 Including Angola, Central African Republic, Chad, Congo, Democratic Republic of Congo, Malawi and Zambia.
8 Including Belize, Costa Rica, Cuba, Dominican Republic, El Salvador, Guatemala, Haiti, Honduras, Mexico, Nicaragua and Panama.
9 Including Burundi, Djibouti, Ethiopia, Kenya, Rwanda, Somali Republic, Sudan, Tanzania and Uganda.
10 Including Bandar Seri Begawan, Burma, China, Hongkong, Indonesia, Kampuchea, Korea, Laos, Malaysia, Nepal, Philippines, Singapore, Taiwan, Thailand and Vietnam.
11 Including Afghanistan, Bangladesh, India, Pakistan and Sri Lanka.
12 Including Iran, Iraq, Kuwait, Persian Gulf States, Republic of North Yemen, Republic of South Yemen, Saudi Arabia and United Arab Emirates.
13 Including Botswana, Lesotho, Mozambique, Namibia, South African Republic, Swaziland and Zimbabwe.
14 Including Argentina, Bolivia, Brazil, Chile, Colombia, Ecuador, Guyana, Paraguay, Peru, Uruguay and Venezuela.
15 Including Benin, Cameroon, Equatorial Guinea, Gabon, Gambia, Ghana, Guinea, Guinea Bissau, Ivory Coast, Liberia, Mali, Mauritania, Niger, Nigeria, Senegal, Sierra Leone, Togo, Upper Volta and Western Sahara.
16 Atlantic Ocean Islands, Indian Ocean Islands and Pacific Ocean Islands and Madeira.
17 Australia & New Zealand, Africa, Caribbean and other areas of the world (excluding USA & Canada).
18 Passengers on pleasure cruises beginning and/or ending at UK seaports (excluding QE2 passengers between Southampton and New York which is included in USA & Canada).

Sources: Department for Transport;
Civil Aviation Authority

14 Retailing

14.1 Index numbers of retail sales[1]

Sales: weekly average 2000=100, seasonally adjusted

| | | | Volume | | | | | | | | | Value | | | | | |
| | | | Predominantly non-food stores | | | | | | | | Predominantly non-food stores | | | | | |
	All retail-ers	Predomi-nantly food stores	Total	Non-special-ised stores	Textile, clothing and footwear stores	House-hold goods stores	Other stores	Non-store and repair	All retail-ers	Predomi-nantly food stores	Total	Non-special-ised stores	Textile, clothing and footwear stores	House-hold goods stores	Other stores	Non-store retaili-ng and repair
Sales in 2000 (£m)	207 149	89 041	106 359	18 781	27 880	27 699	31 999	11 749	207 149	89 041	106 359	18 781	27 880	27 699	31 999	11 749
	EAPS	EAPT	EAPV	EAPU	EAPX	EAPY	EAPW	EAPZ	EAQV	EAQW	EAQY	EAQX	EARA	EARB	EAQZ	EARC
2002	112.2	108.2	116.2	110.5	123.8	117.8	111.6	106.5	110.6	110.4	111.8	107.4	114.9	113.1	110.5	100.5
2003	116.3	111.9	121.3	113.8	129.6	122.3	117.5	105.4	113.7	114.8	114.8	109.2	118.9	113.4	115.6	96.2
2004	123.3	116.5	129.6	118.0	139.2	130.8	127.1	117.1	118.7	119.6	119.8	111.2	124.6	117.2	123.0	102.7
2005	125.8	119.7	131.8	119.3	143.8	131.2	129.2	118.0	119.9	123.6	119.1	110.8	126.4	112.9	123.0	99.1
2006	129.8	122.7	136.6	124.0	151.0	137.4	130.7	122.5	123.3	128.3	121.7	113.9	131.8	114.0	124.3	99.9
2006 Q1	127.1	121.5	132.9	122.1	145.7	132.6	128.4	117.4	120.7	125.8	119.1	112.3	126.9	111.9	122.5	96.4
Q2	129.5	122.5	136.4	125.2	149.4	138.2	130.1	120.7	122.9	127.5	121.8	114.8	130.2	115.3	124.2	98.6
Q3	130.5	123.8	136.9	125.6	151.0	137.5	130.8	122.9	124.3	129.8	122.3	115.2	131.7	114.5	125.1	100.3
Q4	132.3	123.8	139.8	126.3	154.9	141.5	133.2	127.6	125.6	130.7	123.9	115.6	135.0	114.5	127.1	102.9
2006 May	129.5	121.7	136.9	126.0	150.6	137.9	130.6	121.5	122.7	126.5	122.1	115.5	131.3	114.6	124.5	98.9
Jun	130.3	123.7	136.8	124.6	149.4	139.2	130.8	121.6	123.9	129.2	122.2	114.3	130.5	116.0	125.0	99.8
Jul	130.3	124.7	136.2	125.9	151.6	134.9	130.1	119.1	123.9	130.2	121.5	115.3	132.3	111.9	124.1	96.9
Aug	130.9	122.7	138.0	126.5	151.4	140.6	130.9	127.7	124.5	128.8	123.2	116.0	132.1	116.9	125.1	104.5
Sep	130.3	123.9	136.6	124.7	150.1	137.0	131.3	122.2	124.5	130.4	122.2	114.6	130.9	114.6	125.8	99.8
Oct	131.5	123.7	138.8	127.6	153.8	137.9	133.2	124.4	125.1	130.3	123.5	116.7	134.4	113.2	126.8	100.2
Nov	131.8†	123.5	139.2†	125.8†	155.0†	139.2†	133.2†	127.8†	125.4†	130.4	123.8†	115.2†	135.3†	114.2†	127.1†	103.1
Dec	133.2	124.2	141.2	125.6	155.6	146.4	133.2	129.9	126.2	131.3	124.3	115.0	135.3	115.7	127.5	104.9

1 Great Britain only. The motor trades are excluded. Information for periods earlier than those shown is available from ONS Newport (tel. 01633 812713).

Source: Office for National Statistics

14.2 Index numbers of retail sales[1]
Value of retail sales at current prices

Sales: weekly average 2000=100, not seasonally adjusted

| | | Predominantly food stores | | | |
	All retailing	Total value of sales	Non-specialised stores	Specialist food stores	Alcoholic drinks, other beverages and tobacco
Sales in 2000 (£m)	207 149	89 041	76 846	6 393	5 801
	EAFY	EAFS	EAGB	CY3X	CY45
2002	110.6	110.4	112	101	95
2003	113.7	114.8	118	100	87
2004	118.7	119.6	124	101	82
2005	119.9	123.6	129	103	78
2006	123.3	128.3	134	102	76
2006 Q1	110.6	119.7	126	97	67
Q2	118.9	127.2	133	102	76
Q3	119.1	127.0	133	97	77
Q4	144.8	139.3	146	114	84
2006 May	118.2	125.7	132	102	74
Jun	120.2	129.5	136	100	79
Jul	121.1	130.0	136	99	81
Aug	118.4	125.7	132	99	77
Sep	118.1	125.7	132	93	74
Oct	124.1	127.4	133	101	77
Nov	137.1†	134.1	141†	107	78†
Dec	167.5	152.9	159	130	94

14.2 Index numbers of retail sales[1]
Value of retail sales at current prices
continued

Sales: weekly average 2000=100, not seasonally adjusted

Predominantly non-food stores

	Total	Non-specialised stores	Textile, clothing and footwear stores				Household goods stores			
			Total	Retail of textiles	Retail sale of clothing	Retail sale of footwear and leather goods	Total	Retail sale of furniture, lighting, etc	Retail sale of electrical household appliances	Retail sale of hardware, paint and glass
Sales in 2000 (£m)	106 359	18 781	27 880	915	23 725	3 240	27 699	8 706	10 966	8 027
	EAFT	EAGE	EAFU	EAPG	EAGH	EAPH	EAFV	EAPI	EAPJ	EAPK
2002	111.8	107.4	114.9	122	116	101	113.1	113	109	119
2003	114.8	109.2	118.9	114	120	109	113.4	110	107	125
2004	119.8	111.2	124.6	115	127	113	117.2	116	108	132
2005	119.1	110.8	126.4	101	129	114	112.9	111	103	129
2006	121.7	113.9	131.8	95	136	113	114.0	115	104	126
2006 Q1	105.2	95.4	107.0	95	110	91	107.9	116	97	114
Q2	114.6	100.3	124.9	91	128	114	109.5	107	90	139
Q3	115.1	102.7	126.7	87	130	116	108.3	111	93	126
Q4	152.0	157.2	168.7	106	176	133	130.1	126	139	123
2006 May	114.7	100.3	126.9	96	129	117	108.5	107	89	138
Jun	115.0	99.8	126.4	87	129	116	106.5	103	88	135
Jul	117.0	105.5	133.2	89	137	119	105.4	107	88	128
Aug	114.8	103.5	123.3	88	126	114	109.6	112	95	127
Sep	113.9	99.8	124.2	85	127	115	109.6	114	95	125
Oct	122.8	113.2	137.7	97	141	123	114.2	122	101	123
Nov	140.4†	145.3†	153.5†	110†	161	114	123.8†	129	119†	124
Dec	184.8	201.9	205.5	110	216	157	147.8	126	184	121

Predominantly non-food stores

	Other specialised non-food stores						Non-store retail and repair		
	Total	Pharmaceutical medical cosmetic and toilet goods	Retail sale of books, newspapers and periodicals	Retail sale of floor coverings	Photographic, optical & precision equipment, office supplies	Other retail sale in specialist stores nes including secondhand	Total	Retail sale via mail order houses	Non-store retail excepting mail order
Sales in 2000 (£m)	31 999	3 553	5 022	1 788	4 167	17 470	11 749	8 819	2 930
	EAFW	EAPQ	EAPL	EAPM	EAWH	CY4B	EAFX	EAPN	CY4H
2002	110.5	97	106	116	110	114	100.5	97	110
2003	115.6	103	102	110	102	126	96.2	93	105
2004	123.0	105	104	116	131	131	102.7	100	110
2005	123.0	104	105	114	135	130	99.1	98	103
2006	124.3	100	94	126	130	137	99.9	99	104
2006 Q1	107.0	85	90	122	127	110	89.5	90	87
Q2	118.5	95	81	118	125	133	93.8	90	105
Q3	118.2	96	86	127	126	129	95.5	92	107
Q4	153.5	123	120	134	141	174	120.9	122	117
2006 May	117.9	94	80	119	128	131	93.9	93	98
Jun	121.4	101	79	118	124	137	95.7	88	119
Jul	119.6	99	81	120	129	132	91.3	89	99
Aug	118.4	94	84	132	129	129	96.1	91	111
Sep	116.9	94	92	130	122	126	98.3	94	111
Oct	122.7	97	97	139	130	132	110.1	110	110
Nov	140.4†	106†	106†	156†	143†	155†	130.3†	133†	121†
Dec	188.7	157	150	113	149	224	122.0	123	119

1 Great Britain only. The motor trades are excluded. Information for periods earlier than those shown is available from ONS Newport (tel. 01633 812713).

Source: Office for National Statistics

15 External trade in goods

15.1 Values of United Kingdom total trade in goods

£ million BOP basis seasonally adjusted

	Total trade in goods			Total excluding oil			Total excluding oil and erratics[1]		
	Exports	Imports	Balance	Exports	Imports	Balance	Exports	Imports	Balance
	BOKG	BOKH	BOKI	ELBM	ENXP	BQKH	BPBL	BQBG	BPAP
2001	189 093	230 305	−41 212	174 278	220 780	−46 502	162 591	206 231	−43 640
2002	186 524	234 229	−47 705	172 203	225 016	−52 813	160 967	210 840	−49 873
2003	188 320	236 927	−48 607	173 712	225 695	−51 983	161 431	212 703	−51 272
2004	190 877	251 770	−60 893	174 677	236 463	−61 786	162 467	224 251	−61 784
2005	211 616	280 399	−68 783	191 821	258 411	−66 590	179 363	245 401	−66 038
2006	243 857	328 180	−84 323	220 863	301 481	−80 618	208 766	286 176	−77 410
2001 Q4	45 419	55 890	−10 471	42 249	53 577	−11 328	39 425	50 196	−10 771
2002 Q1	46 382	57 754	−11 372	43 106	55 678	−12 572	40 409	52 716	−12 307
Q2	49 102	60 104	−11 002	44 956	57 713	−12 757	41 706	53 888	−12 182
Q3	46 608	58 624	−12 016	43 252	56 216	−12 964	40 443	52 696	−12 253
Q4	44 432	57 747	−13 315	40 889	55 409	−14 520	38 409	51 540	−13 131
2003 Q1	48 666	59 528	−10 862	44 404	56 652	−12 248	41 386	53 304	−11 918
Q2	46 697	58 242	−11 545	43 350	55 674	−12 324	40 430	52 601	−12 171
Q3	46 338	58 640	−12 302	42 843	55 714	−12 871	39 808	52 222	−12 414
Q4	46 619	60 517	−13 898	43 115	57 655	−14 540	39 807	54 576	−14 769
2004 Q1	46 079	60 026	−13 947	42 393	57 040	−14 647	39 419	54 090	−14 671
Q2	47 137	62 384	−15 247	43 397	58 577	−15 180	40 540	55 514	−14 974
Q3	48 218	63 747	−15 529	44 145	59 699	−15 554	40 900	56 514	−15 614
Q4	49 443	65 613	−16 170	44 742	61 147	−16 405	41 608	58 133	−16 525
2005 Q1	48 861	65 004	−16 143	44 358	60 821	−16 463	41 287	57 710	−16 423
Q2	51 723	67 530	−15 807	47 035	62 506	−15 471	44 148	59 505	−15 357
Q3	54 398	72 453	−18 055	49 271	65 799	−16 528	45 971	62 339	−16 368
Q4	56 634	75 412	−18 778	51 157	69 285	−18 128	47 957	65 847	−17 890
2006 Q1	64 288[†]	85 477[†]	−21 189[†]	58 488[†]	78 759[†]	−20 271[†]	55 568[†]	75 006[†]	−19 438[†]
Q2	66 551	89 250	−22 699	60 143	82 441	−22 298	56 872	78 643	−21 771
Q3	57 482	77 336	−19 854	51 562	70 108	−18 546	48 792	66 261	−17 469
Q4	55 536	76 117	−20 581	50 670	70 173	−19 503	47 534	66 266	−18 732
2003 Dec	15 614	20 343	−4 729	14 316	19 420	−5 104	13 196	18 349	−5 153
2004 Jan	15 008	20 307	−5 299	13 748	19 131	−5 383	12 838	18 166	−5 328
Feb	15 177	19 460	−4 283	14 137	18 611	−4 474	13 171	17 737	−4 566
Mar	15 894	20 259	−4 365	14 508	19 298	−4 790	13 410	18 187	−4 777
Apr	15 741	20 791	−5 050	14 524	19 520	−4 996	13 574	18 482	−4 908
May	15 485	20 564	−5 079	14 192	19 319	−5 127	13 285	18 373	−5 088
Jun	15 911	21 029	−5 118	14 681	19 738	−5 057	13 681	18 659	−4 978
Jul	15 919	21 258	−5 339	14 592	19 886	−5 294	13 566	18 840	−5 274
Aug	15 915	21 152	−5 237	14 460	19 981	−5 521	13 379	18 907	−5 528
Sep	16 384	21 337	−4 953	15 093	19 832	−4 739	13 955	18 767	−4 812
Oct	16 239	21 835	−5 596	14 686	20 348	−5 662	13 644	19 245	−5 601
Nov	16 399	21 821	−5 422	14 913	20 196	−5 283	13 933	19 315	−5 382
Dec	16 805	21 957	−5 152	15 143	20 603	−5 460	14 031	19 573	−5 542
2005 Jan	16 296	21 856	−5 560	14 753	20 456	−5 703	13 724	19 398	−5 674
Feb	15 923	21 392	−5 469	14 672	20 068	−5 396	13 623	19 061	−5 438
Mar	16 642	21 756	−5 114	14 933	20 297	−5 364	13 940	19 251	−5 311
Apr	16 932	22 624	−5 692	15 402	20 909	−5 507	14 553	19 878	−5 325
May	16 698	22 140	−5 442	15 154	20 526	−5 372	14 146	19 660	−5 514
Jun	18 093	22 766	−4 673	16 479	21 071	−4 592	15 449	19 967	−4 518
Jul	17 653	23 265	−5 612	15 798	21 315	−5 517	14 664	20 195	−5 531
Aug	18 036	24 365	−6 329	16 543	22 157	−5 614	15 457	20 891	−5 434
Sep	18 709	24 823	−6 114	16 930	22 327	−5 397	15 850	21 253	−5 403
Oct	18 873	24 352	−5 479	16 885	22 120	−5 235	15 866	21 241	−5 375
Nov	18 635	25 224	−6 589	16 960	23 140	−6 180	15 859	21 803	−5 944
Dec	19 126	25 836	−6 710	17 312	24 025	−6 713	16 232	22 803	−6 571
2006 Jan	20 405[†]	27 189[†]	−6 784[†]	18 571[†]	24 727[†]	−6 156[†]	17 593[†]	23 457[†]	−5 864[†]
Feb	21 183	28 695	−7 512	19 425	26 753	−7 328	18 405	25 496	−7 091
Mar	22 700	29 593	−6 893	20 492	27 279	−6 787	19 570	26 053	−6 483
Apr	22 219	29 232	−7 013	20 114	27 230	−7 116	19 044	26 042	−6 998
May	22 498	31 259	−8 761	20 426	28 820	−8 394	19 381	27 451	−8 070
Jun	21 834	28 759	−6 925	19 603	26 391	−6 788	18 447	25 150	−6 703
Jul	19 150	25 919	−6 769	16 936	23 465	−6 529	16 024	22 111	−6 087
Aug	19 502	26 080	−6 578	17 652	23 563	−5 911	16 743	22 314	−5 571
Sep	18 830	25 337	−6 507	16 974	23 080	−6 106	16 025	21 836	−5 811
Oct	18 558	25 126	−6 568	16 788	23 080	−6 292	15 696	21 851	−6 155
Nov	18 602	25 473	−6 871	17 071	23 374	−6 303	15 948	22 026	−6 078
Dec	18 376	25 518	−7 142	16 811	23 719	−6 908	15 890	22 389	−6 499

1 These are defined as ships, North Sea installations, aircraft, precious stones and silver.

Source: Office for National Statistics: 020 7533 6064

15.2 Volume and Price index numbers

Indices 2003=100 BOP basis

	Volume (seasonally adjusted)						Price index (not seasonally adjusted)							
	Total trade in goods		Total excluding oil		Total excluding oil and erratics[1]		Total trade in goods			Total excluding oil			Total excluding oil & erratics[1]	
	Exports	Imports	Exports	Imports	Exports	Imports	Exports	Imports	Terms of trade[2]	Exports	Imports	Terms of trade[2]	Exports	Imports
	BQKU	BQKV	BQKI	BQKJ	BOMA	ELAL	BQKR	BQKS	BQKT	BQKK	BQKL	BQKM	BQAK	ELBA
2001	101.5	93.8	100.8	93.5	103.3	93.1	98.3	103.3	95.2	98.8	104.4	94.6	97.3	103.9
2002	100.3	98.2	99.9	98.6	101.8	98.2	98.2	100.7	97.5	98.7	101.1	97.6	97.7	100.9
2003	100.0	100.0	100.0	100.0	100.0	100.0	100.0	100.0	100.0	100.0	100.0	100.0	100.0	100.0
2004	101.5	106.9	102.0	106.3	102.0	106.8	100.3	99.5	100.8	98.9	98.7	100.2	99.0	99.0
2005	110.9	114.8	112.5	114.6	113.2	115.4	104.7	103.7	101.0	100.6	100.6	100.0	100.8	100.7
2006	127.6	131.2	130.6	131.7	132.9	132.6	107.8	107.4	100.4	101.8	103.0	98.8	102.0	103.1
2001 Q4	99.9	94.0	99.1	92.9	101.1	92.5	95.7	100.1	95.6	97.2	101.8	95.5	96.1†	101.6
2002 Q1	99.7	95.7	99.2	95.8	101.7	96.5	98.4	101.5	96.9	99.7	102.6	97.2	98.6	102.3
Q2	104.8	100.2	103.9	100.6	105.0	100.0	99.3	101.5	97.8	99.6	101.8	97.8	98.7	101.6
Q3	100.5	99.0	100.6	99.5	102.6	99.1	98.1	100.2	97.9	98.2	100.3	97.9	97.3	100.2
Q4	96.1	97.9	95.8	98.4	98.1	97.2	97.1	99.6	97.5	97.2	99.8	97.4	96.3	99.7
2003 Q1	103.2	100.2	102.8	100.8	103.4	100.7	99.9	100.3	99.6	99.0	99.7	99.3	99.0	99.7
Q2	99.2	98.5	99.3	98.3	99.6	98.6	100.3	99.9	100.4	101.0	100.3	100.7	101.0	100.4
Q3	98.0	98.7	98.1	98.4	98.2	97.9	100.4	100.3	100.1	100.6	100.4	100.2	100.5	100.2
Q4	99.5	102.6	99.7	102.5	98.8	102.8	99.3	99.6	99.7	99.4	99.7	99.7	99.5	99.8
2004 Q1	100.1	103.6	100.0	103.4	99.8	103.7	98.1	97.7	100.4	98.0	97.8	100.2	98.1	98.1
Q2	101.4	106.4	102.2	105.5	102.7	105.9	99.5	99.1	100.4	98.6	98.6	100.0	98.7	98.9
Q3	101.8	107.4	102.9	107.1	102.5	107.3	100.9	100.4	100.5	98.8	99.1	99.7	99.0	99.3
Q4	102.7	110.1	103.1	109.4	103.1	110.1	102.7	100.7	102.0	100.3	99.3	101.0	100.4	99.5
2005 Q1	102.4	108.3	102.6	108.3	102.7	108.9	102.7	101.5	101.2	100.6	99.8	100.8	100.8	100.0
Q2	109.2	111.9	110.7	111.5	111.7	112.6	103.3	102.2	101.1	100.0	99.8	100.2	100.3	99.9
Q3	113.8	117.4	116.2	117.1	116.6	117.5	106.2	105.2	101.0	100.4	100.9	99.5	100.7	101.0
Q4	118.3	121.4	120.6	121.5	121.6	122.5	106.6	105.8	100.8	101.3	102.0	99.3	101.5	102.0
2006 Q1	135.1†	136.3†	138.6†	137.0†	142.0†	138.6†	108.3	107.3	100.9	102.6	102.9	99.7	102.7	102.8
Q2	140.1	143.2	143.7	144.6	146.4	146.6	109.4	108.1	101.2†	102.5	103.0	99.5†	102.7	103.0
Q3	119.2	122.6	121.6	122.6	123.9	122.8	108.5	108.0†	100.5†	101.6	103.0†	98.6†	101.9	103.2†
Q4	116.0	122.6	118.3	122.7	119.1	122.6	105.1	106.3	98.9	100.6	103.0	97.7	101.0	103.2
2004 Mar	103.0	104.3	102.4	104.5	101.7	104.2	98.8	98.4	100.4	98.4	98.2	100.2	98.5	98.5
Apr	102.0	106.7	102.6	105.5	103.2	105.8	99.1	98.6	100.5	98.7	98.4	100.3	98.8	98.6
May	99.4	104.7	100.1	104.2	100.8	105.0	100.3	99.8	100.5	99.0	99.0	100.0	99.1	99.3
Jun	102.8	107.7	103.9	106.8	104.2	106.8	99.1	98.9	100.2	98.0	98.4	99.6	98.1	98.7
Jul	102.4	108.8	103.0	107.7	103.0	108.0	99.3	99.1	100.2	98.0	98.5	99.5	98.2	98.8
Aug	100.7	106.9	101.5	107.6	101.0	107.7	101.0	100.6	100.4	98.5	99.0	99.5	98.7	99.2
Sep	102.2	106.6	104.1	106.0	103.5	106.2	102.5	101.4	101.1	100.0	99.8	100.2	100.1	100.0
Oct	100.1	109.1	101.2	109.2	101.0	109.3	104.2	102.2	102.0	100.8	99.9	100.9	101.0	100.1
Nov	101.8	109.4	102.6	108.0	103.2	109.5	103.1	100.9	102.2	100.7	99.6	101.1	100.8	99.7
Dec	106.3	111.8	105.6	110.9	105.2	111.6	100.9	99.1	101.8	99.3	98.5	100.8	99.4	98.8
2005 Jan	102.5	109.6	102.1	109.2	102.3	109.8	102.0	101.1	100.9	100.5	99.8	100.7	100.6	100.0
Feb	100.7	106.5	102.0	106.5	101.8	107.1	102.2	101.3	100.9	100.4	99.8	100.6	100.6	100.0
Mar	103.9	108.9	103.8	109.1	104.1	109.7	103.9	102.2	101.7	100.9	99.8	101.1	101.2	100.0
Apr	106.9	112.8	108.3	112.2	110.0	113.2	103.3	101.7	101.6	100.2	99.5	100.7	100.4	99.6
May	105.5	110.3	106.5	109.7	107.0	111.5	103.4	102.1	101.3	100.6	100.0	100.6	100.8	100.1
Jun	115.2	112.7	117.2	112.7	118.0	113.1	103.3	102.8	100.5	99.3	99.8	99.5	99.6	100.0
Jul	110.0	113.1	111.3	113.1	111.1	113.5	106.0	105.3	100.7	100.8	101.5	99.3	101.0	101.5
Aug	113.4	118.1	116.8	118.0	117.5	117.9	106.5	105.3	101.1	100.4	100.6	99.8	100.7	100.8
Sep	117.9	121.0	120.6	120.1	121.3	121.2	106.1	104.9	101.1	100.0	100.6	99.4	100.3	100.7
Oct	118.4	117.9	120.3	117.3	121.9	119.5	106.9	105.6	101.2	101.1	101.7	99.4	101.4	101.8
Nov	117.1	121.6	119.8	121.5	120.4	121.5	106.5	106.0	100.5	101.3	102.3	99.0	101.5	102.2
Dec	119.5	124.8	121.6	125.7	122.6	126.6	106.4	105.9	100.5	101.4	102.1	99.3	101.6	102.1
2006 Jan	127.9†	129.6†	131.2†	129.4†	133.9†	130.2†	107.8	106.8	100.9	102.0	102.3	99.7	102.2	102.3
Feb	134.3	137.9	138.3	139.4	141.4	141.1	108.1†	107.3	100.7†	102.5	103.0	99.5	102.5†	102.9
Mar	143.1	141.5	146.4	142.3	150.7	144.4	108.9	107.7	101.1	103.2	103.3	99.9	103.3	103.2
Apr	140.2	140.8	144.0	142.9	147.0	145.4	110.5	109.0	101.4	103.3	103.5	99.8	103.4	103.4
May	144.3	150.8	148.6	152.1	151.8	154.0	108.7	107.3	101.3	101.9†	102.3	99.6†	102.1	102.5
Jun	135.7	138.0	138.6	138.7	140.5	140.4	109.0	108.0	100.9	102.2†	103.1	99.1†	102.5	103.2
Jul	117.5	122.6	119.3	122.8	121.6	122.8	110.1	109.2†	100.8	102.4	103.6	98.8	102.7	103.7
Aug	121.3	123.5	124.4	123.5	126.9	123.9	108.6	107.9†	100.6	101.1	102.5†	98.6	101.5	102.7†
Sep	118.7	121.8	121.0	121.4	123.2	121.7	106.7	106.8	99.9	101.2	102.9	98.3	101.5	103.1
Oct	115.2	121.2	116.9	120.9	117.4	121.2	105.4	106.7	98.8	101.0	103.4	97.7	101.3	103.6
Nov	116.6	122.5	119.2	122.1	119.5	121.6	105.0	106.5	98.6	100.7	103.3	97.5	101.1	103.6
Dec	116.1	124.2	118.9	125.0	120.4	125.0	104.9	105.8	99.1	100.2	102.2	98.0	100.5	102.5

1 These are defined as ships, North Sea installations, aircraft, precious stones and silver. 2 Export price index as a percentage of the import price index.

Source: Office for National Statistics: 020 7533 6064

15.3 United Kingdom trade in goods, by commodity group[1]

£ million BOP basis seasonally adjusted

	Food, beverages and tobacco (SITC 0+1)			Basic materials (SITC 2+4)			Fuels (SITC 3)			Semi-manufactures (SITC 5+6)			Finished manufactures (SITC 7+8)		
	Exports	Imports	Balance	Exports	Imports	Balance	Exports	Imports	Balance	Exports	Imports	Balance	Exports	Imports	Balance
	BOPL	BQAR	ELBE	BOPM	BQAS	ELBF	BOPN	BQAT	ELBG	BOPO	BQAU	ELBH	BOPP	BQAV	ELBI
2001	9 630	18 485	−8 855	2 571	6 442	−3 871	16 386	10 795	5 591	50 295	52 910	−2 615	109 188	140 409	−31 221
2002	9 993	19 375	−9 382	2 855	5 958	−3 103	16 000	10 279	5 721	50 223	52 722	−2 499	106 380	144 445	−38 065
2003	10 879	21 187	−10 308	3 335	6 139	−2 804	16 558	12 311	4 247	54 492	56 045	−1 553	102 193	139 641	−37 448
2004	10 578	22 147	−11 569	3 771	6 340	−2 569	17 885	17 547	338	56 466	60 226	−3 760	101 296	143 703	−42 407
2005	10 645	23 696	−13 051	3 980	6 769	−2 789	21 497	25 920	−4 423	59 883	62 677	−2 794	114 490	159 497	−45 007
2006	10 949	25 084	−14 135	4 906	7 898	−2 992	25 173	31 613	−6 440	64 853	69 491	−4 638	136 724	192 044	−55 320
2001 Q4	2 379	4 717	−2 338	633	1 469	−836	3 533	2 597	936	12 240	13 150	−910	26 380	33 620	−7 240
2002 Q1	2 388	4 758	−2 370	647	1 489	−842	3 656	2 329	1 327	12 359	12 919	−560	27 118	35 870	−8 752
Q2	2 469	4 837	−2 368	712	1 512	−800	4 537	2 705	1 832	12 845	13 305	−460	28 222	37 391	−9 169
Q3	2 603	4 802	−2 199	741	1 477	−736	3 722	2 653	1 069	12 578	13 018	−440	26 654	36 323	−9 669
Q4	2 533	4 978	−2 445	755	1 480	−725	4 085	2 592	1 493	12 441	13 480	−1 039	24 386	34 861	−10 475
2003 Q1	2 821	5 129	−2 308	842	1 510	−668	4 714	3 100	1 614	13 240	13 472	−232	26 814	35 944	−9 130
Q2	2 634	5 270	−2 636	813	1 506	−693	3 912	2 837	1 075	13 823	14 194	−371	25 293	34 049	−8 756
Q3	2 759	5 310	−2 551	848	1 491	−643	4 018	3 155	863	13 594	14 204	−610	24 927	34 084	−9 157
Q4	2 665	5 478	−2 813	832	1 632	−800	3 914	3 219	695	13 835	14 175	−340	25 159	35 564	−10 405
2004 Q1	2 563	5 384	−2 821	867	1 572	−705	4 028	3 500	528	13 888	14 361	−473	24 523	34 768	−10 245
Q2	2 741	5 549	−2 808	921	1 563	−642	4 170	4 263	−93	13 829	14 506	−677	25 280	36 051	−10 771
Q3	2 611	5 585	−2 974	995	1 598	−603	4 603	4 626	−23	14 229	15 514	−1 285	25 573	35 998	−10 425
Q4	2 663	5 629	−2 966	988	1 607	−619	5 084	5 158	−74	14 520	15 845	−1 325	25 920	36 886	−10 966
2005 Q1	2 655	5 786	−3 131	957	1 607	−650	4 865	5 111	−246	14 706	15 605	−899	25 465	36 460	−10 995
Q2	2 730	5 963	−3 233	985	1 657	−672	5 198	5 869	−671	14 776	15 471	−695	27 823	38 061	−10 238
Q3	2 664	5 926	−3 262	1 032	1 769	−737	5 558	7 401	−1 843	15 155	15 567	−412	29 708	41 367	−11 659
Q4	2 596	6 021	−3 425	1 006	1 736	−730	5 876	7 539	−1 663	15 246	16 034	−788	31 494	43 609	−12 115
2006 Q1	2 680†	6 204†	−3 524†	1 104†	1 864†	−760†	6 297†	8 383†	−2 086†	15 615†	16 712†	−1 097†	38 234†	51 768†	−13 534†
Q2	2 731	6 172	−3 441	1 214	1 914	−700	7 012	7 778	−766	16 342	17 064	−722	38 964	55 813	−16 849
Q3	2 787	6 276	−3 489	1 227	1 967	−740	6 558	8 325	−1 767	16 325	17 515	−1 190	30 275	42 792	−12 517
Q4	2 751	6 432	−3 681	1 361	2 153	−792	5 306	7 127	−1 821	16 571	18 200	−1 629	29 251	41 671	−12 420
2004 Jan	846	1 780	−934	247	555	−308	1 369	1 345	24	4 543	4 845	−302	7 917	11 650	−3 733
Feb	851	1 802	−951	300	512	−212	1 144	1 026	118	4 713	4 613	100	8 108	11 365	−3 257
Mar	866	1 802	−936	320	505	−185	1 515	1 129	386	4 632	4 903	−271	8 498	11 753	−3 255
Apr	943	1 893	−950	320	547	−227	1 327	1 426	−99	4 610	4 822	−212	8 477	11 946	−3 469
May	883	1 825	−942	305	493	−188	1 433	1 380	53	4 357	4 702	−345	8 440	12 006	−3 566
Jun	915	1 831	−916	296	523	−227	1 410	1 457	−47	4 862	4 982	−120	8 363	12 099	−3 736
Jul	914	1 862	−948	323	549	−226	1 501	1 582	−81	4 656	5 071	−415	8 450	12 038	−3 588
Aug	825	1 830	−1 005	339	526	−187	1 641	1 337	304	4 667	5 266	−599	8 381	12 060	−3 679
Sep	872	1 893	−1 021	333	523	−190	1 461	1 707	−246	4 906	5 177	−271	8 742	11 900	−3 158
Oct	857	1 839	−982	327	516	−189	1 682	1 654	28	4 774	5 400	−626	8 531	12 272	−3 741
Nov	910	1 910	−1 000	340	547	−207	1 617	1 906	−289	4 855	5 130	−275	8 588	12 154	−3 566
Dec	896	1 880	−984	321	544	−223	1 785	1 598	187	4 891	5 315	−424	8 801	12 460	−3 659
2005 Jan	882	1 982	−1 100	296	520	−224	1 647	1 719	−72	4 928	5 009	−81	8 492	12 483	−3 991
Feb	891	1 909	−1 018	307	552	−245	1 348	1 718	−370	4 913	5 301	−388	8 378	11 758	−3 380
Mar	882	1 895	−1 013	354	535	−181	1 870	1 674	196	4 865	5 295	−430	8 595	12 219	−3 624
Apr	898	2 053	−1 155	328	546	−218	1 681	1 947	−266	4 806	5 208	−402	9 146	12 709	−3 563
May	889	1 968	−1 079	325	545	−220	1 750	1 882	−132	4 837	5 053	−216	8 835	12 499	−3 664
Jun	943	1 942	−999	332	566	−234	1 767	2 040	−273	5 133	5 210	−77	9 842	12 853	−3 011
Jul	867	1 971	−1 104	312	570	−258	1 986	2 153	−167	4 886	5 260	−374	9 523	13 150	−3 627
Aug	880	1 992	−1 112	369	599	−230	1 612	2 447	−835	5 189	5 264	−75	9 888	13 927	−4 039
Sep	917	1 963	−1 046	351	600	−249	1 960	2 801	−841	5 080	5 043	37	10 297	14 290	−3 993
Oct	862	2 022	−1 160	320	573	−253	2 124	2 548	−424	4 978	5 018	−40	10 417	14 043	−3 626
Nov	850	1 971	−1 121	321	578	−257	1 786	2 561	−775	5 112	5 247	−135	10 465	14 693	−4 228
Dec	884	2 028	−1 144	365	585	−220	1 966	2 430	−464	5 156	5 769	−613	10 612	14 873	−4 261
2006 Jan	898†	2 110†	−1 212†	326†	610†	−284†	1 989	2 943†	−954†	5 224†	5 522†	−298†	11 870†	15 832†	−3 962†
Feb	885	2 038	−1 153	373	610	−237	1 914†	2 492	−578	5 179	5 690	−511	12 719	17 686	−4 967
Mar	897	2 056	−1 159	405	644	−239	2 394	2 948	−554	5 212	5 500	−288	13 645	18 250	−4 605
Apr	902	2 053	−1 151	385	589	−204	2 287	2 348	−61	5 396	5 690	−294	13 157	18 373	−5 216
May	906	2 057	−1 151	401	657	−256	2 311	2 745	−434	5 515	5 697	−182	13 259	19 941	−6 682
Jun	923	2 062	−1 139	428	668	−240	2 414	2 685	−271	5 431	5 677	−246	12 548	17 499	−4 951
Jul	903	2 069	−1 166	414	626	−212	2 471	2 784	−313	5 129	5 758	−629	10 133	14 519	−4 386
Aug	940	2 118	−1 178	403	671	−268	2 031	2 891	−860	5 672	5 853	−181	10 358	14 392	−4 034
Sep	944	2 089	−1 145	410	670	−260	2 056	2 650	−594	5 524	5 904	−380	9 784	13 881	−4 097
Oct	912	2 127	−1 215	471	715	−244	1 913	2 366	−453	5 584	5 901	−317	9 581	13 836	−4 255
Nov	947	2 174	−1 227	493	695	−202	1 659	2 569	−910	5 586	6 220	−634	9 815	13 641	−3 826
Dec	892	2 131	−1 239	397	743	−346	1 734	2 192	−458	5 401	6 079	−678	9 855	14 194	−4 339

1 More commodity detail is available on a seasonally adjusted BOP basis in tables B1 to B11 inclusive, and C1 to C4 inclusive, of the Monthly Review of External Trade Statistics.

Source: Office for National Statistics: 020 7533 6064

15.4 Volume index numbers, by commodity group[1]

2003=100 BOP basis seasonally adjusted

	Food, beverages and tobacco (SITC 0+1)		Basic materials (SITC 2+4)		Fuels (SITC 3)		Semi-manufactures (SITC 5+6)		Finished manufactures (SITC 7+8)		Total manufactures (SITC 5 to 8)	
	Exports	Imports	Exports	Imports	Exports	Imports	Exports	Imports	Exports	Imports	Exports	Imports
	BPEM	BQBK	BAFB	BQBL	BAFC	BQBM	BAHA	BQBN	BAHY	ELAB	BOGT	ELAJ
2001	92	90	81	109	107	102	91	91	108	94	102	94
2002	96	94	90	101	103	91	94	95	103	101	101	99
2003	100	100	100	100	100	100	100	100	100	100	100	100
2004	99	106	105	100	93	123	104	106	101	106	102	106
2005	98	110	107	102	90	131	108	106	117	118	114	115
2006	99	114	121	110	90	136	116	112	143	142	134	134
2001 Q4	91	91	81	102	108	118	91	94	105	92	100	93
2002 Q1	92	91	83	102	103	94	93	93	104	97	100	96
Q2	95	95	90	103	113	92	96	96	110	103	105	101
Q3	100	94	93	101	96	91	95	94	104	103	101	100
Q4	96	95	94	99	102	89	94	98	96	99	96	99
2003 Q1	106	98	102	100	105	89	98	98	105	103	103	101
Q2	95	100	97	99	100	101	101	100	99	97	99	98
Q3	101	99	101	97	98	104	99	101	97	98	98	99
Q4	98	103	100	104	96	106	102	101	99	103	100	102
2004 Q1	96	104	100	100	98	117	104	104	99	103	100	103
Q2	104	106	103	99	90	125	103	103	102	106	102	105
Q3	98	107	109	101	89	122	105	108	102	106	103	107
Q4	99	108	109	100	93	130	105	109	103	109	104	109
2005 Q1	99	109	102	100	96	125	106	106	102	108	103	107
Q2	101	110	108	100	92	133	108	105	114	114	112	111
Q3	98	109	109	105	83	130	110	106	123	123	118	118
Q4	96	111	107	101	87	136	109	107	130	129	123	122
2006 Q1	97[†]	114	112	106	90[†]	143[†]	111[†]	109	160[†]	151[†]	143[†]	140[†]
Q2	99	112	118[†]	109[†]	95	126	117	110[†]	165	165	148	149
Q3	101	113	121	108	90	138	117	112	127	127	124	123
Q4	100	117	134	116	85	136	119	116	120	125	120	122
2004 Mar	97	105	108	96	107	106	104	106	103	104	103	104
Apr	106	109	107	103	92	134	103	104	102	105	103	105
May	100	105	100	94	89	116	98	100	102	106	100	104
Jun	105	105	101	99	90	126	109	106	101	107	104	107
Jul	104	107	112	105	95	139	104	107	102	107	103	107
Aug	92	105	109	99	92	100	104	111	101	107	102	108
Sep	97	108	107	99	80	126	107	107	104	105	105	106
Oct	95	106	106	96	84	110	104	113	101	108	102	110
Nov	101	111	109	103	88	144	104	105	102	108	103	107
Dec	101	107	112	102	108	136	107	110	106	111	106	111
2005 Jan	98	112	95	99	103	134	106	103	102	110	103	108
Feb	99	107	101	102	82	131	107	108	100	104	103	105
Mar	99	107	110	99	103	110	105	108	104	110	104	109
Apr	101	115	108	100	89	134	105	106	112	114	109	112
May	99	109	106	99	96	132	105	103	108	112	107	109
Jun	104	107	109	101	91	133	113	107	122	115	119	112
Jul	95	108	101	101	91	116	106	107	117	116	113	114
Aug	97	110	115	107	71	125	112	108	123	124	119	119
Sep	101	110	112	108	87	150	111	103	128	128	122	121
Oct	95	112	104	101	91	141	107	100	131	125	123	118
Nov	94	109	104	101	80	140	110	105	129	130	123	122
Dec	98	112	113	101	90	128	111	115	131	132	124	127
2006 Jan	97[†]	116	101[†]	105	85	148	112[†]	109	148[†]	139[†]	135[†]	131
Feb	96	112	114	103	84[†]	127[†]	111	111	160	155	143	143[†]
Mar	97	114	121	110	101	154	111	107	172	160	151	145
Apr	98	113	111	101	92	108	115	110	167	162	149	147
May	98	113[†]	119	112	95	135	120	110[†]	172	177	153	158
Jun	100	111	123	113[†]	98	134	116	109	157	155	143	142
Jul	98	112	120	105	97	130	109	111	128	129	121	124
Aug	102	114	122	111	83	139	122	113	129	128	127	124
Sep	102	114	121	109	91	145	119	113	125	125	123	122
Oct	99	115	136	115	92	134	120	113	118	124	118	121
Nov	103	118	142	112	81	152	121	119	120	122	121	121
Dec	97	117	123	121	82	122	117	116	123	129	121	125

1 Commodity volumes are shown in more detail on a seasonally adjusted BOP basis in tables C1 toC3 inclusive, and D1 to D3 inclusive, of the *Monthly Review of External Trade Statistics.*

Source: Office for National Statistics: 020 7533 6064

15.5 Price index numbers, by commodity group[1]

2003=100 BOP basis not seasonally adjusted

	Food, beverages and tobacco (SITC 0+1)		Basic materials (SITC 2+4)		Fuels (SITC 3)		Semi-manufactures (SITC 5+6)		Finished manufactures (SITC 7+8)		Total manufactures (SITC 5 to 8)	
	Exports	Imports	Exports	Imports	Exports	Imports	Exports	Imports	Exports	Imports	Exports	Imports
	BPAI	ELAN	BPAW	ELAO	BPDU	ELAP	BQAA	ELAQ	BQAB	ELAR	BQAI	ELAY
2001	96	96	94	96	93	85	100	104	99	106	99	105
2002	96	97	95	96	93	92	97	99	100	103	99	102
2003	100	100	100	100	100	100	100	100	100	100	100	100
2004	99	98	108	103	117	118	99	102	98	97	99	98
2005	100	103	115	108	151	168	102	107	99	97	100	100
2006	102	105	123	117	176	199	103	112	100	97	101	102
2001 Q4	96	97	93	93	80	70	97	100	97	104	98	103
2002 Q1	96	98	94	94	85	82	97	99	101	105	100	103
Q2	96	96	96	96	95	95	98	99	101	104	100	103
Q3	96	96	96	95	97	96	98	99	99	102	98	101
Q4	97	98	96	97	96	95	97	98	97	101	97	100
2003 Q1	99	98	98	98	109	112	99	99	99	100	99	100
Q2	101	100	100	100	94	91	101	101	101	100	101	101
Q3	100	101	100	100	99	98	100	100	101	100	101	100
Q4	100	101	101	102	98	99	99	100	99	99	99	100
2004 Q1	99	98	104	103	99	100	98	99	98	97	98	98
Q2	99	99	108	104	112	112	98	101	98	97	98	98
Q3	98	98	108	103	126	128	99	103	98	97	98	99
Q4	100	99	112	103	133	133	101	104	99	97	100	99
2005 Q1	100	101	113	104	126	140	102	106	99	96	100	99
Q2	100	103	114	108	140	153	101	106	99	96	100	99
Q3	100	103	116	108	170	192	101	106	99	97	100	100
Q4	101	103	119	110	169	187	102	108	99	98	100	101
2006 Q1	102	102	120	114	176	199	103	110	100	98	101	102
Q2	103	105	124	115	187	212	103	112	100	97	101	102
Q3	102	106	124	117	185	208	102	113	99	97	100	101
Q4	102	106	122	120	157	178	102	114	98	97	100	102
2004 Jan	100	98	102	102	98	99	99	99	98	97	98	98
Feb	98	97	103	102	96	95	98	99	97	96	97	97
Mar	99	98	107	104	103	105	98	100	98	97	98	98
Apr	99	98	109	105	105	105	98	100	98	97	98	98
May	99	99	109	105	117	118	99	101	98	98	99	99
Jun	98	99	105	103	113	113	98	101	98	97	98	98
Jul	98	98	105	103	115	115	98	102	97	97	98	98
Aug	98	98	109	103	130	135	99	102	98	97	98	99
Sep	99	99	111	103	132	135	100	104	99	98	99	99
Oct	100	99	113	104	145	149	101	104	99	98	100	99
Nov	100	98	113	103	133	132	102	104	99	97	100	99
Dec	99	100	110	102	121	117	100	104	98	95	99	98
2005 Jan	100	101	112	103	119	132	102	106	99	97	100	99
Feb	100	101	112	104	123	135	101	106	99	96	100	99
Mar	100	102	115	105	136	153	102	106	100	96	101	99
Apr	99	102	116	107	138	149	101	106	99	95	100	99
May	100	103	114	108	134	145	102	107	99	96	100	99
Jun	100	103	113	108	147	164	100	105	98	96	99	99
Jul	101	105	114	110	163	185	102	107	99	98	100	100
Aug	100	102	116	108	173	199	101	106	99	97	100	100
Sep	100	102	117	107	174	193	101	106	98	97	99	100
Oct	101	102	120	109	172	186	102	108	99	98	100	101
Nov	101	103	118	111	168	186	102	108	99	98	100	101
Dec	101	104	118	111	167	188	102	108	99	98	100	101
2006 Jan	102	102	118	113	176[†]	201	102	109	100	98	101	101
Feb	102	102	120	114	175	198	103	111	100	98	101	102
Mar	102	103	122	115	176	199	104	111	101	99	102	102
Apr	104	104	124	115	192	222	104	112	101	98	102	102
May	103	104	123	115	185	208	102	112	100	97	101	101
Jun	103	106	125	116	184	207	103	113[†]	100	97	101	102
Jul	103	107	125	117	195	220	103	113	100	98	101	102
Aug	102	106	123	117	192	216	102	113	99	96	100	101
Sep	102	106[†]	123	118	168	187[†]	102	114	99	97	100	101
Oct	102	107	123[†]	121	156	176[†]	102[†]	114	99	97	100[†]	102
Nov	102[†]	106	123[†]	121	155	176	102[†]	114	98[†]	97	100[†]	102
Dec	102	105	121	119	159	183	101	113	98	96	99	101

1 Commodity price indices are shown in more detail on a not seasonally adjusted BOP basis in tables C4 to C6 inclusive, and D4 to D6 inclusive, of the *Monthly Review of External Trade Statistics*.

Source: Office for National Statistics: 020 7533 6064

15.6 United Kingdom exports, by commodity[1]

£ million BOP-consistent basis seasonally adjusted

		2005	2006	2006 Q2	2006 Q3	2006 Q4	2006 Jul	2006 Aug	2006 Sep	2006 Oct	2006 Nov	2006 Dec
0. Food and live animals	BOGG	6 550	6 804	1 679	1 738†	1 708	562†	590	586	565	591	552
Of which:												
01. Meat and meat preparations	BOGS	729	758	183†	201	197	67	66	68	72†	67	58
02. Dairy products and eggs	BQMS	718	724	182	176†	181	57	61	58†	59	63	59
04 & 08. Cereals and animal feeding stuffs	BQMT	1 555	1 605	400†	407	409	130†	138	139	139	138	132
05. Vegetables and fruit	BQMU	514	581	150	147†	145	48	49	50†	48	50	47
1. Beverages and tobacco	BQMZ	4 095	4 145	1 052†	1 049	1 043	341†	350	358	347	356	340
11. Beverages	BQNB	3 481	3 628	916†	919	924	297†	305	317	308	316	300
12. Tobacco	BQOW	614	517	136†	130	119	44†	45	41	39	40	40
2. Crude materials	BQOX	3 745	4 631	1 151†	1 153	1 289	387†	380	386	449	469	371
Of which:												
24. Wood, lumber and cork	BQOY	131	147	39	37	38	13†	11	13	13	12	13
25. Pulp and waste paper	BQOZ	283	336	80	83†	96	28	28	27†	32	34	30
26. Textile fibres	BQPA	516	544	130	135†	131	45	46†	44	45	49	37
28. Metal ores	BQPB	1 713	2 437	620†	602	726	204	195†	203	258	274	194
3. Fuels	BOPN	21 497	25 173	7 012†	6 558	5 306	2 471†	2 031	2 056	1 913	1 659	1 734
33. Petroleum and petroleum products	ELBL	19 795	22 994	6 408†	5 920	4 866	2 214	1 850†	1 856	1 770	1 531	1 565
32, 34 and 35. Coal, gas and electricity	BOQI	1 702	2 179	604†	638	440	257†	181	200	143	128	169
4. Animal and vegetable oils and fats	BQPI	235	275	63†	74	72	27	23†	24	22	24	26
5. Chemicals	ENDG	33 389	37 032	9 353†	9 444	9 263	2 936†	3 244	3 264	3 118	3 139	3 006
Of which:												
51. Organic chemicals	BQPJ	6 703	7 930	2 079†	2 169	1 877	655†	775	739	676	647	554
52. Inorganic chemicals	BQPK	1 555	2 141	601†	514	557	156†	166	192	152	203	202
53. Colouring materials	CSCE	1 635	1 621	416†	401	387	131	140†	130	138	134	115
54. Medicinal products	BQPL	12 321	13 590	3 373†	3 364	3 436	1 042†	1 153	1 169	1 144	1 142	1 150
55. Toilet preparations	CSCF	3 219	3 444	850†	875	870	282†	296	297	300	306	264
57 & 58. Plastics	BQQA	4 298	4 506	1 103†	1 123	1 154	363†	379	381	396	371	387
6. Manufactures classified chiefly by material	BQQB	26 494	27 821	6 989†	6 881	7 308	2 193†	2 428	2 260	2 466	2 447	2 395
Of which:												
63. Wood and cork manufactures	BQQC	255	274	70†	65	64	21	22	22	22†	20	22
64. Paper and paperboard manufactures	BQQD	2 044	2 036	519†	516	515	170†	171	175	174	168	173
65. Textile manufactures	BQQE	2 645	2 697	674†	664	646	223†	220	221	213	227	206
67. Iron and steel	BQQF	5 184	5 181	1 238†	1 284	1 386	421	436†	427	487	463	436
68. Non-ferrous metals	BQQG	3 863	4 868	1 207†	1 238	1 348	374†	408	456	449	458	441
69. Metal manufactures	BQQH	4 067	4 537	1 098†	1 242	1 126	354†	526	362	384	373	369
7. Machinery and transport equipment[2]	BQQI	89 382	110 576	32 403†	23 951	22 638	8 017†	8 214	7 720	7 422	7 575	7 641
71 - 716, 72, 73 & 74. Mechanical machinery	BQQK	25 797	28 320	7 200†	7 138	6 913	2 359†	2 387	2 392	2 306	2 310	2 297
716, 75, 76 & 77. Electrical machinery	BQQL	37 120	55 664	18 368†	10 368	9 329	3 507†	3 677	3 184	2 956	3 023	3 350
78. Road vehicles	BQQM	19 440	19 310	4 901†	4 691	4 615	1 577†	1 588	1 526	1 537	1 546	1 532
79. Other transport equipment	BQQN	7 025	7 282	1 934†	1 754	1 781	574†	562	618	623	696	462
8. Miscellaneous manufactures[2]	BQQO	25 108	26 148	6 561†	6 324	6 613	2 116†	2 144	2 064	2 159	2 240	2 214
Of which:												
84. Clothing	CSCN	2 712	2 883	725†	712	726	240	234†	238	234	245	247
85. Footwear	CSCP	471	522	129†	131	143	43	44	44†	46	50	47
87 & 88. Scientific and photographic	BQQQ	7 245	7 340	1 850†	1 790	1 798	590	593†	607	582	611	605
9. Other commodities and transactions	BOQL	1 121	1 252	288†	310	296	100†	98	112	97	102	97
TOTAL UK EXPORTS	BOKG	211 616	243 857	66 551†	57 482	55 536	19 150†	19 502	18 830	18 558	18 602	18 376

1 The numbers on the left hand side of the table refer to the code numbers of the *Standard International Trade Classification*, Revision 3, which was introduced in January 1988.

2 Sections 7 and 8 are shown by broad economic category in table G2 of the *Monthly Review of External Trade Statistics*.

Source: Office for National Statistics: 020 7533 6064

15.7 United Kingdom imports, by commodity[1]

£ million BOP-consistent basis seasonally adjusted

		2005	2006	2006 Q2	2006 Q3	2006 Q4	2006 Jul	2006 Aug	2006 Sep	2006 Oct	2006 Nov	2006 Dec
0. Food and live animals	BQQR	18 594	19 871	4 874†	4 982	5 133	1 651†	1 667	1 664	1 695	1 740	1 698
Of which:												
01. Meat and meat preparations	BQQS	3 618	3 809	935†	976	991	321†	320	335	331	334	326
02. Dairy products and eggs	BQQT	1 700	1 824	437†	464	476	154†	158	152	163	155	158
04 & 08. Cereals and animal feeding stuffs	BQQU	2 364	2 515	608†	630	656	206	206†	218	219	219	218
05. Vegetables and fruit	BQQV	5 447	5 829	1 409†	1 495	1 511	501	502†	492	496	509	506
1. Beverages and tobacco	BQQW	5 102	5 213	1 298†	1 294	1 299	418†	451	425	432	434	433
11. Beverages	EGAT	3 625	3 715	924†	917	919	292†	320	305	307	305	307
12. Tobacco	EMAI	1 477	1 498	374†	377	380	126†	131	120	125	129	126
2. Crude materials	ENVB	6 128	7 115	1 718†	1 778	1 947	564†	607	607	643	624	680
Of which:												
24. Wood, lumber and cork	ENVC	1 357	1 452	343†	363	408	122†	116	125	132	129	147
25. Pulp and waste paper	EQAH	477	512	135†	126	121	40	44†	42	42	39	40
26. Textile fibres	EQAP	314	297	72	72	80	25	23†	24	29	26	25
28. Metal ores	EHAA	1 999	2 675	622†	684	759	202†	242	240	258	254	247
3. Fuels	BQAT	25 920	31 613	7 778†	8 325	7 127	2 784†	2 891	2 650	2 366	2 569	2 192
33. Petroleum and petroleum products	ENXO	21 988	26 699	6 809†	7 228	5 944	2 454†	2 517	2 257	2 046	2 099	1 799
32, 34 and 35. Coal, gas and electricity	BPBI	3 932	4 914	969†	1 097	1 183	330†	374	393	320	470	393
4. Animal and vegetable oils and fats	EHAB	641	783	196†	189	206	62	64†	63	72	71	63
5. Chemicals	ENGA	29 208	31 872	7 719†	7 850	8 452	2 500†	2 660	2 690	2 751	2 856	2 845
Of which:												
51. Organic chemicals	EHAC	7 183	7 737	1 782†	1 817	2 240	607†	558	652	729	763	748
52. Inorganic chemicals	EHAE	1 507	2 141	555†	461	639	118†	152	191	193	261	185
53. Colouring materials	CSCR	1 073	1 121	273	284†	288	92	94	98†	96	94	98
54. Medicinal products	EHAF	8 503	9 169	2 263†	2 307	2 337	720†	831	756	756	762	819
55. Toilet preparations	CSCS	3 035	3 313	798†	831	811	278†	276	277	268	272	271
57 & 58. Plastics	EHAG	5 037	5 449	1 331†	1 400	1 398	446†	494	460	462	454	482
6. Manufactures classified chiefly by material	EHAH	33 469	37 619	9 345†	9 665	9 748	3 258†	3 193	3 214	3 150	3 364	3 234
Of which:												
63. Wood and cork manufactures	EHAI	1 506	1 583	385†	403	418	130	134†	139	137	136	145
64. Paper and paperboard manufactures	EHAJ	4 820	5 041	1 219†	1 257	1 298	422†	410	425	430	429	439
65. Textile manufactures	EHAK	3 844	4 034	992†	1 014	1 023	333†	338	343	342	350	331
67. Iron and steel	EHAL	4 402	4 958	1 142†	1 308	1 400	423†	431	454	446	464	490
68. Non-ferrous metals	EHAM	3 923	6 195	1 697†	1 703	1 561	606†	552	545	490	550	521
69. Metal manufactures	EHAN	5 355	5 831	1 447†	1 500	1 495	494†	508	498	490	507	498
7. Machinery and transport equipment[2]	EHAO	117 321	147 031	44 609†	31 604	30 298	10 801†	10 664	10 139	10 018	9 905	10 375
71 - 716, 72, 73 & 74. Mechanical machinery	EHAQ	21 851	22 745	5 622†	5 668	5 765	1 887†	1 932	1 849	1 925	1 928	1 912
716, 75, 76 & 77. Electrical machinery	EHAR	55 736	81 908	28 606†	15 308	13 730	5 421†	5 143	4 744	4 533	4 434	4 763
78. Road vehicles	EHAS	31 434	33 039	8 260†	8 261	8 319	2 711†	2 805	2 745	2 760	2 735	2 824
79. Other transport equipment	EHAT	8 300	9 339	2 121†	2 367	2 484	782†	784	801	800	808	876
8. Miscellaneous manufactures[2]	EHAU	42 176	45 013	11 204†	11 188	11 373	3 718†	3 728	3 742	3 818	3 736	3 819
Of which:												
84. Clothing	CSDR	11 297	11 839	2 975†	2 906	2 963	982†	950	974	979	992	992
85. Footwear	CSDS	2 563	2 701	681	663†	666	217	217	229†	218	224	224
87 & 88. Scientific and photographic	EHAW	7 415	7 704	1 918†	1 887	1 933	632†	637	618	649	648	636
9. Other commodities and transactions	BQAW	1 840	2 050	509	461	534	163	155	143	181	174†	179
TOTAL UK IMPORTS	BOKH	280 399	328 180	89 250†	77 336	76 117	25 919†	26 080	25 337	25 126	25 473	25 518

1 The numbers on the left hand side of the table refer to the code numbers of the *Standard International Trade Classification,* Revision 3, which was introduced in January 1988.

2 Sections 7 and 8 are shown by broad economic category in table G2 of the *Monthly Review of External Trade Statistics.*

Source: Office for National Statistics: 020 7533 6064

15.8 United Kingdom exports, by area

£ million BOP-consistent basis seasonally adjusted

		2005	2006	2006 Q2	2006 Q3	2006 Q4	2006 Jul	2006 Aug	2006 Sep	2006 Oct	2006 Nov	2006 Dec
European Union:[1]	LGCK	120 623	152 066	42 707†	35 426	33 683	12 034†	12 062	11 330	11 228	11 111	11 344
EMU members:	QAKW	108 742	134 857	37 120†	31 788	30 191	10 789†	10 844	10 155	10 053	9 913	10 225
Austria	CHMY	1 332	1 689	478†	371	368	120†	140	111	130	128	110
Belgium & Luxembourg	CHNQ	11 394	14 933	3 977†	3 704	3 443	1 131†	1 479	1 094	1 171	1 112	1 160
Finland	CHMZ	1 514	1 871	457†	481	454	149†	149	183	152	156	146
France	ENYL	19 932	29 025	8 960†	6 171	5 766	2 107†	2 088	1 976	1 968	1 846	1 952
Germany	ENYO	23 024	27 473	7 356†	6 688	6 741	2 386†	2 119	2 183	2 130	2 173	2 438
Greece	CHNT	1 366	1 481	350†	374	368	124†	127	123	125	126	117
Irish Republic	CHNS	16 297	17 376	4 169†	4 440	4 363	1 431†	1 571	1 438	1 468	1 496	1 399
Italy	CHNO	8 791	9 554	2 350†	2 377	2 282	816†	791	770	747	756	779
Netherlands	CHNP	12 718	16 624	4 569†	3 835	3 592	1 348†	1 274	1 213	1 220	1 175	1 197
Portugal	CHNU	1 697	2 349	738†	569	387	232†	151	186	118	152	117
Spain	CHNV	10 677	12 482	3 716†	2 778	2 427	945†	955	878	824	793	810
Non-EMU members:[1]	BQIA	11 881	17 209	5 587†	3 638	3 492	1 245†	1 218	1 175	1 175	1 198	1 119
Of which:												
Czech Rep	FKML	1 080	1 571	469†	353	341	128†	115	110	111	120	110
Denmark	CHNR	2 314	3 889	1 346†	818	647	273†	299	246	206	225	216
Hungary	QALC	834	850	206†	210	217	68†	70	72	72	72	73
Poland	ERDR	1 653	2 789	791†	543	584	178†	183	182	189	210	185
Sweden	CHNA	4 588	5 244	1 426†	1 278	1 236	460†	401	417	404	417	415
Other Western Europe:	HCJD	9 730	9 068	2 397†	2 011	2 129	688†	658	665	752	689	688
Of which:												
Iceland	EPLW	179	187	51	45	45	17	13	15	14	17†	14
Norway	EPLX	2 211	2 070	568	490	504	167	144	179	167	172†	165
Switzerland	EPLV	4 985	4 098	1 054	820	918	283	282	255	355	276†	287
Turkey	EOBA	2 160	2 418	623†	580	592	190†	191	199	204	191	197
North America:	HBZQ	35 010	36 884	9 825†	9 071	8 687	2 757†	3 123	3 191	2 861	2 955	2 871
Of which:												
Canada	EOBC	3 278	3 892	1 106†	860	899	219†	316	325	295	300	304
Mexico	EPJX	638	750	177	195	219	57	68	70	71	79†	69
USA	EOBB	30 915	32 052	8 483†	7 975	7 523	2 466†	2 725	2 784	2 481	2 557	2 485
Other OECD countries:	HCII	8 577	8 718	2 189†	2 149	2 223	679†	699	771	714	785	724
Of which:												
Australia	EPMA	2 580	2 482	618†	607	652	206†	205	196	206	210	236
Japan	EOBD	3 900	4 115	1 040†	985	1 006	310†	311	364	311	360	335
New Zealand	EPMB	415	373	102†	91	84	29†	29	33	29	29	26
South Korea	ERDM	1 677	1 748	429	466	481	134	154	178	168	186†	127
Oil exporting countries:	HDII	10 852	9 066	2 233†	1 994	2 086	673†	718	603	626	769	691
Of which:												
Brunei	QALF	43	79	14	41	12	22	17	2	7	3	2
Dubai	QALI	4 658	2 831	720†	452	418	163†	167	122	119	173	126
Indonesia	FKMR	366	312	93	64	74	24	19	21	21	31†	22
Kuwait	QATB	426	439	118	109	116	40	35	34	30	33†	53
Nigeria	QATE	799	820	219†	199	207	63†	64	72	63	70	74
Saudi Arabia	ERDI	1 559	1 647	438	414	440	146	145	123	124	160†	156
Rest of the World	HCHW	26 824	28 055	7 200†	6 831	6 728	2 319†	2 242	2 270	2 377	2 293	2 058
Of which:												
Brazil	FKMO	836	916	214†	194	252	47	77†	70	95	80	77
China	ERDN	2 811	3 258	818	816	896	277	279	260	306	299†	291
Egypt	QALL	543	576	130	139	177	55	41	43	48	61†	68
Hong Kong	ERDG	3 090	2 855	752	671†	701	222	201	248†	230	226	245
India	ERDJ	2 800	2 688	664†	584	786	161†	213	210	231	298	257
Israel	ERDL	1 352	1 285	357†	256	339	87†	74	95	112	121	106
Malaysia	ERDK	1 089	875	224	209	233	64	72	73	83†	76†	74
Pakistan	FKMU	461	487	142†	115	101	30	37	48	33†	36	32
Philippines	FKMX	279	241	54	64	59	20	22	22	20	22†	17
Russia	ERDQ	1 869	2 046	446†	522	565	168†	179	175	191	208	166
Singapore	ERDH	2 080	2 315	616	499	597	167	181	151	209	181†	207
South Africa	EPME	2 073	2 177	564	488	526	183	161	144	185†	172	169
Taiwan	ERDP	939	909	228†	219	232	70	74†	78	78	81	73
Thailand	ERDO	638	566	133	147	152	47	45	55	47	60†	45

1 Includes the ten countries which joined the EU on 1 May 2004; Cyprus, Czech Republic, Estonia, Hungary, Latvia, Lithuania, Malta, Poland, Slovakia, Slovenia.

Source: Office for National Statistics: 020 7533 6064

15.9 United Kingdom imports, by area

£ million BOP-consistent basis seasonally adjusted

		2005	2006	2006 Q2	2006 Q3	2006 Q4	2006 Jul	2006 Aug	2006 Sep	2006 Oct	2006 Nov	2006 Dec
European Union:[1]	LGDC	157 453	190 520	54 884†	42 886	41 561	14 498†	14 427	13 961	13 756	13 627	14 178
EMU members	QAKX	139 144	163 997	45 954†	37 553	36 401	12 676†	12 704	12 173	12 005	11 902	12 494
Austria	CHNB	2 464	2 929	848†	696	607	247†	230	219	215	192	200
Belgium & Luxembourg	CHNY	15 160	18 927	5 102†	4 584	3 885	1 430†	1 765	1 389	1 320	1 263	1 302
Finland	CHNC	2 431	2 891	721	755†	722	275†	237	243	236	239	247
France	ENYP	22 194	30 651	9 499†	6 412	5 970	2 231†	2 127	2 054	2 011	1 945	2 014
Germany	ENYS	39 185	42 965	11 148†	10 414	10 643	3 461†	3 491	3 462	3 358	3 464	3 821
Greece	CHOB	698	749	228†	162	155	52†	47	63	60	56	39
Irish Republic	CHOA	10 414	10 713	2 717†	2 577	2 734	817†	890	870	925	910	899
Italy	CHNW	12 679	13 251	3 446†	3 199	3 158	1 050†	1 088	1 061	1 037	1 054	1 067
Netherlands	CHNX	20 443	23 717	6 464†	5 417	5 409	1 833†	1 805	1 779	1 804	1 794	1 811
Portugal	CHOC	2 020	3 592	1 456†	663	511	306†	192	165	174	157	180
Spain	CHOD	11 456	13 612	4 325†	2 674	2 607	974†	832	868	865	828	914
Non-EMU members:[1]	BQIB	18 309	26 523	8 930†	5 333	5 160	1 822†	1 723	1 788	1 751	1 725	1 684
Of which:												
Czech Rep	FKMM	1 884	2 598	655	621†	676	212	195†	214	212	198	266
Denmark	CHNZ	4 397	6 215	2 297†	1 141	974	400†	374	367	336	327	311
Hungary	QALD	1 861	2 128	581†	462	548	148†	160	154	170	179	199
Poland	ERED	2 320	4 018	1 281†	744	702	260†	230	254	263	256	183
Sweden	CHND	5 464	6 408	1 750†	1 518	1 519	516†	500	502	491	509	519
Other Western Europe:	HBTS	20 066	23 382	6 047†	6 137	5 238	2 023†	2 301	1 813	1 653	1 814	1 771
Of which:												
Iceland	EPMW	346	402	123	98	96	41	26	31	35	32	29
Norway	EPMX	12 074	14 417	3 765†	3 904	2 987	1 262†	1 543	1 099	879	1 065	1 043
Switzerland	EPMV	3 881	4 372	1 135†	1 077	1 089	341	391†	345	392	384	313
Turkey	EOBU	3 510	3 947	965†	993	1 007	355†	320	318	325	314	368
North America:	HCRB	27 124	31 501	7 854†	7 544	8 239	2 360†	2 575	2 609	2 710	2 783	2 746
Of which:												
Canada	EOBW	4 155	4 996	1 231†	1 192	1 243	380†	402	410	418	413	412
Mexico	EPJY	446	450	110	96	108	34	33	29	35	47	26
USA	EOBV	22 181	25 775	6 447†	6 175	6 826	1 925†	2 109	2 141	2 233	2 299	2 294
Other OECD countries:	HDJQ	14 419	13 720	3 473†	3 395	3 362	1 162	1 102†	1 131	1 099	1 148	1 115
Of which:												
Australia	EPNA	2 100	2 121	562	523	531	164	178	181	158	212	161
Japan	EOBX	8 665	7 907	1 981†	1 951	1 904	665†	639	647	643	626	635
New Zealand	EPNB	591	605	148	158†	159	55	49†	54	58	53	48
South Korea	ERDY	3 063	3 087	782†	763	768	278†	236	249	240	257	271
Oil exporting countries:	HCPC	6 017	7 022	1 715†	1 835	1 678	637†	620	578	567	619	492
Of which:												
Brunei	QALG	25	71	18	30	12	12	18	–	5	–	7
Dubai	QALJ	643	682	162†	150	165	55†	51	44	52	53	60
Indonesia	FKMS	839	962	231†	225	242	82†	71	72	80	79	83
Kuwait	QATC	367	745	233	225	121	48	78	99	56	42	23
Nigeria	QATF	152	207	68†	72	25	13	53†	6	4	15	6
Saudi Arabia	ERDU	1 713	1 236	283	360	249	211	90	59	97	78	74
Rest of the World	HCIF	55 320	62 035	15 277†	15 539	16 039	5 239†	5 055	5 245	5 341	5 482	5 216
Of which:												
Brazil	FKMP	1 739	1 896	435	456†	523	139	150	167†	170	183	170
China	ERDZ	12 963	15 270	3 798†	3 717	4 007	1 237†	1 219	1 261	1 321	1 342	1 344
Egypt	QALM	350	661	171	139	200	33	50	56	42	75	83
Hong Kong	ERDS	6 602	7 355	1 879†	1 789	1 830	605†	583	601	611	624	595
India	ERDV	2 783	3 129	822†	781	753	269	253†	259	249	250	254
Israel	ERDX	1 003	967	242†	239	238	79†	81	79	82	78	78
Malaysia	ERDW	1 815	1 899	503†	455	479	146†	155	154	161	153	165
Pakistan	FKMV	487	513	121†	128	123	43	42†	43	43	39	41
Philippines	FKMY	712	740	186†	181	166	64†	61	56	49	60	57
Russia	EREC	5 011	5 750	1 531	1 614	1 264	522	534	558	437	476	351
Singapore	ERDT	3 830	3 765	738	744	1 187	304	201	239	405	458†	324
South Africa	EPNE	3 941	3 917	1 179	799	839	263	268	268	218	345	276
Taiwan	EREB	2 227	2 348	615†	573	588	194	193†	186	195	198	195
Thailand	EREA	1 718	1 930	483†	474	471	161	157†	156	159	158	154

1 Includes the ten countries which joined the EU on 1 May 2004; Cyprus, Czech Republic, Estonia, Hungary, Latvia, Lithuania, Malta, Poland, Slovakia, Slovenia.

Source: Office for National Statistics: 020 7533 6064

15.10 Import penetration and export sales ratios for products of manufacturing industry[1]

Standard Industrial Classification 1992

Per cent

		SIC Division	2002	2003	2004	2003 Q1	2003 Q2	2003 Q3	2003 Q4	2004 Q1	2004 Q2	2004 Q3	2004 Q4
Ratio 1 Imports/Home Demand													
Description													
Total of divisions below	BAZI			80	86	87	80	88	89	87	83	85	89	91
Textiles	BAZJ	17	70	73	77	68	71	78	76	73	73	81	80
Wearing apparel; dressing and dyeing of fur	BAZK	18	93	97	100	95	97	100	97	101	96	104	99
Chemicals and chemical products	BAZL	24	80	86	87	79	91	90	87	83	85	89	93
Ratio 2 Imports/Home Demand plus exports													
Description													
Total of divisions below	BAZM			48	50	51	48	49	50	50	49	49	52	52
Textiles	BAZN	17	52	53	55	50	51	56	55	53	53	58	56
Wearing apparel; dressing and dyeing of fur	BAZO	18	76	78	80	78	78	78	78	80	80	81	81
Chemicals and chemical products	BAZP	24	42	44	45	43	45	44	45	44	44	46	46
Ratio 3 Exports/Sales													
Description													
Total of divisions below	BAZQ			77	84	85	76	87	88	84	81	82	87	90
Textiles	BAZR	17	53	59	63	53	57	64	62	58	60	69	67
Wearing apparel; dressing and dyeing of fur	BAZS	18	77	90	100	82	88	101	89	103	85	115	96
Chemicals and chemical products	BAZT	24	81	88	88	80	92	91	88	83	86	89	94
Ratio 4 Exports/Sales plus imports													
Description													
Total of divisions below	BAZU			40	42	42	39	44	44	42	41	42	42	43
Textiles	BAZV	17	26	28	28	26	28	28	28	27	28	29	29
Wearing apparel; dressing and dyeing of fur	BAZW	18	19	20	20	18	19	22	19	21	17	22	19
Chemicals and chemical products	BAZX	24	47	49	48	46	51	51	49	47	48	48	50

1 As from the end of 2004, quarterly data will not be collected for these industries. Annual results for the years 2005 onwards will be published on the ONS website at http://www.statistics.gov.uk/ as usual with various omissions.

Source: Office for National Statistics: 01633 812921

16 UK Balance of payments

16.1 Balance of payments
Summary

£ million

	Seasonally adjusted (balances)					Not seasonally adjusted			
	Trade in goods and services	Income	Current transfers	Current balance	Capital balance	Current balance	Capital balance	Net financial transactions	Net errors and omissions[1]
	IKBJ	HBOJ	IKBP	HBOP	FNVQ	HBOG	FKMJ	HBNT	HHDH
1996	−2 518	556	−4 755	−6 717	1 260	−6 717	1 260	2 811	2 646
1997	1 764	3 314	−5 918	−840	958	−840	958	−8 771	8 653
1998	−7 141	12 320	−8 374	−3 195	489	−3 195	489	9 922	−7 216
1999	−15 454	1 270	−7 533	−21 717	747	−21 717	747	21 416	−446
2000	−19 361	4 540	−10 012	−24 833	1 703	−24 833	1 703	12 604	10 526
2001	−26 789	11 664	−6 759	−21 884	1 318	−21 884	1 318	17 503	3 063
2002	−30 875	23 443	−9 081	−16 513	932	−16 513	932	7 202	8 379
2003	−29 445	24 646	−10 122	−14 921	1 466	−14 921	1 466	20 507	−7 052
2004	−34 975	26 596	−10 949	−19 328	2 063	−19 328	2 063	5 641	11 624
2005	−44 593	27 159	−12 082	−29 516	2 633	−29 516	2 633	30 736	−3 853
1996 Q2	−1 031	1 066	−1 528	−1 493	291	−2 864	291	1 667	906
Q3	−320	−433	−893	−1 646	430	−1 849	430	3 716	−2 297
Q4	173	−197	−1 270	−1 294	279	−211	279	2 453	−2 521
1997 Q1	1 253	362	−1 726	−111	369	−112	335	−3 363	3 140
Q2	333	1 741	−1 354	720	59	−255	80	747	−572
Q3	759	1 686	−1 719	726	193	505	205	1 979	−2 689
Q4	−581	−475	−1 119	−2 175	337	−978	338	−8 134	8 774
1998 Q1	−1 066	1 700	−2 190	−1 556	30	−2 649	−2	828	1 823
Q2	−1 362	1 841	−1 600	−1 121	−65	−2 029	−39	2 502	−434
Q3	−1 776	4 854	−1 788	1 290	219	1 919	230	−1 610	−539
Q4	−2 937	3 925	−2 796	−1 808	305	−436	300	8 202	−8 066
1999 Q1	−4 467	−70	−2 039	−6 576	−1	−6 836	−29	1 188	5 677
Q2	−2 994	−110	−1 644	−4 748	205	−6 066	229	1 620	4 217
Q3	−3 110	792	−2 020	−4 338	244	−3 550	254	4 751	−1 455
Q4	−4 883	658	−1 830	−6 055	299	−5 265	293	13 857	−8 885
2000 Q1	−4 516	1 407	−1 863	−4 972	248	−4 539	222	4 677	−360
Q2	−3 889	464	−2 247	−5 672	656	−6 373	677	1 686	4 010
Q3	−5 420	2 545	−2 791	−5 666	401	−5 473	417	−3 054	8 110
Q4	−5 536	124	−3 111	−8 523	398	−8 448	387	9 295	−1 234
2001 Q1	−4 931	2 545	−1 867	−4 253	240	−2 913	221	−2 437	5 129
Q2	−6 502	3 074	−2 720	−6 148	569	−7 780	594	3 118	4 068
Q3	−9 205	3 620	26	−5 559	212	−4 822	225	3 538	1 059
Q4	−6 151	2 425	−2 198	−5 924	297	−6 369	278	13 284	−7 193
2002 Q1	−7 426	5 283	−2 298	−4 441	154	−2 990	130	−10 393	13 253
Q2	−7 368	4 270	−2 557	−5 655	158	−7 917	184	−64	7 797
Q3	−7 237	6 924	−1 519	−1 832	160	−1 305	174	3 423	−2 292
Q4	−8 844	6 966	−2 707	−4 585	460	−4 301	444	14 236	−10 379
2003 Q1	−6 132	7 932	−2 364	−564	236	535	202	−7 851	7 114
Q2	−6 582	5 098	−2 926	−4 410	208	−6 719	237	11 603	−5 121
Q3	−7 600	4 688	−2 479	−5 391	329	−4 636	348	4 628	−340
Q4	−9 131	6 928	−2 353	−4 556	693	−4 101	679	12 127	−8 705
2004 Q1	−8 067	5 825	−2 686	−4 928	735	−4 045	697	−5 552	8 900
Q2	−8 407	6 377	−2 439	−4 469	601	−6 699	634	9 386	−3 321
Q3	−9 036	4 954	−2 807	−6 889	266	−7 432	283	1 662	5 487
Q4	−9 465	9 440	−3 017	−3 042	461	−1 152	449	145	558
2005 Q1	−10 100	7 132	−3 422	−6 390	773	−7 011	732	1 816	4 463
Q2	−9 124	9 485	−2 575	−2 214	666	−3 463	700	3 263	−500
Q3	−13 586	6 567	−3 024	−10 043	356	−10 455	373	17 252	−7 170
Q4	−11 783	3 975	−3 061	−10 869	838	−8 587	828	8 405	−646
2006 Q1	−13 817	6 632	−3 057	−10 242	553	−11 018	493	2 465	8 060
Q2	−14 327	8 806	−2 743	−8 264	−1 904	−8 864	−1 867	16 099	−5 368
Q3	−13 330	6 556	−2 655	−9 429	466	−9 832	489	9 750	−407

1 This series represents net errors and omissions in the balance of payments accounts. It is the converse of the current and capital balances (HBOG and FKMJ) and net financial account transactions (HBNT) and is required to balance these three accounts, not seasonally adjusted.

Source: Office for National Statistics: 020 7533 6078

99

16.2 Balance of payments
Current account balances (seasonally adjusted)

£ million

	Trade in goods and services			Income			Current transfers				
	Trade in goods	Trade in services	Total trade	Compensation of employees	Investment income	Total income	Central government	Other sectors	Total current transfers	Current balance	Current balance as % of GDP [1]
	BOKI	IKBD	IKBJ	IJAJ	HBOM	HBOJ	FNSV	FNTC	IKBP	HBOP	AA6H
1996	−13 722	11 204	−2 518	93	463	556	−2 469	−2 286	−4 755	−6 717	−0.9
1997	−12 342	14 106	1 764	83	3 231	3 314	−3 087	−2 831	−5 918	−840	−0.1
1998	−21 813	14 672	−7 141	−10	12 330	12 320	−5 020	−3 354	−8 374	−3 195	−0.4
1999	−29 051	13 597	−15 454	201	1 069	1 270	−3 940	−3 593	−7 533	−21 717	−2.4
2000	−32 976	13 615	−19 361	150	4 390	4 540	−5 550	−4 462	−10 012	−24 833	−2.6
2001	−41 212	14 423	−26 789	66	11 598	11 664	−2 593	−4 166	−6 759	−21 884	−2.2
2002	−47 705	16 830	−30 875	67	23 376	23 443	−5 633	−3 448	−9 081	−16 513	−1.6
2003	−48 607	19 162	−29 445	59	24 587	24 646	−6 976	−3 146	−10 122	−14 921	−1.3
2004	−60 893	25 918	−34 975	71	26 525	26 596	−8 304	−2 645	−10 949	−19 328	−1.6
2005	−68 783	24 190	−44 593	65	27 094	27 159	−9 481	−2 601	−12 082	−29 516	−2.4
1996 Q2	−3 717	2 686	−1 031	21	1 045	1 066	−405	−1 123	−1 528	−1 493	−0.8
Q3	−3 089	2 769	−320	−16	−417	−433	−486	−407	−893	−1 646	−0.9
Q4	−3 041	3 214	173	−26	−171	−197	−539	−731	−1 270	−1 294	−0.7
1997 Q1	−2 303	3 556	1 253	1	361	362	−806	−920	−1 726	−111	−0.1
Q2	−3 140	3 473	333	18	1 723	1 741	−1 088	−266	−1 354	720	0.4
Q3	−2 777	3 536	759	22	1 664	1 686	−843	−876	−1 719	726	0.4
Q4	−4 122	3 541	−581	42	−517	−475	−350	−769	−1 119	−2 175	−1.0
1998 Q1	−4 734	3 668	−1 066	75	1 625	1 700	−1 319	−871	−2 190	−1 556	−0.7
Q2	−4 977	3 615	−1 362	−27	1 868	1 841	−843	−757	−1 600	−1 121	−0.5
Q3	−5 782	4 006	−1 776	−29	4 883	4 854	−1 279	−509	−1 788	1 290	0.6
Q4	−6 320	3 383	−2 937	−29	3 954	3 925	−1 579	−1 217	−2 796	−1 808	−0.8
1999 Q1	−7 934	3 467	−4 467	33	−103	−70	−1 022	−1 017	−2 039	−6 576	−3.0
Q2	−6 598	3 604	−2 994	89	−199	−110	−824	−820	−1 644	−4 748	−2.1
Q3	−6 598	3 488	−3 110	47	745	792	−948	−1 072	−2 020	−4 338	−1.9
Q4	−7 921	3 038	−4 883	32	626	658	−1 146	−684	−1 830	−6 055	−2.6
2000 Q1	−7 480	2 964	−4 516	13	1 394	1 407	−1 276	−587	−1 863	−4 972	−2.1
Q2	−7 405	3 516	−3 889	82	382	464	−1 227	−1 020	−2 247	−5 672	−2.4
Q3	−8 844	3 424	−5 420	30	2 515	2 545	−1 219	−1 572	−2 791	−5 666	−2.4
Q4	−9 247	3 711	−5 536	25	99	124	−1 828	−1 283	−3 111	−8 523	−3.5
2001 Q1	−9 180	4 249	−4 931	−53	2 598	2 545	−1 037	−830	−1 867	−4 253	−1.7
Q2	−11 080	4 578	−6 502	65	3 009	3 074	−1 379	−1 341	−2 720	−6 148	−2.5
Q3	−10 481	1 276	−9 205	29	3 591	3 620	967	−941	26	−5 559	−2.2
Q4	−10 471	4 320	−6 151	25	2 400	2 425	−1 144	−1 054	−2 198	−5 924	−2.3
2002 Q1	−11 372	3 946	−7 426	8	5 275	5 283	−1 065	−1 233	−2 298	−4 441	−1.7
Q2	−11 002	3 634	−7 368	19	4 251	4 270	−1 310	−1 247	−2 557	−5 655	−2.2
Q3	−12 016	4 779	−7 237	23	6 901	6 924	−1 317	−202	−1 519	−1 832	−0.7
Q4	−13 315	4 471	−8 844	17	6 949	6 966	−1 941	−766	−2 707	−4 585	−1.7
2003 Q1	−10 862	4 730	−6 132	15	7 917	7 932	−1 600	−764	−2 364	−564	−0.2
Q2	−11 545	4 963	−6 582	23	5 075	5 098	−1 960	−966	−2 926	−4 410	−1.6
Q3	−12 302	4 702	−7 600	11	4 677	4 688	−1 639	−840	−2 479	−5 391	−1.9
Q4	−13 898	4 767	−9 131	10	6 918	6 928	−1 777	−576	−2 353	−4 556	−1.6
2004 Q1	−13 947	5 880	−8 067	15	5 810	5 825	−1 962	−724	−2 686	−4 928	−1.7
Q2	−15 247	6 840	−8 407	32	6 345	6 377	−1 906	−533	−2 439	−4 469	−1.5
Q3	−15 529	6 493	−9 036	17	4 937	4 954	−2 151	−656	−2 807	−6 889	−2.3
Q4	−16 170	6 705	−9 465	7	9 433	9 440	−2 285	−732	−3 017	−3 042	−1.0
2005 Q1	−16 143	6 043	−10 100	6	7 126	7 132	−2 622	−800	−3 422	−6 390	−2.1
Q2	−15 807	6 683	−9 124	30	9 455	9 485	−2 259	−316	−2 575	−2 214	−0.7
Q3	−18 055	4 469	−13 586	18	6 549	6 567	−2 271	−753	−3 024	−10 043	−3.3
Q4	−18 778	6 995	−11 783	11	3 964	3 975	−2 329	−732	−3 061	−10 869	−3.5
2006 Q1	−21 475	7 658	−13 817	1	6 631	6 632	−2 373	−684	−3 057	−10 242	−3.3
Q2	−20 929	6 602	−14 327	23	8 783	8 806	−2 137	−606	−2 743	−8 264	−2.6
Q3	−20 499	7 169	−13 330	6	6 550	6 556	−2 481	−174	−2 655	−9 429	−2.9

1 Using series YBHA: GDP at current market prices.

Source: Office for National Statistics: 020 7533 6078

16.3 Balance of payments
Summary of financial account

£ million

	Investment in the UK				UK investment abroad						Net transactions				
	Direct investment	Portfolio investment	Other investment	Total	Direct investment	Portfolio investment	Financial derivatives	Other investment	Reserve assets	Total	Direct investment	Portfolio investment	Other investment	Reserve assets	Total
	HJYU	HHZF	XBMN	HBNS	-HJYP	-HHZC	-ZPNN	-XBMM	-LTCV	-HBNR	HJYV	HHZD	HHYR	LTCV	HBNT
1996	17 564	42 996	160 723	221 283	23 516	59 760	−963	136 669	−510	218 472	−5 952	−16 764	24 054	510	2 811
1997	22 900	26 786	196 670	246 356	37 302	51 941	−1 156	169 420	−2 380	255 127	−14 402	−25 155	27 250	2 380	−8 771
1998	45 054	20 853	67 640	133 547	73 786	32 073	3 043	14 887	−164	123 625	−28 732	−11 220	52 753	164	9 922
1999	55 066	114 106	55 469	224 641	125 602	21 390	−2 685	59 557	−639	203 225	−70 536	92 716	−4 088	639	21 416
2000	80 566	164 543	267 030	512 139	155 582	65 563	−1 553	276 028	3 915	499 535	−75 016	98 980	−8 998	−3 915	12 604
2001	37 348	48 148	223 969	309 465	42 827	86 551	−8 417	174 086	−3 085	291 962	−5 479	−38 403	49 883	3 085	17 503
2002	16 782	51 010	71 187	138 979	35 041	1 011	−1 001	97 185	−459	131 777	−18 259	49 999	−25 998	459	7 202
2003	16 776	95 222	245 370	357 368	40 889	36 267	5 401	255 863	−1 559	336 861	−24 113	58 955	−10 493	1 559	20 507
2004	42 416	87 247	404 321	533 984	53 831	140 853	7 875	325 588	196	528 343	−11 415	−53 606	78 733	−196	5 641
2005	107 794	125 528	510 035	743 357	50 002	166 692	2 451	492 820	656	712 621	57 792	−41 164	17 215	−656	30 736
1996 Q2	5 041	6 667	45 773	57 481	4 839	22 177	−71	28 634	235	55 814	202	−15 510	17 139	−235	1 667
Q3	5 259	9 048	28 891	43 198	2 966	24 541	−393	13 279	−911	39 482	2 293	−15 493	15 612	911	3 716
Q4	3 921	19 044	51 288	74 253	6 196	13 913	−332	50 523	1 500	71 800	−2 275	5 131	765	−1 500	2 453
1997 Q1	8 798	8 031	76 114	92 943	8 815	13 891	−490	75 548	−1 458	96 306	−17	−5 860	566	1 458	−3 363
Q2	4 803	9 191	49 657	63 651	4 304	36 374	70	21 931	225	62 904	499	−27 183	27 726	−225	747
Q3	3 430	9 499	20 085	33 014	18 189	−2 663	−232	15 405	336	31 035	−14 759	12 162	4 680	−336	1 979
Q4	5 869	65	50 814	56 748	5 994	4 339	−504	56 536	−1 483	64 882	−125	−4 274	−5 722	1 483	−8 134
1998 Q1	11 004	−911	41 997	52 090	5 484	23 977	−626	23 425	−998	51 262	5 520	−24 888	18 572	998	828
Q2	8 055	−9 451	49 236	47 840	7 107	9 160	595	28 167	309	45 338	948	−18 611	21 069	−309	2 502
Q3	13 199	841	19 228	33 268	20 212	−17 831	1 531	30 653	313	34 878	−7 013	18 672	−11 425	−313	−1 610
Q4	12 796	30 374	−42 821	349	40 983	16 767	1 543	−67 358	212	−7 853	−28 187	13 607	24 537	−212	8 202
1999 Q1	12 832	19 069	68 198	100 099	8 149	15 377	−1 519	77 741	−837	98 911	4 683	3 692	−9 543	837	1 188
Q2	5 262	84 215	61 300	150 777	84 661	12 107	441	51 746	202	149 157	−79 399	72 108	9 554	−202	1 620
Q3	12 863	6 224	−29 530	−10 443	11 589	9 965	535	−36 524	−759	−15 194	1 274	−3 741	6 994	759	4 751
Q4	24 109	4 598	−44 499	−15 792	21 203	−16 059	−2 142	−33 406	755	−29 649	2 906	20 657	−11 093	−755	13 857
2000 Q1	14 601	91 753	139 095	245 449	112 991	−13 178	492	142 932	−2 465	240 772	−98 390	104 931	−3 837	2 465	4 677
Q2	22 735	29 873	60 168	112 776	34 299	45 177	−926	31 994	546	111 090	−11 564	−15 304	28 174	−546	1 686
Q3	43 931	12 932	29 864	86 727	5 849	19 518	−526	63 410	1 530	89 781	38 082	−6 586	−33 546	−1 530	−3 054
Q4	−701	29 985	37 903	67 187	2 443	14 046	−593	37 692	4 304	57 892	−3 144	15 939	211	−4 304	9 295
2001 Q1	16 426	20 888	214 849	252 163	21 156	37 299	−2 331	201 075	−2 599	254 600	−4 730	−16 411	13 774	2 599	−2 437
Q2	12 210	7 969	−20 847	−668	11 835	30 083	1 473	−47 214	37	−3 786	375	−22 114	26 367	−37	3 118
Q3	5 518	8 173	−549	13 142	8 648	11 046	−5 843	−3 749	−498	9 604	−3 130	−2 873	3 200	498	3 538
Q4	3 194	11 118	30 516	44 828	1 188	8 123	−1 716	23 974	−25	31 544	2 006	2 995	6 542	25	13 284
2002 Q1	−6 050	9 049	24 262	27 261	18 143	−6 575	−340	26 954	−528	37 654	−24 193	15 624	−2 692	528	−10 393
Q2	15 376	22 727	−7 367	30 736	16 315	42 400	−1 968	−25 969	22	30 800	−939	−19 673	18 602	−22	−64
Q3	1 061	2 789	−5 335	−1 485	15 331	−36 798	1 855	14 022	682	−4 908	−14 270	39 587	−19 357	−682	3 423
Q4	6 395	16 445	59 627	82 467	−14 748	1 984	−548	82 178	−635	68 231	21 143	14 461	−22 551	635	14 236
2003 Q1	3 656	16 431	112 428	132 515	20 994	15 526	7 677	97 795	−1 626	140 366	−17 338	905	14 633	1 626	−7 851
Q2	5 771	12 864	122 701	141 336	14 213	25 426	−2 302	93 043	−647	129 733	−8 442	−12 562	29 658	647	11 603
Q3	1 186	41 115	−5 094	37 207	7 323	−3 958	1 348	26 563	1 303	32 579	−6 137	45 073	−31 657	−1 303	4 628
Q4	6 163	24 812	15 335	46 310	−1 641	−727	−1 322	38 462	−589	34 183	7 804	25 539	−23 127	589	12 127
2004 Q1	9 292	49 901	208 542	267 735	17 558	45 902	4 504	205 848	−525	273 287	−8 266	3 999	2 694	525	−5 552
Q2	7 558	12 011	71 679	91 248	13 609	−198	3 967	65 079	−595	81 862	−6 051	12 209	6 600	595	9 386
Q3	9 141	16 471	63 115	88 727	19 421	47 649	1 278	18 771	−54	87 065	−10 280	−31 178	44 344	54	1 662
Q4	16 425	8 864	60 985	86 274	3 243	47 500	−1 874	35 890	1 370	86 129	13 182	−38 636	25 095	−1 370	145
2005 Q1	19 365	54 286	199 030	272 681	24 064	28 299	−269	219 304	−533	270 865	−4 699	25 987	−20 274	533	1 816
Q2	16 660	27 434	129 329	173 423	3 899	54 574	1 155	110 005	527	170 160	12 761	−27 140	19 324	−527	3 263
Q3	71 249	8 221	115 689	195 159	11 673	49 521	1 574	114 763	376	177 907	59 576	−41 300	926	−376	17 252
Q4	520	35 587	65 987	102 094	10 366	34 298	−9	48 748	286	93 689	−9 846	1 289	17 239	−286	8 405
2006 Q1	31 494	39 258	306 119	376 871	15 342	44 990	5 968	308 571	−465	374 406	16 152	−5 732	−2 452	465	2 465
Q2	19 018	25 920	12 377	57 315	10 421	42 181	2 915	−14 091	−210	41 216	8 597	−16 261	26 468	210	16 099
Q3	13 357	54 156	43 928	111 441	−1 498	25 315	4 035	73 454	385	101 691	14 855	28 841	−29 526	−385	9 750

Source: Office for National Statistics: 020 7533 6078

17 Government finance

17.1 Public sector finances

£ millions[1]

	Public sector surplus on current budget	Public sector net investment	Net Borrowing					Public sector net debt[2]	Public sector net debt as percentage of GDP
			Central government	Local government	General government	Public corporations	Public sector		
	ANMU	-ANNW	-NMFJ	-NMOE	-NNBK	-CPCM	-ANNX	RUTN	RUTO
2000	20 736[†]	3 747[†]	-16 727[†]	654[†]	-16 073[†]	-916[†]	-17 472	321.9	33.0
2001	18 423	8 780	-9 307	-497	-9 804	161	-10 083	324.2	31.8
2002	-8 180	10 173	19 153	-1 866	17 287	1 066	18 761	351.6[†]	32.7
2003	-21 457	15 620	39 351	-3 067	36 284	793	36 837	383.3	33.5
2004	-21 611	16 126	37 232	526	37 758	-21	37 726	425.8	35.4[†]
2005	-19 334	21 490	36 068	2 032	38 100	2 724	39 134	467.1	37.3
2006	-9 977	27 835	38 119	-1 589	36 530	1 282	37 812	504.1	38.1
1999/00	20 029	4 787	-15 320	1 347	-13 973	-1 269	-16 251	345.4	36.6
2000/01	22 991	3 328	-18 999	-351	-19 350	-313	-19 896	312.4	31.7
2001/02	9 637	10 718	1 073	-167	906	175	921	317.6	30.7
2002/03	-12 717	12 075	25 459	-2 325	23 134	1 658	24 916	349.1	32.0
2003/04	-19 865	14 596	36 302	-1 818	34 484	-23	34 094	384.4	33.1
2004/05	-20 423	19 378	38 648	706	39 354	447	39 068	424.0	34.9
2005/06	-15 670	21 929	32 892	2 984	35 876	1 723	36 887	463.6	36.5
2000 Q4	2 138[†]	1 102[†]	-2 040[†]	1 162[†]	-878[†]	-158[†]	-1 229	321.9	33.0
2001 Q1	18 116	2 780	-14 753	-838	-15 591	255	-15 425	312.4	31.7
Q2	-3 001	1 196	5 060	-796	4 264	-67	4 005	319.7	32.1
Q3	4 705	1 954	-3 444	818	-2 626	-125	-2 858	313.6	31.1
Q4	-1 397	2 850	3 830	319	4 149	98	4 195	324.2	31.8
2002 Q1	9 330	4 718	-4 373	-508	-4 881	269	-4 421	317.6[†]	30.7[†]
Q2	-9 395	715	10 877	-680	10 197	-87	10 216	324.2	30.9
Q3	-742	2 430	3 939	-640	3 299	-127	3 221	328.2	30.9
Q4	-7 373	2 310	8 710	-38	8 672	1 011	9 745	351.6	32.7
2003 Q1	4 793	6 620	1 933	-967	966	861	1 734	349.1	32.0
Q2	-12 010	2 200	16 096	-1 880	14 216	-6	14 168	357.7	32.2
Q3	-4 054	3 337	7 255	139	7 394	-3	7 377	362.9	32.2
Q4	-10 186	3 463	14 067	-359	13 708	-59	13 558	383.3	33.5
2004 Q1	6 385	5 596	-1 116	282	-834	45	-1 009	384.4	33.1
Q2	-11 858	2 835	16 150	-1 421	14 729	-36	14 963	397.4	33.8
Q3	-5 939	3 345	8 613	725	9 338	-54	9 252	403.4	33.9
Q4	-10 199	4 350	13 585	940	14 525	24	14 520	425.8	35.4
2005 Q1	7 573	8 848	300	462	762	513	333	424.0	34.9
Q2	-11 294	2 520	11 566	-1 508	10 058	3 756	13 437	439.4	35.9
Q3	-3 566	4 803	7 383	1 810	9 193	-824	8 296	447.1	36.1
Q4	-12 047	5 319	16 819	1 268	18 087	-721	17 068	467.1	37.3
2006 Q1	11 237	9 287	-2 876	1 414	-1 462	-488	-1 914	463.6	36.5
Q2	-10 947	5 735	19 887	-5 162	14 725	1 957	16 289[†]	486.3	37.8
Q3	-1 322	5 086	6 062	614	6 676	-268	5 859	489.6	37.5
Q4	-10 149	7 429	16 013	1 680	17 693	-115	17 578	504.1	38.1
2004 Dec	-5 149	1 648	6 367	105	6 472	325	6 797	425.8[†]	35.4[†]
2005 Jan	11 841	2 578	-10 497	766	-9 731	468	-9 263	409.3	33.9
Feb	1 577	3 066	3 444	-1 831	1 613	-124	1 489	410.1	33.9
Mar	-5 216	2 891	6 763	1 106	7 869	238	8 107	424.0	34.9
Apr	1 481	409	-4 568	-453	-5 021	3 949	-1 072	422.7	34.7
May	-7 823	1 070	11 093	-2 093	9 000	-107	8 893	427.8	35.0
Jun	-4 622	994	4 785	841	5 626	-10	5 616	439.4	35.9
Jul	3 725	1 494	-3 592	1 343	-2 249	18	-2 231	431.1	35.1
Aug	-3 765	1 732	6 067	28	6 095	-598	5 497	436.8	35.4
Sep	-3 285	1 745	4 789	410	5 199	-169	5 030	447.1	36.1
Oct	2 492	1 641	-1 627	1 032	-595	-256	-851	443.0	35.7
Nov	-8 236	2 000	11 079	-353	10 726	-490	10 236	451.6	36.2
Dec	-6 058	1 625	7 203	381	7 584	99	7 683	467.1	37.3
2006 Jan	12 756	2 822	-12 677	1 665	-11 012	1 078	-9 934	445.1	35.4
Feb	563	3 221	3 715	-1 349	2 366	292	2 658	447.3	35.4
Mar	-1 921	3 441	6 024	1 127	7 151	-1 789	5 362	463.6	36.5
Apr	-698[†]	1 170	3 035[†]	-3 609	-574[†]	2 442	1 868[†]	462.9	36.3
May	-6 606	1 502	9 836	-1 418	8 418	-310	8 108	470.4	36.7
Jun	-3 225	3 088	6 664	-237	6 427	-114	6 313	486.3	37.8
Jul	8 505	1 682	-6 578	22	-6 556	-267	-6 823	473.1	36.6
Aug	-4 952	2 008	6 437	698	7 135	-175	6 960	477.5	36.8
Sep	-4 250	1 472[†]	5 650	-168[†]	5 482	240[†]	5 722	489.6	37.5
Oct	3 440	2 583[†]	-2 275	1 030[†]	-1 245	388[†]	-857	481.7	36.8
Nov	-8 908	2 310	11 500	310	11 810	-592	11 218	490.0	37.2
Dec	-4 681	2 536	6 788	340	7 128	89	7 217	504.1	38.1

1 Unless otherwise stated.
2 £ billion

Source: Office for National Statistics: 020 7533 5991

17.2 Central government transactions and fiscal balances

£ million

	Current receipts									
	Taxes on production	of which	Taxes on income and wealth				Compulsory social contributions	Interest and dividends	Other receipts[3]	Total
	Total	VAT	Total	Income and capital gains tax[1]	Other[2]	Other taxes				
	NMBY	NZGF	NMCU	LIBR	LIBP	LIQR	AIIH	LIQP	LIQQ	ANBV
1999/00	124 955	58 676	133 994	98 819	35 175	8 852	56 935	7 542	6 058	338 336
2000/01	129 123	60 736	144 263	110 324	33 939	8 559	62 068	8 923	7 187	360 123
2001/02	132 882	64 730	145 179	111 688	33 491	9 419	63 162	7 665	7 468	365 775
2002/03	139 651	69 081	143 290	112 373	30 917	9 527	63 529	7 770	7 640	371 407
2003/04	148 625	76 627	145 586	115 233	30 353	10 106	75 148	7 539	7 494	394 498
2004/05	154 784	79 960	161 218	124 977	36 126	10 701	80 209	7 025	7 824	421 646
2005/06	159 324	81 549	180 423	135 272	45 151	11 357	85 404	7 110	8 037	451 655
2005 Mar	12 658	6 689	12 482	11 185	1 297	838	8 083	1 017	670	35 748
Apr	13 176	6 779	15 493	9 369	6 124	1 011	6 588	505	664	37 437
May	13 029	6 696	9 855	8 962	893	907	6 686	601	667	31 745
Jun	13 005	6 661	10 246	8 942	1 304	849	6 995	528	669	32 292
Jul	13 503	6 965	20 644	13 184	7 460	1 009	6 936	446	652	43 190
Aug	13 258	6 677	11 998	10 927	1 071	1 012	6 663	504	657	34 092
Sep	13 677	7 082	11 529	9 192	2 337	1 054	6 942	747	661	34 610
Oct	13 772	7 209	17 886	9 449	8 437	928	6 813	557	676	40 632
Nov	13 698	6 985	10 160	9 028	1 132	813	6 858	664	682	32 875
Dec	13 496	6 882	11 402	9 260	2 142	875	7 348	536	681	34 338
2006 Jan	12 849	6 593	30 814	19 820	10 994	908	7 461	575	673	53 280
Feb	12 633	6 221	16 375	14 924	1 451	901	7 689	480	676	38 754
Mar	13 228	6 799	14 021	12 215	1 806	1 090	8 425	967	679	38 410
Apr	13 848†	7 162	15 857	10 014	5 843	943	6 963†	505	701	38 817†
May	13 732	7 041	10 640†	9 245†	1 395	980	6 951	626	697	33 626
Jun	13 991	7 073	11 034	9 652	1 382	1 066	7 117	547	699	34 454
Jul	13 954	7 218	24 820	14 706	10 114	936	7 095	503	700	48 008
Aug	14 111	7 116	12 707	11 206	1 501	1 029	6 808	489	705	35 849
Sep	14 627	7 800†	12 470	10 013	2 457†	1 023	7 185	581	703	36 589
Oct	14 747	7 728	19 631	9 666	9 965	995	7 014	744	707	43 838
Nov	14 263	7 189	10 643	9 174	1 469	922	7 024	652†	711†	34 215
Dec	14 265	7 117	13 389	10 748	2 641	912	7 665	505	711	37 447

	Current expenditure				Saving, gross plus capital taxes	Depreciation	Surplus on current budget	Net investment	Net borrowing
	Interest	Net Social Benefits	Other	Total					
	NMFX	GZSJ	LIQS	ANLP	ANPM	NSRN	ANLV	-ANNS	-NMFJ
1999/00	25 015	93 246	189 621	307 882	30 454	5 432	25 417	9 000	−15 320
2000/01	25 995	96 016	205 390	327 401	32 722	5 491	27 589	8 212	−18 999
2001/02	22 095	105 586	220 330	348 011	17 764	5 503	12 642	13 104	1 073
2002/03	20 943	109 206	244 717	374 866	−3 459	5 696	−8 746	16 396	25 459
2003/04	22 324	117 260	267 250	406 834	−12 336	5 976	−17 860	18 810	36 302
2004/05	23 974	122 944	287 071	433 908	−12 262	6 327	−18 044	20 199	38 648
2005/06	25 821	128 147	305 574	459 448	−7 793	6 828	−13 912	18 379	32 892
2005 Mar	1 592	10 030	27 538	39 160	−3 412	504	−3 916	2 847	6 763
Apr	2 231	10 097	22 541	34 869	2 568	499	2 069	−2 499	−4 568
May	2 224	10 123	28 588	40 935	−9 190	502	−9 692	1 401	11 093
Jun	1 874	10 245	23 778	35 897	−3 605	504	−4 109	676	4 785
Jul	2 506	10 515	24 635	37 656	5 534	493	5 041	1 449	−3 592
Aug	2 450	10 629	24 910	37 989	−3 897	497	−4 394	1 673	6 067
Sep	1 203	10 581	25 433	37 217	−2 607	500	−3 107	1 682	4 789
Oct	2 350	10 542	23 696	36 588	4 044	516	3 528	1 901	−1 627
Nov	2 457	13 477	25 465	41 399	−8 524	520	−9 044	2 035	11 079
Dec	2 124	10 739	26 281	39 144	−4 806	522	−5 328	1 875	7 203
2006 Jan	2 618	10 477	24 491	37 586	15 694	520	15 174	2 497	−12 677
Feb	2 186	9 986	26 946	39 118	−364	522	−886	2 829	3 715
Mar	1 598	10 642	28 810	41 050	−2 640	524	−3 164	2 860	6 024
Apr	2 222	10 276†	27 306	39 804†	−987†	533	−1 520†	1 515	3 035†
May	2 253	10 800	28 112	41 165	−7 539	533	−8 072	1 764	9 836
Jun	1 826	10 723	26 899	39 448	−4 994	534	−5 528	1 136	6 664
Jul	2 667	10 948	25 749	39 364	8 644	541	8 103	1 525	−6 578
Aug	2 396	11 096	26 261	39 753	−3 904	544	−4 448	1 989	6 437
Sep	1 353	11 192	27 426	39 971	−3 382	545	−3 927	1 723	5 650
Oct	2 489	11 202	25 692†	39 383	4 455	542	3 913	1 638†	−2 275
Nov	2 660	13 077	26 954	42 691	−8 476	544	−9 020	2 480	11 500
Dec	2 285	11 193	27 557	41 035	−3 588	546	−4 134	2 654	6 788

1 Includes capital gains tax paid by households. Includes income tax and capital gains tax paid by corporations.
2 Mainly comprises corporation tax and petroleum revenue tax.
3 Includes receipts from the spectrum.

Source: Office for National Statistics: 020 7533 5991

17.3 Public sector aggregates[1]

£ millions, not seasonally adjusted

	Surplus on current budget[2]		Net investment[3]		Net borrowing[4]		Net cash requirement[5]	
	General Government	Public Sector	General Government	Public Sector	General Government	Public Sector	General Government	Public Sector
Calendar years	ANLW	ANMU	-ANNV	-ANNW	-NNBK	-ANNX	RUUS	RURQ
2000	22 444†	20 736†	6 371†	3 747†	−16 073†	−17 472	−38 840	−36 870
2001	19 577	18 423	9 773	8 780	−9 804	−10 083	−3 768	−1 928
2002	−6 571	−8 180	10 716	10 173	17 287	18 761	16 421	19 310
2003	−18 330	−21 457	17 954	15 620	36 284	36 837	38 214	38 521
2004	−18 848	−21 611	18 910	16 126	37 758	37 726	41 321	42 324
2005	−21 336	−19 334	16 764	21 490	38 100	39 134	41 870	40 951
2006	..	−9 977	..	27 835	36 530	37 812	36 799	33 186
Financial years								
1999/00	21 704	20 029	7 731	4 787	−13 973	−16 251	−9 685	−7 973
2000/01	24 819	22 991	5 469	3 328	−19 350	−19 896	−37 862	−36 323
2001/02	10 268	9 637	11 174	10 718	906	921	2 943	3 637
2002/03	−10 137	−12 717	12 997	12 075	23 134	24 916	21 499	25 221
2003/04	−16 947	−19 865	17 537	14 596	34 484	34 094	40 005	39 732
2004/05	−18 718	−20 423	20 636	19 378	39 354	39 068	38 737	38 571
2005/06	−18 292	−15 670	17 584	21 929	35 876	36 887	40 075	40 031
Quarterly								
1999 Q1	12 835†	12 286†	4 010†	3 656†	−8 825†	−8 960	−6 457	−4 796
Q2	−5 150	−5 328	493	−	5 643	5 170	5 318	5 306
Q3	4 975	4 613	1 435	767	−3 540	−4 086	−3 154	−3 215
Q4	5 447	4 883	1 685	821	−3 762	−4 334	2 223	2 167
2000 Q1	16 432	15 861	4 118	3 199	−12 314	−13 001	−14 072	−12 231
Q2	−1 733	−2 118	−296	−880	1 437	1 394	−12 221	−11 819
Q3	5 174	4 855	856	326	−4 318	−4 636	−16 734	−16 486
Q4	2 571	2 138	1 693	1 102	−878	−1 229	4 187	3 666
2001 Q1	18 807	18 116	3 216	2 780	−15 591	−15 425	−13 094	−11 684
Q2	−2 815	−3 001	1 449	1 196	4 264	4 005	6 246	6 352
Q3	4 790	4 705	2 164	1 954	−2 626	−2 858	−6 322	−6 101
Q4	−1 205	−1 397	2 944	2 850	4 149	4 195	9 402	9 505
2002 Q1	9 498	9 330	4 617	4 718	−4 881	−4 421	−6 383	−6 119
Q2	−8 996	−9 395	1 201	715	10 197	10 216	7 126	7 045
Q3	−412	−742	2 887	2 430	3 299	3 221	−145	1 329
Q4	−6 661	−7 373	2 011	2 310	8 672	9 745	15 823	17 055
2003 Q1	5 932	4 793	6 898	6 620	966	1 734	−1 305	−208
Q2	−11 189	−12 010	3 027	2 200	14 216	14 168	16 404	16 266
Q3	−3 491	−4 054	3 903	3 337	7 394	7 377	6 036	5 903
Q4	−9 582	−10 186	4 126	3 463	13 708	13 558	17 079	16 560
2004 Q1	7 315	6 385	6 481	5 596	−834	−1 009	486	1 003
Q2	−11 096	−11 858	3 633	2 835	14 729	14 963	11 577	11 690
Q3	−5 286	−5 939	4 052	3 345	9 338	9 252	6 968	7 370
Q4	−9 781	−10 199	4 744	4 350	14 525	14 520	22 290	22 261
2005 Q1	7 445	7 573	8 207	8 848	762	333	−2 098	−2 750
Q2	−11 363	−11 294	−1 305	2 520	10 058	13 437	15 944	16 254
Q3	−4 546	−3 566	4 647	4 803	9 193	8 296	8 463	8 181
Q4	−12 872	−12 047	5 215	5 319	18 087	17 068	19 561	19 266
2006 Q1	10 489	11 237	9 027	9 287	−1 462	−1 914	−3 893	−3 670
Q2	−11 156	−10 947	3 569	5 735	14 725	16 289†	19 252	19 022
Q3	−1 720	−1 322	4 956	5 086	6 676	5 859	5 811†	5 531†
Q4	..	−10 149	..	7 429	17 693	17 578	15 629	12 303

1 National accounts entities as defined under the European System of Accounts 1995 (ESA95).
2 Net saving, plus capital taxes.
3 Gross capital formation, plus payments less receipts, of investment grants less depreciation.

4 Net borrowing = net investment minus surplus on current budget
5 Previously called Public Sector Borrowing Requirement (PSBR).

Source: Office for National Statistics: 020 7533 5984

17.4 Selected financial statistics[1]

£ million

	Building societies		Unit trusts[3]	Net equity of households in life assurance and pension funds' reserves
	Advances			
	Not seasonally adjusted	Seasonally adjusted		

Amount outstanding as at 31 Dec

	AHIF		AGXB	
2005	215 997		347 420	

Transactions

	AAMN	AHHU	AGXE	NBYD
2002	14 961	14 932[†]	7 978	46 302
2003	23 816	23 712	9 753	34 654
2004	22 078	21 833	5 718	40 582
2005	20 419	20 402	12 030	55 994
2005 Q4	4 924	5 276[†]	4 342	16 594
2006 Q1	5 021	6 172	7 562	19 130
Q2	7 205	7 190	4 191	13 627
Q3	7 514	5 910	4 719[†]	12 912
2005 Dec	1 896	1 899[†]	2 052	..
2006 Jan	1 798	2 160	1 862	..
Feb	1 103	1 823	2 264	..
Mar	2 120	2 189	3 436	..
Apr	1 212	1 797	2 339	..
May	1 781	1 909	298	..
Jun	4 212	3 484	1 554	..
Jul	1 495	951	454	..
Aug	3 383	2 495	1 652	..
Sep	2 636	2 464	2 613[†]	..
Oct	4 627	5 444	1 665	..
Nov	3 152	2 600	632	..

	Banks[4]				Consumer credit[5]		of which Credit cards[5]	
	UK private sector deposits		Lending to the private sector					
	Sterling (Not seasonally adjusted)	Other currencies	Sterling (Not seasonally adjusted)	Other currencies	Not seasonally adjusted	Seasonally adjusted	Not seasonally adjusted	Seasonally adjusted

Amount outstanding as at 31 Dec

	AEAS	AGAK	AECE	AECK	VZRD	VZRI	VZRE	VZRJ
2006	..	..	..	..	212 650	212 369	55 797	54 903

Transactions

					Net lending	Net lending	Net lending	Net lending
	AEAT	AEAZ	AECF	AECL	VZQC	RLMH	VZQS	VZQX
2003	60 629	40 328	103 591	..	22 401	22 510[†]	8 710	8 926[†]
2004	86 098	27 697	133 774	..	25 337	25 421	9 998	9 935
2005	137 264[†]	39 719[†]	137 554	..	19 666[†]	19 696	6 166	6 151
2006	..	..	..	..	12 556	12 556	2 117	2 164
2006 Q1	32 018[†]	33 898[†]	48 216	..	2 369[†]	3 439[†]	−830	935[†]
Q2	51 074	−8 568	75 483	..	4 107	3 185	1 178	496
Q3	36 307	23 434	42 412[†]	..	3 050	2 838	364	254
Q4	..	..	..	..	3 030	3 094	1 405	479
2006 Jan	−9 591[†]	−3 427[†]	7 041	..	1 933[†]	1 343[†]	−148	526[†]
Feb	13 406	14 661	15 624	..	250	1 423	−150	364
Mar	28 203	22 664	25 551	..	186	666	−532	128
Apr	12 208	6 388	25 295	..	1 578	974	716	167
May	6 575	−2 407	14 806	..	1 759	1 229	285	165
Jun	32 291	−12 549	35 382	..	770	789	177	33
Jul	−2 902	15 867	16 413[†]	..	1 194	1 097	114	124
Aug	6 900	7 167	11 612	..	622	827	−122	−303
Sep	32 309	400	14 387	..	1 234	1 076	371[†]	397
Oct	7 593	6 791	2 164	..	1 185	1 187	−132[†]	194
Nov	7 981	7 773	16 504	..	771	1 134	369	172
Dec	..	..	..	..	1 075	1 029	1 168	254

1 For further details see *Financial Statistics,* Tables 1.2E, 3.2B, 4.2A, 4.3A, 4.3B, 5.2D, 6.2A, 10.5D.
2 Total administered by the Department for National Savings.
3 Including open ended investment companies (OEICs).
4 Monthly figures relate to calendar months.
5 Data have been revised back to February 2003 due to the inclusion of some additional other specialist lenders and the removal of some non-resident based securitisation vehicles.

Sources: Office for National Statistics;
Department for National Savings;
Building Societies Commission;
Association of Unit Trusts and Investment Funds;
Bank of England;
Department of Trade and Industry

17.5 Monetary aggregates

£ million

	Amount outstanding					
	'Narrow' money		'Broad' money			
	M0-the wide monetary base[1]		Retail deposits and cash in M4		M4	
	Not seasonally adjusted	Seasonally adjusted	Not seasonally adjusted	Seasonally adjusted	Not seasonally adjusted	Seasonally adjusted
	AVAD	AVAE	VQXV	VQWU	AUYM	AUYN
2002	39 540	37 237	703 920	703 186	1 008 751	1 009 319[†]
2003	42 317	40 000	777 347	777 045[†]	1 081 299	1 081 407
2004	44 466	42 284	845 654	845 837	1 179 192	1 178 781
2005	47 093	44 274	922 713	923 519	1 328 344[†]	1 327 708
2004 Q2	41 109	41 408	810 989	808 973[†]	1 133 432	1 128 046[†]
Q3	41 748	41 810	827 059	828 363	1 148 480	1 153 241
Q4	44 466	42 284	845 654	845 837	1 179 192	1 178 781
2005 Q1	42 395	42 634	862 152	862 591	1 216 916[†]	1 218 043
Q2	42 656	42 967	885 455	883 270	1 250 591	1 244 738
Q3	43 969	44 076	903 230	904 697	1 277 158	1 282 633
Q4	47 093	44 274	922 713	923 519	1 328 344	1 327 708
2006 Q1	44 669	45 501	945 342[†]	945 087	1 365 278	1 366 260
Q2	..	..	962 917	960 534	1 419 761	1 413 463
Q3	..	..	973 789	975 663	1 459 035	1 465 317
2005 Nov	44 644	44 412	913 946	912 665[†]	1 308 244[†]	1 307 601[†]
Dec	47 093	44 274	922 713	918 873	1 328 344	1 322 757
2006 Jan	45 567	45 274	915 855[†]	925 161	1 319 877	1 332 292
Feb	44 367	45 251	924 506	934 880	1 335 157	1 346 410
Mar	44 669	45 501	945 342	941 776	1 365 278	1 358 892
Apr	45 939	45 878	954 339	948 922	1 378 952	1 375 688
May	..	..	953 269	952 547	1 385 873	1 384 360
Jun	..	..	962 917	957 341	1 419 761	1 405 047
Jul	..	..	961 211	961 928	1 417 739	1 419 165
Aug	..	..	965 851	966 687	1 426 145	1 433 209
Sep	..	..	973 789	972 212	1 459 035	1 456 886
Oct	..	..	976 281	978 239	1 468 109	1 469 826
Nov	..	..	988 012	985 588	1 478 127	1 477 060

1 No longer supplied by The Bank of England.

Source: Bank of England

17.6 Selected interest rates, exchange rates and security prices[1]

	Selected retail banks' base rate	Average discount rate for 91 day Treasury bills	Inter bank 3 months bid rate	Inter bank 3 months offer rate	British government securities 20 years yield[2]	Exchange rate US spot
	ZCMG	AJNB	HSAJ	HSAK	AJLX	LUSS
2005 Jun	4.75	4.52	4.69	4.73	4.29	1.7925
Jul	4.75	4.43	4.54	4.56	4.33	1.7607
Aug	4.50[†]	4.38	4.52	4.54	4.34	1.7990
Sep	4.50	4.40	4.52	4.55	4.26	1.7688
Oct	4.50	4.42	4.54	4.56	4.36	1.7700
Nov	4.50	4.40	4.55	4.58	4.25	1.7304
Dec	4.50	4.43	4.57	4.59	4.14	1.7166
2006 Jan	4.50	4.40	4.52	4.54	3.81	1.7775
Feb	4.50	4.39	4.51	4.53	3.96	1.7511
Mar	4.50	4.41	4.54	4.56	4.15	1.7345
Apr	4.50	4.45	4.60	4.63	4.32	1.8179
May	4.50	4.51	4.66	4.68	4.43	1.8712
Jun	4.50	4.54	4.71	4.73	4.46	1.8494
Jul	4.50	4.58	4.73	4.74	4.45	1.8671
Aug	4.75	4.77	4.94	4.95	4.42	1.9018
Sep	4.75	4.87	5.02	5.05	4.29	1.8682
Oct	4.75	4.98	5.14	5.16	4.35	1.9073
Nov	5.00	5.04	5.20	5.22	4.27	1.9670
Dec	5.00	5.11	5.26	5.29	4.33	1.9570
2007 Jan	5.25	..	..	..	..	..

1 As from December 2003 *The Financial Times* Actuaries indices have been removed as The Bank of England are no longer able to provide these data
2 Average of working days.

Source: Bank of England

18 Prices and wages

18.1 Consumer Prices Index[1]: Detailed figures by division[2]

	Food and non-alcoholic beverages	Alcoholic beverages and tobacco	Clothing and footwear	Housing, water, electricity, gas & other fuels	Furniture, household equipment & routine maintenance	Health[3]	Transport	Communication	Recreation and culture	Education[3]	Restaurants and hotels	Miscellaneous goods and services[3]	CPI (overall index)
Index level (2005=100)													
COICOP Division	01	02	03	04	05	06	07	08	09	10	11	12	
Weights 2006	102	44	65	108	73	24	155	25	147	17	134	106	1000
	D7BU	D7BV	D7BW	D7BX	D7BY	D7BZ	D7C2	D7C3	D7C4	D7C5	D7C6	D7C7	D7BT
2004 Dec	99.0	97.8	104.5	96.6	102.7	98.0	98.0	100.5	101.2	98.7	98.0	97.4	99.1
2005 Jan	99.2	98.8	100.7	97.2	98.6	98.7	96.4	100.3	100.7	98.7	98.2	98.5	98.6
Feb	99.9	99.2	100.5	97.4	99.0	98.8	97.3	100.5	100.2	98.7	98.4	98.6	98.8
Mar	100.8	98.7	101.3	97.7	100.7	98.9	98.0	100.4	100.2	98.7	98.7	98.8	99.3
Apr	99.9	99.8	100.9	99.6	99.3	99.5	98.5	100.5	100.7	98.7	99.5	99.3	99.7
May	100.6	100.2	101.0	99.8	100.1	99.7	99.5	100.1	100.4	98.7	99.8	99.4	100.0
Jun	100.6	100.3	100.6	100.0	100.4	99.8	99.6	100.3	100.3	98.7	99.9	99.6	100.0
Jul	99.4	100.6	96.5	100.4	99.4	100.6	102.1	99.5	99.8	98.7	100.3	100.4	100.1
Aug	99.6	100.5	98.2	100.4	99.6	100.8	103.3	99.9	99.3	98.7	100.6	100.6	100.4
Sep	99.7	100.3	99.9	100.7	100.2	100.7	102.2	99.8	99.6	100.8	100.8	100.8	100.6
Oct	99.5	100.7	99.9	101.7	99.4	101.0	102.0	99.5	99.7	103.3	101.2	101.2	100.7
Nov	100.1	100.7	100.5	102.3	100.5	101.0	100.4	99.6	99.5	103.3	101.2	101.4	100.7
Dec	100.7	100.2	100.1	102.8	102.8	100.3	100.7	99.4	99.5	103.3	101.4	101.5	101.0
2006 Jan	100.4	101.0	96.0	103.3	97.8	101.0	101.2	100.9	98.6	103.3	101.5	102.0	100.5
Feb	101.0	100.8	95.9	103.6	98.5	101.1	101.4	101.0	99.4	103.3	101.9	102.1	100.9
Mar	100.4	101.1	96.5	104.5	100.3	101.0	101.4	100.9	98.9	103.3	102.2	102.4	101.1
Apr	100.2	102.3	96.5	107.3	98.5	102.2	102.9	100.9	99.1	103.3	102.5	102.3	101.7
May	101.7	102.5	97.2	108.8	99.3	102.6	103.5	99.7	98.9	103.3	103.0	103.4	102.2
Jun	102.4	103.6	96.7	109.7	100.2	102.5	103.5	100.0	98.7	103.3	103.1	103.9	102.5
Jul	102.6	103.4	92.2	110.5	98.1	103.0	105.5	99.8	98.4	103.3	103.5	104.0	102.5
Aug	103.0	103.8	94.4	110.9	99.1	103.4	105.8	99.2	98.4	103.3	103.6	104.5	102.9
Sep	103.6	103.7	96.4	111.5	100.6	103.6	102.9	99.6	98.6	107.9	103.8	104.7	103.0
Oct	104.2	103.9	96.6	112.7	99.0	104.2	101.5	100.4	98.6	117.8	104.2	105.0	103.2
Nov	105.1	103.4	97.2	113.7	100.0	104.1	101.1	100.3	98.7	117.8	104.5	105.0	103.4
Dec	105.5	103.0	96.0	114.5	103.3	104.2	102.8	99.9	99.2	117.8	104.7	104.9	104.0
Percentage change on a year earlier													
	D7G8	D7G9	D7GA	D7GB	D7GC	D7GD	D7GE	D7GF	D7GG	D7GH	D7GI	D7GJ	D7G7
2004 Dec	−0.1	2.1	−5.4	5.4	0.7	1.9	4.0	−2.2	−0.6	5.0	2.9	3.2	1.7
2005 Jan	0.4	2.5	−5.9	5.8	−0.5	2.4	2.7	−2.3	−0.4	5.0	2.9	3.6	1.6
Feb	0.8	2.9	−5.8	5.9	−1.2	2.7	3.2	−2.7	−0.7	5.0	3.1	3.4	1.7
Mar	1.7	2.2	−5.1	5.8	–	2.6	4.0	−2.9	−0.7	5.0	2.8	3.6	1.9
Apr	1.0	2.0	−5.3	6.5	−1.0	2.7	3.8	−3.2	−0.6	5.0	3.3	4.3	1.9
May	1.4	2.3	−5.5	6.4	−1.0	2.8	3.3	−3.6	−0.5	5.0	3.3	4.3	1.9
Jun	2.2	2.3	−4.8	6.4	−0.6	2.7	3.4	−2.9	−1.2	5.0	3.2	4.6	2.0
Jul	1.7	2.2	−4.8	6.7	0.6	3.4	4.6	−2.5	−1.5	5.0	3.4	5.2	2.3
Aug	2.2	1.8	−4.4	6.3	0.2	3.7	5.4	−2.1	−2.0	5.0	3.5	5.1	2.4
Sep	2.0	1.4	−5.3	6.5	−0.2	3.1	6.0	−1.2	−1.6	4.7	3.7	5.1	2.5
Oct	1.5	1.9	−5.3	6.5	−0.2	3.0	5.8	−1.6	−1.5	4.7	3.7	4.2	2.3
Nov	1.7	2.5	−5.1	6.5	0.1	2.9	4.1	−1.2	−1.5	4.7	3.5	4.3	2.1
Dec	1.7	2.5	−4.2	6.4	–	2.4	2.8	−1.0	−1.7	4.7	3.5	4.2	1.9
2006 Jan	1.2	2.3	−4.7	6.3	−0.8	2.3	5.1	0.5	−2.1	4.7	3.4	3.5	1.9
Feb	1.1	1.6	−4.7	6.4	−0.5	2.3	4.2	0.4	−0.8	4.7	3.5	3.6	2.0
Mar	−0.4	2.5	−4.7	7.0	−0.4	2.1	3.5	0.5	−1.4	4.7	3.6	3.7	1.8
Apr	0.3	2.5	−4.4	7.7	−0.8	2.7	4.4	0.3	−1.6	4.7	3.0	4.0	2.0
May	1.1	2.2	−3.7	9.0	−0.8	2.9	4.0	−0.4	−1.6	4.7	3.2	4.1	2.2
Jun	1.8	3.3	−3.9	9.8	−0.1	2.7	3.9	−0.3	−1.6	4.7	3.2	4.4	2.5
Jul	3.2	2.8	−4.5	10.0	−1.3	2.4	3.3	0.3	−1.4	4.7	3.2	3.6	2.4
Aug	3.4	3.3	−3.9	10.5	−0.5	2.6	2.4	−0.7	−0.9	4.7	3.0	3.9	2.5
Sep	4.0	3.4	−3.5	10.7	0.3	2.8	0.6	−0.2	−1.0	7.1	2.9	3.9	2.4
Oct	4.7	3.2	−3.3	10.8	−0.4	3.1	−0.5	0.9	−1.1	14.0	3.0	3.8	2.4
Nov	5.0	2.7	−3.2	11.1	−0.5	3.1	0.8	0.7	−0.7	14.0	3.2	3.5	2.7
Dec	4.6	2.7	−4.1	11.4	0.6	3.9	2.1	0.5	−0.3	14.0	3.2	3.3	3.0

Note: Further information on the consumer prices index is available from the National Statistics website: www.statistics.gov.uk/cpi

1 Prior to 10 December 2003, the consumer prices index (CPI) was published in the UK as the harmonised index of consumer prices (HICP).

2 Inflation rates prior to 1997 and index levels prior to 1996 are estimated. Further details are given in *Economic Trends No. 541 December 1998.* These details are also available on the National Statistics website: http://www.statistics.gov.uk/cci/article.asp?ID=31&Pos=3&Col-Rank=2&Rank=720

3 The coverage of these categories was extended in January 2000; further extensions to coverage came into effect in January 2001 for health and miscellaneous goods and services; the coverage of miscellaneous goods and services was further extended with effect from January 2002 (details are given in a series of Economic Trends articles available on the National Statistics website:www.statistics.gov.uk/cpi)

Source: Office for National Statistics: 020 7533 5874

18.2 Consumer Prices Index[1]: Detailed figures by divisions, groups and classes

	Weights	Index (2005=100)						Percentage change over 12 months					
	2006	2006 Jul	2006 Aug	2006 Sep	2006 Oct	2006 Nov	2006 Dec	2006 Jul	2006 Aug	2006 Sep	2006 Oct	2006 Nov	2006 Dec
CPI (overall index)	*1 000.0*	102.5	102.9	103.0	103.2	103.4	104.0	*2.4*	*2.5*	*2.4*	*2.4*	*2.7*	*3.0*
01 Food and non-alcoholic beverages	*102.0*	102.6	103.0	103.6	104.2	105.1	105.4	*3.2*	*3.4*	*4.0*	*4.7*	*5.0*	*4.6*
02 Alcoholic beverages and tobacco	*44.0*	103.4	103.8	103.7	103.9	103.4	103.0	*2.8*	*3.3*	*3.4*	*3.2*	*2.7*	*2.7*
03 Clothing and footwear	*65.0*	92.2	94.4	96.4	96.6	97.2	96.0	*−4.5*	*−3.9*	*−3.5*	*−3.3*	*−3.2*	*−4.1*
04 Housing, water, electricity, gas and other fuels	*108.0*	110.5	110.9	111.5	112.7	113.7	114.5	*10.0*	*10.5*	*10.7*	*10.8*	*11.1*	*11.4*
05 Furniture, household equipment and maintenance	*73.0*	98.1	99.1	100.6	99.0	100.0	103.3	*−1.3*	*−0.5*	*0.3*	*−0.4*	*−0.5*	*0.6*
06 Health	*24.0*	103.0	103.4	103.6	104.2	104.1	104.2	*2.4*	*2.6*	*2.8*	*3.1*	*3.1*	*3.9*
07 Transport	*155.0*	105.5	105.8	102.9	101.5	101.1	102.8	*3.3*	*2.4*	*0.6*	*−0.5*	*0.8*	*2.1*
08 Communication	*25.0*	99.8	99.2	99.6	100.4	100.3	99.9	*0.3*	*−0.7*	*−0.2*	*0.9*	*0.7*	*0.5*
09 Recreation and culture	*147.0*	98.4	98.4	98.6	98.6	98.7	99.2	*−1.4*	*−0.9*	*−1.0*	*−1.1*	*−0.7*	*−0.3*
10 Education	*17.0*	103.3	103.3	107.9	117.8	117.8	117.8	*4.7*	*4.7*	*7.1*	*14.0*	*14.0*	*14.0*
11 Restaurants and hotels	*134.0*	103.5	103.6	103.8	104.2	104.5	104.7	*3.2*	*3.0*	*2.9*	*3.0*	*3.2*	*3.2*
12 Miscellaneous goods and services	*106.0*	104.0	104.5	104.7	105.0	105.0	104.9	*3.6*	*3.9*	*3.9*	*3.8*	*3.5*	*3.3*
All goods	*554*	101.1	101.8	102.1	101.8	102.3	103.0	*1.8*	*2.2*	*1.8*	*1.5*	*1.8*	*2.3*
All services	*446*	104.1	104.2	104.1	104.8	104.8	105.4	*3.2*	*2.9*	*3.2*	*3.6*	*3.7*	*3.8*
01.1 Food	*90.0*	102.4	102.9	103.3	103.9	105.0	105.2	*3.2*	*3.4*	*3.7*	*4.5*	*4.9*	*4.5*
01.1.1 Bread and cereals	*15.0*	101.5	101.9	102.5	102.5	103.3	104.0	*1.8*	*1.5*	*2.3*	*2.4*	*2.7*	*2.9*
01.1.2 Meat	*21.0*	102.8	103.2	102.7	103.5	103.7	104.3	*3.1*	*3.1*	*2.6*	*4.0*	*2.6*	*3.6*
01.1.3 Fish	*4.0*	108.9	110.9	111.5	113.7	113.0	114.9	*9.1*	*11.1*	*11.1*	*12.2*	*11.4*	*12.2*
01.1.4 Milk, cheese and eggs	*13.0*	102.2	102.8	102.7	102.8	103.0	103.3	*2.3*	*2.4*	*1.9*	*1.5*	*1.6*	*1.8*
01.1.5 Oils and fats	*2.0*	107.4	108.3	107.0	107.6	108.4	108.3	*8.1*	*9.0*	*8.2*	*8.9*	*10.4*	*9.9*
01.1.6 Fruit	*8.0*	97.5	99.8	102.6	104.3	108.0	107.2	*1.1*	*0.9*	*4.0*	*4.5*	*5.5*	*4.9*
01.1.7 Vegetables including potatoes and tubers	*14.0*	102.5	100.8	102.5	103.1	107.1	106.4	*4.7*	*5.3*	*5.9*	*9.3*	*13.3*	*8.1*
01.1.8 Sugar, jam, syrups, chocolate and confectionery	*11.0*	103.9	104.9	104.7	104.5	104.5	104.5	*3.5*	*4.4*	*4.2*	*3.4*	*3.3*	*3.4*
01.1.9 Food products nec[2]	*2.0*	98.6	99.2	99.1	100.6	99.6	100.0	*−1.1*	*−0.4*	*−0.1*	*1.3*	*0.4*	*1.2*
01.2 Non-alcoholic beverages	*12.0*	103.8	104.4	105.9	106.3	105.7	106.2	*3.1*	*3.5*	*5.8*	*6.4*	*5.8*	*5.2*
01.2.1 Coffee, tea and cocoa	*3.0*	103.7	108.3	109.9	110.1	108.4	110.1	*2.1*	*6.2*	*8.2*	*9.6*	*6.9*	*8.7*
01.2.2 Mineral waters, soft drinks and juices	*9.0*	103.9	103.0	104.5	105.0	104.8	105.0	*3.5*	*2.6*	*4.9*	*5.3*	*5.4*	*4.1*
02.1 Alcoholic beverages	*18.0*	100.3	101.1	101.0	101.3	100.1	99.0	*−0.7*	*0.2*	*0.6*	*1.1*	*0.7*	*0.8*
02.1.1 Spirits	*5.0*	99.1	99.9	100.3	101.1	98.4	96.9	*−1.7*	*0.1*	*−0.1*	*0.7*	*0.1*	*0.3*
02.1.2 Wine	*9.0*	101.2	102.5	102.0	101.8	101.3	99.8	*−0.6*	*0.9*	*1.5*	*1.3*	*1.1*	*0.7*
02.1.3 Beer	*4.0*	99.6	99.3	99.4	100.3	99.2	99.6	*−0.2*	*−1.1*	*−0.9*	*0.7*	*0.3*	*1.5*
02.2 Tobacco	*26.0*	105.6	105.7	105.8	105.8	105.8	105.9	*5.4*	*5.5*	*5.6*	*4.7*	*4.0*	*4.1*
03.1 Clothing	*56.0*	91.9	94.3	96.4	96.6	97.3	96.2	*−4.6*	*−3.9*	*−3.5*	*−3.3*	*−3.2*	*−4.2*
03.1.2 Garments	*52.0*	91.4	93.8	95.9	96.2	96.9	95.7	*−5.0*	*−4.3*	*−3.8*	*−3.7*	*−3.6*	*−4.7*
03.1.3 Other clothing and clothing accessories	*3.0*	97.2	99.2	101.1	101.4	102.4	102.3	*−1.6*	*−0.6*	*0.0*	*1.2*	*1.3*	*1.1*
03.1.4 Cleaning, repair and hire of clothing	*1.0*	104.6	105.0	105.5	105.7	105.8	106.1	*4.7*	*4.7*	*4.7*	*4.6*	*4.3*	*4.3*
03.2 Footwear including repairs	*9.0*	94.2	94.8	96.5	96.5	96.6	94.7	*−3.8*	*−3.9*	*−3.6*	*−3.1*	*−3.0*	*−3.4*
04.1 Actual rentals for housing	*47.0*	103.8	103.9	103.9	104.0	104.2	104.3	*3.1*	*3.1*	*3.0*	*3.0*	*3.2*	*3.2*
04.3 Regular maintenance and repair of the dwelling	*19.0*	101.4	102.1	102.5	102.4	102.7	103.5	*0.5*	*2.5*	*2.9*	*2.3*	*2.8*	*3.3*
04.3.1 Materials for maintenance and repair	*11.0*	99.3	100.2	100.6	100.2	100.6	101.7	*−2.0*	*1.4*	*2.0*	*1.1*	*1.9*	*2.6*
04.3.2 Services for maintenance and repair	*8.0*	104.6	104.9	105.2	..	..	..	*4.2*	*4.2*	*4.2*	*..*	*..*	*..*
04.4 Water supply and misc. services for the dwelling	*10.0*	108.2	108.2	108.2	108.2	108.2	108.2	*5.5*	*5.5*	*5.5*	*5.5*	*5.5*	*5.5*
04.4.1 Water supply	*5.0*	109.0	109.0	109.0	109.0	109.0	109.0	*5.7*	*5.7*	*5.7*	*5.7*	*5.7*	*5.7*
04.4.3 Sewerage collection	*5.0*	107.5	107.5	107.5	107.5	107.5	107.5	*5.2*	*5.2*	*5.2*	*5.2*	*5.2*	*5.2*
04.5 Electricity, gas and other fuels	*32.0*	127.8	128.8	130.5	134.6	137.6	139.9	*29.2*	*29.6*	*29.7*	*29.9*	*30.0*	*30.2*
04.5.1 Electricity	*15.0*	124.6	125.6	127.1	130.5	133.0	134.6	*25.9*	*26.5*	*26.7*	*27.3*	*27.0*	*27.3*
04.5.2 Gas	*14.0*	134.0	135.5	138.7	145.4	150.1	152.9	*36.3*	*37.8*	*39.0*	*40.6*	*39.9*	*39.8*
04.5.3 Liquid fuels	*2.0*	122.9	120.2	113.9	104.5	99.8	104.5	*16.7*	*7.5*	*−0.3*	*−12.0*	*−5.7*	*−4.7*
04.5.4 Solid fuels	*1.0*	104.8	105.7	106.9	112.4	113.7	113.7	*7.6*	*7.1*	*7.1*	*8.0*	*8.3*	*8.0*
05.1 Furniture, furnishings and carpets	*32.0*	98.5	99.3	101.5	99.3	100.7	108.3	*−1.8*	*0.2*	*1.7*	*0.2*	*−0.3*	*1.9*
05.1.1 Furniture and furnishings	*25.0*	98.2	98.8	101.9	98.9	101.0	109.8	*−2.4*	*−0.1*	*2.0*	*0.3*	*0.1*	*2.3*
05.1.2 Carpets and other floor coverings	*7.0*	99.5	101.2	100.2	101.0	99.9	102.9	*0.2*	*1.5*	*0.5*	*0.3*	*−1.5*	*−0.1*
05.2 Household textiles	*8.0*	93.3	94.8	95.1	93.8	95.8	95.8	*−3.4*	*−4.1*	*−5.7*	*−4.1*	*−4.0*	*−4.2*
05.3 Household appliances, fitting and repairs	*9.0*	95.5	96.8	99.6	95.1	95.0	95.0	*−4.5*	*−4.1*	*−1.2*	*−4.9*	*−4.8*	*−4.7*
05.3.1/2 Major appliances and small electric goods	*8.0*	94.8	96.3	99.4	94.3	94.2	94.2	*−5.2*	*−4.8*	*−1.5*	*−5.5*	*−5.5*	*−5.4*
05.3.3 Repair of household appliances	*1.0*	100.9	101.0	101.3	101.4	101.6	101.8	*0.9*	*0.8*	*1.0*	*0.7*	*1.1*	*1.2*
05.4 Glassware, tableware and household utensils	*7.0*	96.2	97.7	98.8	98.8	99.7	99.9	*−1.4*	*−1.8*	*−1.2*	*−0.9*	*−1.0*	*−1.0*
05.5 Tools and equipment for house and garden	*6.0*	100.2	100.7	101.5	101.9	101.7	101.9	*0.2*	*1.4*	*1.8*	*2.6*	*2.5*	*2.3*
05.6 Goods and services for routine maintenance	*11.0*	104.1	104.4	104.5	105.2	105.3	105.6	*4.0*	*4.0*	*3.6*	*4.1*	*4.5*	*3.7*
05.6.1 Non-durable household goods	*5.0*	103.3	103.7	103.6	104.0	104.5	104.6	*3.3*	*3.4*	*2.7*	*3.0*	*4.4*	*2.9*
05.6.2 Domestic services and household services	*6.0*	104.7	105.1	105.3	106.2	105.9	106.4	*4.6*	*4.6*	*4.4*	*5.0*	*4.4*	*4.3*
06.1 Medical products, appliances and equipment	*11.0*	98.9	99.5	99.8	100.0	99.7	99.9	*−1.2*	*−0.8*	*−0.2*	*−0.4*	*−0.4*	*1.2*
06.1.1 Pharmaceutical products	*5.0*	99.1	100.5	101.0	101.1	100.7	101.3	*−1.1*	*0.2*	*1.3*	*0.6*	*0.8*	*3.2*
06.1.2/3 Other medical and therapeutic equipment	*6.0*	98.8	98.8	98.8	99.2	98.9	98.7	*−1.1*	*−1.5*	*−1.4*	*−1.0*	*−1.4*	*−0.5*

18.2 Consumer Prices Index[1]: Detailed figures by divisions, groups and classes
continued

	Weights	Index (2005=100)						Percentage change over 12 months					
	2006	2006 Jul	2006 Aug	2006 Sep	2006 Oct	2006 Nov	2006 Dec	2006 Jul	2006 Aug	2006 Sep	2006 Oct	2006 Nov	2006 Dec
06.2 Out-patient services	4.0	104.9	105.3	105.4	105.7	106.0	106.1	4.3	4.5	4.5	4.5	4.6	4.3
06.2.1/3 Medical services and paramedical services	2.0	103.3	103.6	103.8	104.2	104.4	104.2	2.9	3.0	3.1	3.4	3.3	2.6
06.2.2 Dental services	2.0	106.5	107.1	107.0	107.2	107.6	107.9	5.6	5.9	5.8	5.6	5.9	5.9
06.3 Hospital services	9.0	107.7	107.8	107.8	109.0	109.1	109.2	6.1	6.0	6.0	6.9	6.9	7.0
07.1 Purchase of vehicles	52.0	99.5	99.5	99.4	99.4	99.4	99.4	-0.4	-0.1	-0.2	-0.2	0.1	0.2
07.1.1A New cars	31.0	101.0	101.1	101.1	101.2	101.5	101.5	0.7	0.8	0.7	0.7	1.4	1.3
07.1.1B Second hand cars	18.0	96.9	96.5	96.4	96.1	95.8	95.4	-2.5	-1.7	-1.7	-1.9	-2.1	-1.8
07.1.2/3 Motorcycles and bicycles	3.0	98.7	99.0	99.4	99.3	98.9	99.7	-1.6	-0.7	-0.2	-0.1	-0.5	0.4
07.2 Operation of personal transport equipment	71.0	108.6	109.0	105.7	103.2	103.3	104.6	7.9	6.4	0.9	-1.1	0.3	2.9
07.2.1 Spare parts and accessories	6.0	103.2	103.0	103.3	103.2	103.4	103.3	3.7	2.8	2.6	2.5	2.3	2.2
07.2.2 Fuels and lubricants	35.0	111.6	112.2	105.3	99.8	99.3	101.6	10.4	7.5	-3.8	-7.6	-4.6	0.8
07.2.3 Maintenance and repairs	24.0	106.1	106.6	107.0	107.7	108.6	109.0	6.1	5.8	5.9	6.1	6.2	6.2
07.2.4 Other services	6.0	104.4	104.6	104.7	104.7	105.0	105.3	2.8	2.7	2.5	2.1	2.2	2.1
07.3 Transport services	32.0	108.4	109.4	102.4	101.1	99.1	104.6	0.3	-1.3	1.2	0.3	2.4	3.2
07.3.1 Passenger transport by railway	8.0	104.8	105.5	103.7	103.6	104.6	104.1	4.7	5.2	3.7	3.2	3.8	3.6
07.3.2 Passenger transport by road	15.0	101.4	101.8	102.4	103.0	103.6	103.9	1.1	1.3	1.4	1.3	2.0	1.8
07.3.3 Passenger transport by air	7.0	116.2	118.4	92.3	90.5	83.3	103.6	-5.2	-9.2	-9.6	-10.0	-5.2	0.5
07.3.4 Passenger transport by sea and inland waterway	2.0	127.0	129.2	124.0	107.8	98.5	103.6	3.8	-0.1	20.5	10.6	6.7	13.2
08.1 Postal services	1.0	107.9	107.9	114.3	114.3	114.3	114.3	6.9	6.9	13.1	13.1	13.1	13.1
08.2/3 Telephone and telefax equipment and services	24.0	99.4	98.7	98.9	99.8	99.7	99.3	0.0	-1.1	-0.8	0.4	0.2	0.0
09.1 Audio-visual equipment and related products	27.0	88.6	87.0	87.6	86.7	87.2	88.0	-10.6	-10.5	-9.2	-10.0	-7.8	-7.5
09.1.1 Reception and reproduction of sound and pictures	6.0	84.8	83.7	84.8	84.1	84.6	84.3	-15.1	-13.8	-12.3	-9.8	-8.8	-8.9
09.1.2 Photographic, cinematographic and optical equipment	5.0	75.0	72.9	74.8	74.6	73.2	71.7	-23.2	-25.0	-20.9	-19.5	-14.5	-17.0
09.1.3 Data processing equipment	5.0	85.7	87.1	85.9	81.9	82.6	83.1	-10.6	-8.2	-9.5	-14.5	-13.0	-13.0
09.1.4 Recording media	10.0	97.3	94.0	94.6	94.6	96.3	99.4	-3.2	-4.0	-2.9	-4.7	-2.2	-0.1
09.1.5 Repair of audio-visual equipment & related products	1.0	104.3	104.7	104.6	104.9	104.8	105.0	3.9	3.8	3.4	3.1	3.0	3.2
09.2 Other major durables for recreation and culture	9.0	100.8	100.8	100.8	101.0	100.9	100.9	0.8	0.7	0.7	0.8	0.5	0.5
09.2.1/2 Major durables for in/outdoor recreation	9.0	100.8	100.8	100.8	101.0	100.9	100.9	0.8	0.7	0.7	0.8	0.5	0.5
09.3 Other recreational items, gardens and pets	39.0	96.4	97.8	97.0	97.2	97.0	97.5	-2.5	-1.3	-2.8	-2.9	-2.9	-1.9
09.3.1 Games, toys and hobbies	23.0	93.4	95.7	94.0	94.5	93.9	94.5	-4.9	-3.7	-6.1	-5.8	-6.1	-4.1
09.3.2 Equipment for sport and open-air recreation	4.0	97.2	99.0	98.6	98.5	98.2	98.0	-2.5	1.5	-0.4	-1.1	0.3	-0.8
09.3.3 Gardens, plants and flowers	5.0	98.8	98.6	98.9	98.3	99.8	100.5	0.1	-0.4	0.6	-0.8	0.3	0.3
09.3.4/5 Pets, related products and services	7.0	103.8	103.3	104.2	104.4	104.4	104.6	3.6	3.5	3.4	3.5	3.3	3.3
09.4 Recreational and cultural services	28.0	104.9	104.5	105.9	106.6	106.5	107.0	4.2	4.3	4.5	5.1	4.5	4.4
09.4.1 Recreational and sporting services	8.0	104.5	104.8	107.2	107.2	107.3	107.3	5.2	5.0	4.8	4.8	4.6	4.5
09.4.2 Cultural services	20.0	105.1	104.3	105.3	106.4	106.2	106.9	3.9	4.1	4.4	5.2	4.5	4.4
09.5 Books, newspapers and stationery	17.0	103.9	103.6	104.6	104.1	104.7	105.5	3.4	3.8	4.2	3.6	3.6	4.7
09.5.1 Books	5.0	103.7	103.1	103.8	102.6	103.8	105.8	1.5	2.5	2.3	1.8	3.4	7.2
09.5.2 Newspapers and periodicals	7.0	106.3	106.6	108.4	108.1	108.6	108.8	7.0	7.4	8.1	7.3	5.8	5.5
09.5.3/4 Misc. printed matter, stationery, drawing materials	5.0	100.9	100.0	100.1	100.1	100.2	100.6	0.4	0.1	0.7	0.2	0.6	0.9
09.6 Package holidays	27.0	99.3	99.5	99.2	99.3	99.7	99.7	-0.7	-0.3	-0.2	0.0	0.7	0.8
10.0 Education	17.0	103.3	103.3	107.9	117.8	117.8	117.8	4.7	4.7	7.1	14.0	14.0	14.0
11.1 Catering services	116.0	103.2	103.4	103.5	103.9	104.3	104.6	3.0	2.9	2.9	2.9	3.2	3.2
11.1.1 Restaurants & cafes	103.0	103.3	103.5	103.6	104.0	104.3	104.6	3.0	3.0	3.0	3.0	3.1	3.2
11.1.2 Canteens	13.0	102.6	102.6	103.0	103.4	104.9	104.9	3.0	1.5	2.0	2.4	3.5	3.3
11.2 Accommodation services	18.0	105.1	105.1	105.5	106.2	105.6	105.5	4.4	4.1	3.3	3.6	3.6	3.1
12.1 Personal care	32.0	101.6	102.5	102.7	102.8	102.7	102.2	2.0	2.9	3.2	2.8	2.7	2.1
12.1.1 Hairdressing and personal grooming establishments	8.0	103.9	104.1	104.1	104.5	104.8	105.1	3.6	3.7	3.4	3.4	3.6	3.5
12.1.2/3 Appliances and products for personal care	24.0	100.8	101.9	102.1	102.1	101.9	101.1	1.3	2.6	3.1	2.6	2.4	1.7
12.3 Personal effects nec[2]	11.0	101.8	103.8	104.4	104.3	104.3	104.2	3.6	3.7	3.4	3.9	3.5	3.4
12.3.1 Jewellery, clocks and watches	7.0	104.6	106.1	106.7	106.7	106.8	106.4	6.2	5.6	5.9	6.1	6.1	5.9
12.3.2 Other personal effects	4.0	97.1	100.1	100.8	100.4	100.2	100.6	-0.3	0.8	-0.8	0.4	-0.8	-0.8
12.4 Social protection	12.0	105.9	106.1	106.8	107.2	107.7	108.0	5.6	5.2	5.3	5.2	5.4	5.6
12.5 Insurance	9.0	103.4	102.6	102.5	102.4	103.5	103.0	1.9	2.9	2.7	2.3	2.7	2.8
12.5.2 House contents insurance	2.0	102.3	102.6	102.5	103.3	102.9	102.5	1.8	2.0	1.6	2.8	2.2	1.5
12.5.3 Health insurance	2.0	108.9	108.9	108.9	110.0	110.0	110.0	7.6	7.6	7.6	8.3	8.3	8.3
12.5.4 Transport insurance	5.0	101.7	100.3	100.1	99.3	101.3	100.5	0.0	1.3	1.1	-0.2	0.8	1.1
12.6 Financial services nec[2]	29.0	105.2	105.1	105.2	105.7	105.0	105.0	3.5	3.4	3.5	3.2	2.5	2.5
12.6.2 Other financial services nec[2]	29.0	105.2	105.1	105.2	105.7	105.0	105.0	3.5	3.4	3.5	3.2	2.5	2.5
12.7 Other services nec[2]	13.0	106.6	106.8	107.1	108.0	108.2	108.6	5.9	5.8	5.8	6.2	5.9	5.8

1 Prior to 10 December 2003, the consumer prices index (CPI) was published in the UK as the harmonised index of consumer prices (HICP).

2 nec - not elsewhere covered

Source: Office for National Statistics: 020 7533 5874

18.3 Retail Prices Index

13 January 1987=100

| | ALL ITEMS (RPI) | mortgage interest payments (RPIX) | All items excluding | | | | Food and catering | Alcohol and tobacco | Housing and household expenditure | Personal expenditure | Travel and leisure | Consumer durables | All items excluding mortgage interest payments & indirect taxes (RPIY)[3] |
			mortgage interest payments and depreciation[1]	housing	food	seasonal food[2]							
Weights													
	CZGU	CZGY	DOGZ	CZGX	CZGV	CZGW	CBVV	CBVW	CBVX	CBVY	CBVZ	CBWA	
1997	1 000	961	932	814	864	981	185	114	351	96	254	122	
1998	1 000	955	923	803	870	982	178	105	359	95	263	121	
1999	1 000	958	928	807	872	980	179	100	358	95	268	127	
2000	1 000	960	924	805	882	982	170	95	355	101	279	126	
2001	1 000	954	914	795	884	982	169	97	362	96	276	125	
2002	1 000	964	924	801	886	980	166	99	363	94	278	126	
2003	1 000	961	919	797	891	983	160	98	365	92	285	126	
2004	1 000	961	914	791	889	981	160	97	367	93	283	121	
2005	1 000	950	901	776	890	981	159	96	387	89	269	122	
2006	1 000	950	906	778	895	983	155	96	392	90	267	117	
Annual averages													
	CHAW	CHMK	CHON	CHAZ	CHAY	CHAX	CHBS	CHBT	CHBU	CHBV	CHBW	CHBY	CBZW
1997	157.5	156.5	156.4	152.9	160.5	158.5	150.4	183.2	158.4	137.7	159.0	117.3	151.5
1998	162.9	160.6	160.3	156.2	166.5	163.8	153.4	192.3	166.2	139.9	162.8	115.9	154.5
1999	165.4	164.3	163.6	158.9	169.4	166.5	155.4	202.6	167.7	139.6	165.6	112.3	157.1
2000	170.3	167.7	166.4	161.3	175.1	171.4	156.7	210.3	176.2	137.2	170.3	108.0	159.9
2001	173.3	171.3	169.5	163.7	178.0	174.3	162.2	216.9	180.0	135.7	172.0	105.0	163.7
2002	176.2	175.1	172.5	166.0	181.1	177.2	164.8	222.3	184.6	133.2	174.2	101.9	167.5
2003	181.3	180.0	176.2	168.7	186.7	182.4	167.9	228.0	194.3	133.2	177.0	99.8	172.0
2004	186.7	184.0	179.1	170.9	192.8	187.9	170.0	233.6	207.4	131.5	178.1	97.7	175.5
2005	192.0	188.2	182.6	173.7	198.7	193.3	172.9	239.8	219.4	131.0	179.2	95.3	179.4
2006	198.1	193.7	187.8	178.3	205.2	199.5	176.9	247.1	231.8	131.7	181.1	94.0	184.8
Monthly figures													
2003 Dec	183.5	181.8	177.6	169.9	189.0	184.5	169.9	229.1	199.4	133.4	176.7	100.3	173.5
2004 Jan	183.1	181.4	177.1	169.3	188.6	184.2	169.7	229.8	198.9	130.9	177.1	97.0	173.2
Feb	183.8	182.0	177.7	170.0	189.3	184.9	170.1	230.0	200.0	131.6	177.6	98.0	173.9
Mar	184.6	182.5	178.1	170.4	190.2	185.7	170.2	231.2	201.9	132.1	177.4	98.5	174.3
Apr	185.7	183.6	179.1	170.8	191.6	186.9	170.2	233.4	204.5	132.2	177.8	98.2	174.9
May	186.5	184.3	179.7	171.4	192.4	187.6	170.6	233.8	205.8	132.3	178.5	98.6	175.6
Jun	186.8	184.2	179.5	171.2	193.0	188.1	169.8	234.2	207.4	131.7	178.5	98.5	175.6
Jul	186.8	183.8	178.9	170.5	193.1	188.2	169.2	234.7	208.3	129.0	178.8	95.6	175.1
Aug	187.4	184.3	179.3	170.9	193.8	188.8	169.1	235.2	209.3	130.1	179.1	96.4	175.7
Sep	188.1	184.7	179.4	171.1	194.6	189.5	169.3	235.3	211.3	132.0	178.1	97.7	176.1
Oct	188.6	185.1	179.8	171.3	195.1	189.9	169.9	235.5	212.5	132.2	177.9	97.2	176.6
Nov	189.0	185.4	180.1	171.6	195.5	190.3	170.4	235.0	213.4	132.6	177.9	97.6	176.9
Dec	189.9	186.4	180.9	172.5	196.4	191.2	171.2	234.7	215.6	131.8	178.6	99.1	177.9
2005 Jan	188.9	185.2	179.8	171.2	195.2	190.1	171.6	236.0	214.3	129.4	177.1	94.5	176.7
Feb	189.6	185.9	180.4	171.9	195.9	190.8	172.4	236.9	214.9	130.2	177.6	95.0	177.4
Mar	190.5	186.8	181.4	173.0	196.8	191.6	173.4	236.8	216.3	131.4	178.1	96.7	178.3
Apr	191.6	187.8	182.4	173.3	198.2	192.9	172.7	239.4	218.3	131.4	179.3	95.6	179.0
May	192.0	188.2	182.7	173.7	198.6	193.2	173.7	240.2	219.0	131.4	179.1	95.9	179.4
Jun	192.2	188.3	182.8	173.8	198.8	193.4	173.6	240.5	219.7	131.4	178.9	95.8	179.5
Jul	192.2	188.3	182.7	173.5	199.1	193.7	172.4	241.0	220.2	128.8	180.2	94.0	179.5
Aug	192.6	188.6	183.0	173.8	199.5	194.1	172.7	241.0	220.5	130.3	180.2	94.1	179.8
Sep	193.1	189.3	183.7	174.6	200.0	194.5	172.7	241.1	220.7	131.6	181.0	95.1	180.5
Oct	193.3	189.5	183.8	174.7	200.4	194.8	172.7	241.8	221.4	131.8	180.8	94.7	180.7
Nov	193.6	189.7	184.0	174.9	200.5	195.0	173.4	241.9	222.5	132.2	179.6	95.4	180.9
Dec	194.1	190.2	184.5	175.5	201.0	195.5	174.1	241.6	224.5	131.9	179.0	97.0	181.5
2006 Jan	193.4	189.4	183.7	174.5	200.3	194.8	174.1	242.5	223.0	129.1	179.4	92.4	180.7
Feb	194.2	190.1	184.4	175.2	201.0	195.6	174.9	242.8	224.0	130.0	179.9	93.5	181.4
Mar	195.0	190.8	185.2	176.0	202.0	196.4	174.3	243.8	225.8	131.1	180.0	95.1	182.2
Apr	196.5	192.3	186.7	177.0	203.8	198.0	174.2	245.8	228.3	131.7	181.6	93.6	183.2
May	197.7	193.6	187.8	178.2	204.9	199.1	176.1	246.8	230.0	132.7	182.1	94.3	184.5
Jun	198.5	194.2	188.4	178.9	205.7	199.8	176.8	248.3	231.6	132.6	181.9	94.7	185.2
Jul	198.5	194.2	188.3	178.7	205.6	199.9	177.1	248.3	231.5	129.4	183.3	91.8	185.2
Aug	199.2	194.9	188.9	179.3	206.4	200.7	177.6	249.1	232.6	131.3	183.3	93.0	186.0
Sep	200.1	195.3	189.2	179.6	207.4	201.5	178.1	249.2	235.9	133.0	181.2	94.8	186.4
Oct	200.4	195.5	189.3	179.7	207.5	201.7	179.1	249.7	237.3	133.4	179.6	93.7	186.7
Nov	201.1	196.2	190.0	180.4	208.2	202.4	180.2	249.6	238.7	133.7	179.8	94.5	187.5
Dec	202.7	197.4	191.2	181.7	210.1	204.1	180.6	249.4	242.7	132.9	181.0	96.7	188.6

Note: Further information on the RPI is available from the National Statistics Website: www.statistics.gov.uk/rpi.

1 This series has been constructed using the index for all items excluding mortgage interest payments prior to February 1995.

2 Seasonal food is defined as items of food the prices of which show significant seasonal variations. These are fresh fruit and vegetables, fresh fish, eggs and home-killed lamb.

3 There are no weights available for RPIY.

Source: Office for National Statistics: 020 7533 5874

18.4 Retail Prices Index[1]
Detailed figures for various groups, sub-groups and sections

13 January 1987=100

		Group and sub-group weights in 2006	2006 Jan	2006 Feb	2006 Mar	2006 Apr	2006 May	2006 Jun	2006 Jul	2006 Aug	2006 Sep	2006 Oct	2006 Nov	2006 Dec
ALL ITEMS (RPI)	CHAW	1000	193.4	194.2	195.0	196.5	197.7	198.5	198.5	199.2	200.1	200.4	201.1	202.7
All items excluding:														
mortgage interest payments (RPIX)	CHMK	950	189.4	190.1	190.8	192.3	193.6	194.2	194.2	194.9	195.3	195.5	196.2	197.4
mortgage interest payments and depreciation	CHON	906	183.7	184.4	185.2	186.7	187.8	188.4	188.3	188.9	189.2	189.3	190.0	191.2
mortgage interest payments and council tax	DQAD	911	186.7	187.4	188.2	189.3	190.6	191.3	191.2	192.0	192.4	192.6	193.3	194.5
housing	CHAZ	778	174.5	175.2	176.0	177.0	178.2	178.9	178.7	179.3	179.6	179.7	180.4	181.7
food	CHAY	895	200.3	201.0	202.0	203.8	204.9	205.7	205.6	206.4	207.4	207.5	208.2	210.1
seasonal food	CHAX	983	194.8	195.6	196.4	198.0	199.1	199.8	199.9	200.7	201.5	201.7	202.4	204.1
All items excluding mortgage interest payments and indirect taxes (RPIY)[2]	CBZW		180.7	181.4	182.2	183.2	184.5	185.2	185.2	186.0	186.4	186.7	187.5	188.6
Food	CHBA	105	154.2	155.2	154.4	154.1	156.3	157.1	157.3	157.7	158.4	159.3	160.6	160.8
Bread	DOAA	4	159.7	162.9	162.2	163.7	163.3	162.9	163.6	165.1	166.8	166.8	169.4	170.5
Cereals	DOAB	3	145.5	144.5	144.1	143.5	142.5	143.7	144.3	144.0	144.9	143.7	144.8	145.8
Biscuits and cakes	DOAC	6	165.1	166.7	167.8	167.4	168.9	169.0	168.8	169.5	169.0	169.0	170.4	171.4
Beef	DOAD	4	133.3	134.2	132.0	134.3	137.1	137.3	142.1	140.1	141.7	143.3	142.4	141.7
Lamb	DOAE	2	180.8	182.6	180.7	183.9	188.9	190.0	190.0	189.9	189.6	188.9	188.3	190.4
of which home-killed lamb	DOAF	1	181.5	183.9	180.9	185.0	193.7	196.7	193.5	195.3	191.8	189.5	191.3	194.4
Pork	DOAG	1	158.5	159.0	155.8	155.2	157.1	157.7	159.9	158.9	155.6	159.7	156.1	156.8
Bacon	DOAH	2	177.3	175.2	175.5	174.7	177.2	180.4	183.2	185.1	184.6	187.1	189.8	190.0
Poultry	DOAI	4	109.1	110.3	108.2	108.5	108.7	109.9	111.4	110.5	106.8	109.6	108.5	109.8
Other meat	DOAJ	7	145.7	146.4	144.8	144.6	146.1	145.9	146.4	148.0	148.0	147.9	148.4	149.8
Fish	DOAK	4	158.8	159.1	159.0	160.5	159.2	162.4	164.8	166.9	167.6	169.1	168.9	169.9
of which fresh fish	DOAL	2	164.7	167.6	166.5	167.5	167.2	172.5	176.8	178.6	180.6	183.0	182.1	180.6
Butter	DOAM	1	171.6	171.0	169.9	168.1	168.6	169.0	169.6	173.8	173.5	173.7	173.8	174.1
Oils and fats	DOAN	1	130.1	130.6	135.4	138.6	143.7	144.9	144.9	145.6	143.3	144.8	145.8	145.7
Cheese	DOAO	3	176.8	175.1	175.0	175.7	176.0	176.0	175.1	176.4	175.9	175.7	176.0	176.4
Eggs	DOAP	1	160.7	157.8	162.4	158.2	158.9	162.0	162.3	166.8	168.1	169.6	170.5	179.0
Milk, fresh	DOAQ	5	186.8	191.1	178.0	174.9	173.2	178.2	187.9	188.4	188.2	188.9	188.8	189.0
Milk products	DOAR	4	143.6	144.2	144.4	144.1	146.0	145.5	147.2	147.5	147.8	147.1	147.8	147.1
Tea	DOAS	1	145.5	149.9	145.6	146.9	154.6	158.7	157.9	161.1	162.4	162.5	161.4	161.5
Coffee and other hot drinks	DOAT	1	116.9	117.0	115.6	116.4	116.5	116.0	116.1	123.3	125.0	125.2	123.6	125.9
Soft drinks	DOAU	10	188.5	190.1	190.2	191.2	192.1	192.3	192.2	191.0	193.6	193.9	193.7	193.9
Sugar and preserves	DOAV	1	155.9	156.3	155.3	154.8	156.1	156.3	156.4	156.8	158.4	158.5	159.9	160.6
Sweets and chocolates	DOAW	10	186.5	187.4	188.3	190.7	192.3	192.7	192.1	190.8	193.9	193.3	193.1	193.3
Potatoes	DOAX	4	149.7	152.1	152.3	151.5	153.2	156.9	153.7	151.0	151.7	154.7	161.1	162.0
of which unprocessed potatoes	DOAY	1	136.0	137.3	138.6	139.2	147.1	151.0	149.4	138.7	141.4	144.3	150.7	152.0
Vegetables other than potatoes	DOAZ	8	132.8	135.3	135.0	129.8	137.5	138.8	137.5	135.0	136.3	138.8	144.1	142.4
of which fresh vegetables	DOBA	6	119.8	122.9	122.5	116.4	125.5	126.9	125.4	121.9	122.4	124.1	129.8	127.6
Fruit	DOBB	7	142.7	142.7	142.1	140.5	149.0	148.5	142.3	144.7	150.1	152.0	158.2	157.1
of which fresh fruit	DOBC	6	139.1	139.0	138.2	136.3	146.1	145.5	138.6	141.5	147.4	148.9	156.6	155.3
Other foods	DOBD	11	150.0	149.7	149.5	149.3	150.4	150.4	150.0	151.3	150.7	152.5	151.7	151.6
Catering	CHBC	50	242.7	243.0	243.4	243.7	244.7	245.3	245.9	246.4	246.6	247.8	248.5	249.2
Restaurant meals	DOBE	27	238.0	238.0	238.5	238.7	239.7	240.2	240.7	241.1	240.8	242.1	242.8	243.6
Canteen meals	DOBF	4	296.9	296.4	297.2	297.8	298.2	298.2	298.5	298.3	301.6	302.7	305.0	304.7
Take-aways and snacks	DOBG	19	231.8	232.5	232.7	233.2	234.3	234.9	235.8	236.5	237.0	237.9	238.3	239.1
Alcoholic drink	CHBD	67	209.3	209.7	210.9	211.7	212.6	213.2	213.2	214.0	214.1	214.7	214.5	214.3
Beer	DOBH	36	227.4	228.0	229.2	230.5	232.1	232.2	232.7	233.1	233.3	234.0	234.6	235.1
Beer on sales	DOBI	31	245.8	246.8	248.2	249.4	251.1	251.5	252.1	252.6	252.8	253.5	254.3	255.0
Beer off sales	DOBJ	5	146.8	145.8	146.4	148.1	148.8	147.7	147.8	147.8	148.1	149.1	148.4	148.4
Wines and spirits	DOBK	31	185.4	185.5	186.6	186.9	187.2	188.2	187.7	189.0	189.0	189.4	188.4	187.6
Wines and spirits on sales	DOBL	17	230.9	231.9	232.9	233.9	235.2	235.5	235.8	236.6	237.1	237.7	238.3	238.8
Wines and spirits off sales	DOBM	14	156.0	155.6	156.8	156.5	155.9	157.6	156.4	158.2	157.8	157.9	155.8	153.6
Tobacco	CHBE	29	334.7	334.6	334.8	340.9	342.2	346.9	347.1	347.5	347.7	347.9	348.0	348.1
Cigarettes	DOBN	26	342.1	341.9	342.1	348.5	349.7	354.5	354.7	355.2	355.4	355.6	355.6	355.8
Other tobacco	DOBO	3	266.6	266.9	267.0	271.1	272.1	276.5	276.5	276.7	276.7	276.9	276.8	276.9

18.4 Retail Prices Index[1]
Detailed figures for various groups, sub-groups and sections

continued

13 January 1987 = 100

		Group and sub-group weights in 2006	2006 Jan	2006 Feb	2006 Mar	2006 Apr	2006 May	2006 Jun	2006 Jul	2006 Aug	2006 Sep	2006 Oct	2006 Nov	2006 Dec
Housing	CHBF	222	292.4	293.2	293.7	298.7	299.8	300.7	302.1	303.6	307.8	308.9	310.0	313.8
Rent	DOBP	45	273.6	273.8	273.8	278.0	278.3	278.5	280.5	280.6	280.8	280.8	281.3	281.5
Mortgage interest payments	DOBQ	50	298.2	300.1	302.4	304.3	306.3	308.1	310.1	312.5	327.8	330.3	332.8	347.6
Depreciation (Jan 1995 = 100)	CHOO	44	261.3	262.5	261.9	263.4	265.8	267.9	269.8	273.0	276.2	278.3	279.5	281.1
Council tax and Rates	DOBR	39	268.1	268.1	268.1	280.6	280.7	280.7	280.7	280.7	280.7	280.7	280.7	280.7
Water and other charges	DOBS	12	335.7	335.7	335.7	354.1	354.1	354.1	354.1	354.1	354.1	354.1	354.1	354.1
Repairs and maintenance charges	DOBT	12	307.9	308.3	308.7	310.5	311.3	311.9	313.4	314.3	315.4	317.1	317.6	318.9
Do-it-yourself materials	DOBU	13	157.9	158.0	158.4	158.5	157.9	158.2	158.4	159.7	161.0	160.7	161.1	162.4
Dwelling insurance and ground rent	DOBV	7	275.8	280.3	280.3	280.7	281.4	278.6	278.5	280.0	280.5	279.2	280.1	279.5
Fuel and light	CHBG	33	173.3	174.9	179.5	188.4	197.2	201.8	204.1	205.6	208.2	214.5	219.0	222.7
Coal and solid fuels	DOBW	1	185.0	185.2	185.2	185.7	187.6	183.9	183.7	185.3	187.6	197.3	199.5	199.5
Electricity	DOBX	15	162.6	164.5	168.8	176.1	183.5	187.9	189.9	191.3	193.7	198.8	202.6	205.0
Gas	DOBY	14	174.3	175.1	180.8	192.5	204.4	210.4	212.6	215.1	220.3	231.1	238.7	243.3
Oil and other fuels	DOBZ	3	261.6	266.2	268.5	275.4	277.0	276.6	285.4	280.0	267.3	249.3	240.0	249.4
Household goods	CHBH	71	142.9	144.7	148.4	145.1	146.5	148.4	145.2	146.2	148.9	146.1	147.9	154.6
Furniture	DOCA	26	158.5	159.7	171.6	162.1	164.5	169.9	162.4	163.6	169.4	164.0	168.5	187.7
Furnishings	DOCB	11	153.8	158.0	158.4	157.8	160.4	159.6	155.9	159.1	158.2	158.1	158.4	162.5
Electrical appliances	DOCC	8	75.9	77.6	75.3	74.1	74.5	74.4	75.6	76.1	79.0	74.7	74.2	73.3
Other household equipment	DOCD	5	138.0	139.7	140.3	140.8	141.6	142.5	137.4	140.0	141.5	140.3	142.5	141.8
Household consumables	DOCE	14	157.6	158.6	159.4	159.8	159.8	160.8	159.8	159.6	160.2	160.1	160.8	161.2
Pet care	DOCF	7	166.8	167.3	167.6	168.9	169.9	170.0	170.5	169.6	171.2	171.4	171.4	171.8
Household services	CHBI	66	187.8	187.5	187.9	189.0	188.4	189.3	189.5	189.5	191.5	196.6	196.2	196.3
Postage	DOCG	1	177.0	177.0	177.0	188.9	188.9	188.9	188.9	188.9	200.2	200.2	200.2	200.2
Telephones, telemessages, etc	DOCH	24	89.7	89.7	89.7	89.7	88.4	88.6	88.6	88.5	87.9	87.8	88.6	88.2
Domestic services	DOCI	12	283.6	284.2	284.7	286.1	287.3	289.1	290.9	291.9	293.7	295.0	295.7	296.6
Fees and subscriptions	DOCJ	29	266.4	265.0	266.2	268.6	269.1	270.9	271.2	272.2	277.8	291.8	290.7	291.1
Clothing and footwear	CHBJ	49	92.6	93.3	94.6	94.8	95.7	95.3	91.5	93.6	95.5	95.9	96.3	95.0
Men's outerwear	DOCK	10	93.8	95.4	96.5	96.7	97.5	96.6	93.5	94.8	97.8	97.5	98.2	97.5
Women's outerwear	DOCL	17	67.8	68.4	70.0	70.1	70.8	70.4	65.2	67.8	69.8	70.6	70.6	69.0
Children's outerwear	DOCM	6	86.9	85.8	86.5	87.6	89.3	89.4	88.0	89.4	89.0	89.7	90.5	90.8
Other clothing	DOCN	7	146.7	147.2	147.7	147.5	148.4	148.3	146.1	148.9	150.4	149.6	151.1	150.9
Footwear	DOCO	9	106.8	108.1	108.7	108.8	109.7	109.3	107.5	108.3	110.3	110.3	110.4	108.3
Personal goods and services	CHBQ	41	204.8	206.1	206.6	208.3	209.3	210.0	208.6	209.9	210.8	211.2	211.3	211.9
Personal articles	DOCP	12	129.9	132.3	133.2	133.4	134.6	135.9	134.0	135.8	137.2	136.9	137.0	138.1
Chemists goods	DOCQ	16	189.1	189.4	189.3	190.8	190.7	190.6	189.4	190.2	190.7	191.3	191.1	190.8
Personal services	DOCR	13	347.0	347.4	347.9	353.3	356.0	356.6	356.6	357.4	357.6	359.1	360.0	360.7
Motoring expenditure	CHBK	140	185.5	185.1	185.8	188.1	189.5	189.2	190.5	190.2	186.6	183.5	183.8	184.6
Purchase of motor vehicles	DOCS	56	107.7	107.4	107.1	106.9	106.8	106.4	106.1	105.6	105.5	105.2	104.9	104.5
Maintenance of motor vehicles	DOCT	20	288.2	288.6	289.6	290.4	291.5	292.9	294.4	295.2	296.4	298.0	300.3	301.4
Petrol and oil	DOCU	40	261.5	263.1	262.9	273.1	282.1	279.6	284.7	286.2	268.6	254.4	253.2	259.0
Vehicle tax and insurance	DOCV	24	279.3	275.3	282.0	285.0	280.8	283.5	286.9	283.7	283.3	282.3	286.9	285.3
Fares and other travel costs	CHBR	19	225.3	225.4	225.3	230.2	227.6	228.9	238.4	240.2	229.0	228.1	226.1	234.4
Rail fares	DOCW	5	245.1	248.0	247.9	252.0	251.3	250.5	251.8	254.1	248.0	247.6	250.9	249.4
Bus and coach fares	DOCX	4	267.8	267.3	268.8	256.8	253.4	253.6	254.0	254.8	257.3	259.3	261.4	262.3
Other travel costs	DOCY	10	193.4	192.6	192.0	201.8	198.9	201.3	216.2	218.0	201.3	199.4	194.2	208.2
Leisure goods	CHBL	41	92.6	94.2	93.4	92.4	91.9	91.9	91.5	91.6	91.9	91.2	91.6	92.2
Audio-visual equipment	DOCZ	9	18.4	18.6	18.0	17.7	17.3	17.2	17.1	17.2	17.2	16.8	16.9	16.9
CDs and tapes	DODA	4	102.9	99.2	100.9	99.8	100.9	100.1	99.3	96.4	97.3	97.4	99.2	101.8
Toys, photographic and sports goods	DODB	12	90.3	93.0	91.9	90.8	90.3	90.0	89.1	90.1	89.7	89.6	89.1	89.4
Books and newspapers	DODC	10	238.2	247.0	246.6	245.9	246.4	247.4	249.4	249.0	252.6	251.9	253.2	254.5
Gardening products	DODD	6	144.9	146.4	146.2	143.8	143.9	145.4	144.1	143.9	143.6	142.7	144.2	145.3
Leisure services	CHBM	67	261.9	262.7	262.8	265.8	266.2	265.8	267.3	267.4	269.1	270.5	270.8	271.4
Television licences and rentals	DODE	12	161.7	161.7	161.7	164.0	164.0	164.1	164.7	164.7	164.7	166.1	166.1	166.2
Entertainment and other recreation	DODF	17	342.1	343.1	342.6	349.9	348.6	348.7	351.4	348.4	358.7	361.3	360.1	362.3
Foreign holidays (Jan 1993 = 100)	CHMQ	30	166.3	166.7	166.8	168.0	168.5	167.9	168.7	169.3	169.1	169.5	170.3	170.4
UK holidays (Jan 1994 = 100)	CHMS	8	159.5	161.3	161.8	162.4	163.3	163.6	164.9	165.8	165.5	166.2	166.3	166.8

Note: Indices are given to one decimal place to provide as much information as is available but precision is greater at higher levels of aggregation, ie at sub-group and group levels. Further information on the RPI is available from the National Statistics Website: www.statistics.gov.uk/rpi.

2 The taxes excluded are council tax, VAT, duties, vehicle excise duty, insurance tax and airport tax. There are no weights available for RPIY.

Source: Office for National Statistics: 020 7533 5874

1 *Retail Prices Index 1914-1990* contains group and sub-group indices and weights back to 1956, group indices back to 1947, together with cost of living indices as far back as 1914.

18.5 Retail Prices Index (All Items)

	Annual average	Jan	Feb	Mar	Apr	May	Jun	Jul	Aug	Sep	Oct	Nov	Dec
January 1962=100													
1962	101.6	100.0	100.1	100.5	101.9	102.2	102.9	102.5	101.6	101.5	101.4	101.8	102.3
1963	103.6	102.7	103.6	103.7	104.0	103.9	103.9	103.3	103.0	103.3	103.7	104.0	104.2
1964	107.0	104.7	104.8	105.2	106.1	107.0	107.4	107.4	107.8	107.8	107.9	108.8	109.2
1965	112.1	109.5	109.5	109.9	112.0	112.4	112.7	112.7	112.9	113.0	113.1	113.6	114.1
1966	116.5	114.3	114.4	114.6	116.0	116.8	117.1	116.6	117.3	117.1	117.4	118.1	118.3
1967	119.4	118.5	118.6	118.6	119.5	119.4	119.9	119.2	118.9	118.8	119.7	120.4	121.2
1968	125.0	121.6	122.2	122.6	124.8	124.9	125.4	125.5	125.7	125.8	126.4	126.7	128.4
1969	131.8	129.1	129.8	130.3	131.7	131.5	132.1	132.1	131.8	132.2	133.2	133.5	134.4
1970	140.2	135.5	136.2	137.0	139.1	139.5	139.9	140.9	140.8	141.5	143.0	144.0	145.0
1971	153.4	147.0	147.8	149.0	152.2	153.2	154.3	155.2	155.3	155.5	156.4	157.3	158.1
1972	164.3	159.0	159.8	160.3	161.8	162.6	163.7	164.2	165.5	166.4	168.7	169.3	170.2
1973	179.4	171.3	172.4	173.4	176.7	178.0	178.9	179.7	180.2	181.8	185.4	186.8	188.2
1974	..	191.8	..	..	..	..	..	..	..	..	..	..	..
January 1974=100													
1974	108.5	100.0	101.7	102.6	106.1	107.6	108.7	109.7	109.8	111.0	113.2	115.2	116.9
1975	134.8	119.9	121.9	124.3	129.1	134.5	137.1	138.5	139.3	140.5	142.5	144.2	146.0
1976	157.1	147.9	149.8	150.6	153.5	155.2	156.0	156.3	158.5	160.6	163.5	165.8	168.0
1977	182.0	172.4	174.1	175.8	180.3	181.7	183.6	183.8	184.7	185.7	186.5	187.4	188.4
1978	197.1	189.5	190.6	191.8	194.6	195.7	197.2	198.1	199.4	200.2	201.1	202.5	204.2
1979	223.5	207.2	208.9	210.6	214.2	215.9	219.6	229.1	230.9	233.2	235.6	237.7	239.4
1980	263.7	245.3	248.8	252.2	260.8	263.2	265.7	267.9	268.5	270.2	271.9	274.1	275.6
1981	295.0	277.3	279.8	284.0	292.2	294.1	295.8	297.1	299.3	301.0	303.7	306.9	308.8
1982	320.4	310.6	310.7	313.4	319.7	322.0	322.9	323.0	323.1	322.9	324.5	326.1	325.5
1983	335.1	325.9	327.3	327.9	332.5	333.9	334.7	336.5	338.0	339.5	340.7	341.9	342.8
1984	351.8	342.6	344.0	345.1	349.7	351.0	351.9	351.5	354.8	355.5	357.7	358.8	358.5
1985	373.2	359.8	362.7	366.1	373.9	375.6	376.4	375.7	376.7	376.5	377.1	378.4	378.9
1986	385.9	379.7	381.1	381.6	385.3	386.0	385.8	384.7	385.9	387.8	388.4	391.7	393.0
1987	..	394.5	..	..	..	..	..	..	..	..	..	..	..
January 1987=100													
1990	126.1	119.5	120.2	121.4	125.1	126.2	126.7	126.8	128.1	129.3	130.3	130.0	129.9
1991	133.5	130.2	130.9	131.4	133.1	133.5	134.1	133.8	134.1	134.6	135.1	135.6	135.7
1992	138.5	135.6	136.3	136.7	138.8	139.3	139.3	138.8	138.9	139.4	139.9	139.7	139.2
1993	140.7	137.9	138.8	139.3	140.6	141.1	141.0	140.7	141.3	141.9	141.8	141.6	141.9
1994	144.1	141.3	142.1	142.5	144.2	144.7	144.7	144.0	144.7	145.0	145.2	145.3	146.0
1995	149.1	146.0	146.9	147.5	149.0	149.6	149.8	149.1	149.9	150.6	149.8	149.8	150.7
1996	152.7	150.2	150.9	151.5	152.6	152.9	153.0	152.4	153.1	153.8	153.8	153.9	154.4
1997	157.5	154.4	155.0	155.4	156.3	156.9	157.5	157.5	158.5	159.3	159.5	159.6	160.0
1998	162.9	159.5	160.3	160.8	162.6	163.5	163.4	163.0	163.7	164.4	164.5	164.4	164.4
1999	165.4	163.4	163.7	164.1	165.2	165.6	165.6	165.1	165.5	166.2	166.5	166.7	167.3
2000	170.3	166.6	167.5	168.4	170.1	170.7	171.1	170.5	170.5	171.7	171.6	172.1	172.2
2001	173.3	171.1	172.0	172.2	173.1	174.2	174.4	173.3	174.0	174.6	174.3	173.6	173.4
2002	176.2	173.3	173.8	174.5	175.7	176.2	176.2	175.9	176.4	177.6	177.9	178.2	178.5
2003	181.3	178.4	179.3	179.9	181.2	181.5	181.3	181.3	181.6	182.5	182.6	182.7	183.5
2004	186.7	183.1	183.8	184.6	185.7	186.5	186.8	186.8	187.4	188.1	188.6	189.0	189.9
2005	192.0	188.9	189.6	190.5	191.6	192.0	192.2	192.2	192.6	193.1	193.3	193.6	194.1
2006	198.1	193.4	194.2	195.0	196.5	197.7	198.5	198.5	199.2	200.1	200.4	201.1	202.7

Note: Further information on the RPI is available from the National Statistics Website: www.statistics.gov.uk/rpi.

Source: Office for National Statistics: 020 7533 5874

18.6 Harmonised Indices of Consumer Prices (HICPs) - International comparisons : EU countries

percentage changes over 12 months

Per cent

		2004	2005	2006	2005 Dec	2006 Jan	2006 Feb	2006 Mar	2006 Apr	2006 May	2006 Jun	2006 Jul	2006 Aug	2006 Sep	2006 Oct	2006 Nov	2006 Dec
European Union countries																	
United Kingdom[1]	D7G7	1.3	2.1	2.3	1.9	1.9	2.0	1.8	2.0	2.2	2.5	2.4	2.5	2.4	2.4	2.7	3.0
Austria	D7SK	2.0	2.1	..	1.6	1.5	1.5	1.3	2.1	2.1	1.9	2.0	2.1	1.3	1.3†	1.6	..
Belgium	D7SL	1.9	2.5	..	2.8	2.8	2.8	2.2	2.6	2.8	2.5	2.4	2.3	1.9	1.7	2.0	..
Cyprus	D7RO	1.9	2.0	..	1.4	2.0	2.3	2.6	2.5	2.5	2.6	2.8	2.7	2.2	1.7	1.3	..
Czech Republic	D7RP	2.6	1.6	..	1.9	2.4	2.4	2.4	2.3	2.8	2.3	2.4	2.6	2.2	0.8	1.0	..
Denmark	D7SM	0.9	1.7	..	2.2	2.0	2.1	1.8	1.8	2.1	2.1	2.0	1.9	1.5	1.4	1.8	..
Estonia	D7RQ	3.0	4.1	..	3.6	4.7	4.5	4.0	4.3	4.6	4.4	4.5	5.0	3.8	3.8	4.7	..
Finland	D7SN	0.1	0.8	..	1.1	1.2	1.3	1.2	1.5	1.7	1.5	1.4	1.3	0.8	0.9	1.3	..
France	D7SO	2.3	1.9	..	1.8	2.3	2.0	1.7	2.0	2.4	2.2	2.2	2.1	1.5	1.2	1.6	..
Germany	D7SP	1.8	1.9	..	2.1	2.1	2.1	1.9	2.3	2.1	2.0	2.1	1.8	1.0	1.1	1.5	..
Greece	D7SQ	3.0	3.5	..	3.5	3.0	3.1	3.3	3.5	3.3	3.4	3.9	3.4	3.1	3.1	3.2	..
Hungary	D7RR	6.8	3.5	..	3.3	2.5	2.3	2.4	2.4	2.9	2.9	3.2	4.7	5.9	6.3	6.4	..
Ireland	D7SS	2.3	2.2	..	1.9	2.5	2.7	2.8	2.7	3.0	2.9	2.9	3.2	2.2	2.2	2.4	..
Italy	D7ST	2.3	2.2	..	2.1	2.2	2.2	2.2	2.3	2.3	2.4	2.3	2.3	2.4	1.9	2.0	..
Latvia	D7RS	6.2	6.9	..	7.1	7.6	7.0	6.6	6.1	7.1	6.3	6.9	6.8	5.9	5.6	6.3	..
Lithuania	D7RT	1.2	2.7	..	3.0	3.5	3.4	3.1	3.4	3.6	3.7	4.4	4.3	3.3	3.7	4.4	..
Luxembourg	D7SU	3.2	3.8	..	3.4	4.1	3.9	3.7	3.5	3.6	3.9	3.4	3.1	2.0	0.6	1.8	..
Malta	D7RU	2.7	2.5	..	3.4	2.4	2.3	2.9	3.5	3.5	3.3	3.6	3.0	3.1	1.7	0.9	..
Netherlands	D7SV	1.4	1.5	..	2.0	1.8	1.4	1.4	1.8	1.8	1.8	1.7	1.9	1.5	1.3	1.6	..
Poland	D7RV	3.6	2.2	..	0.8	0.9	0.9	0.9	1.2	1.5	1.5	1.4	1.7	1.4	1.1	1.3	..
Portugal	D7SX	2.5	2.1	..	2.5	2.7	3.0	3.8	3.7	3.7	3.5	3.0	2.7	3.0	2.6	2.4	..
Slovakia	D7RW	7.5	2.8	..	3.9	4.1	4.3	4.3	4.4	4.8	4.5	5.0	5.0	4.5	3.1	3.7	..
Slovenia	D7RX	3.7	2.5	..	2.4	2.6	2.3	2.0	2.8	3.4	3.0	1.9	3.1	2.5	1.5	2.4	..
Spain	D7SY	3.1	3.4	..	3.7	4.2	4.1	3.9	3.9	4.1	4.0	4.0	3.8	2.9	2.6	2.7	..
Sweden	D7SZ	1.0	0.8	..	1.3	1.1	1.1	1.5	1.8	1.9	1.9	1.8	1.6	1.2	1.2	1.5	..
EICP[2] EU 15 average	CLNX	..	..	..	..	..	..	..	..	..	..	..	..	..	..	..	..
EICP[2] EU 25 average[3]	D7RY	2.0	2.2	..	2.1	2.3	2.2	2.1	2.3	2.4	2.4	2.4	2.3	1.9	1.8	2.1	..

Note: Further information on HICP is available from the National Statistics Website: www.statistics.gov.uk/hicp.

1 Published as the Consumer Prices Index (CPI) in the UK. (UK 2005=100, others 1996=100)
2 The EICP (European Index of Consumer Prices) is the official EU aggregate It covers 15 member states until April 2004 and 25 member states from May 2004, the new member states being integrated using a chain index

formula. The EU 25 annual average for 2004 is calculated from the EU 15 average from January to April and the EU 25 average from May to December.
3 The coverage of the European Union was extended to include Cyprus, Czech Republic, Estonia, Hungary, Latvia, Lithuania, Malta, Poland, Slovakia and Slovenia with effect from 1 May 2004. Data for the EU 25 average is only available from May 2004. Note: April 2006 data for 'Austria' and 'EU 25 average' is Provisional.

Source: Statistical Office of the European Communities (Eurostat)

18.7 Internal purchasing power of the pound (based on RPI)[1]

Pence

	Year in which purchasing power was 100p																			
	1986	1987	1988	1989	1990	1991	1992	1993	1994	1995	1996	1997	1998	1999	2000	2001	2002	2003	2004	2005
	BAMS	BAMT	BAMU	BAMV	BAMW	BASX	CZVM	CBXX	DOFX	DOHR	DOLM	DTUL	CDQG	JKZZ	ZMHO	IKHI	FAUI	SEZH	C687	E9AO
1986	100	104	109	118	129	136	142	144	147	152	156	161	167	169	174	177	180	185	191	196
1987	96	100	105	113	124	131	136	138	141	146	150	155	160	162	167	170	173	178	183	188
1988	92	95	100	108	118	125	130	132	135	139	143	147	152	155	159	162	165	170	175	180
1989	85	88	93	100	109	116	120	122	125	129	133	137	141	144	148	150	153	157	162	167
1990	78	81	85	91	100	106	110	112	114	118	121	125	129	131	135	137	140	144	148	152
1991	73	76	80	86	94	100	104	105	108	112	114	118	122	124	128	130	132	136	140	144
1992	71	74	77	83	91	96	100	102	104	108	110	114	118	119	123	125	127	131	135	139
1993	70	72	76	82	90	95	98	100	102	106	109	112	116	118	121	123	125	129	133	136
1994	68	71	74	80	88	93	96	98	100	103	106	109	113	115	118	120	122	126	130	133
1995	66	68	72	77	85	90	93	94	97	100	102	106	109	111	114	116	118	122	125	129
1996	64	67	70	75	83	87	91	92	94	98	100	103	107	108	112	113	115	119	122	126
1997	62	65	68	73	80	85	88	89	92	95	97	100	103	105	108	110	112	115	119	122
1998	60	63	66	71	77	82	85	86	88	92	94	97	100	102	105	106	108	111	115	118
1999	59	62	65	70	76	81	84	85	87	90	92	95	98	100	103	105	107	110	113	116
2000	57	60	63	68	74	78	81	83	85	88	90	92	96	97	100	102	103	106	110	113
2001	56	59	62	66	73	77	80	81	83	86	88	91	94	95	98	100	102	105	108	111
2002	56	58	61	65	72	76	79	80	82	85	87	89	92	94	97	98	100	103	106	109
2003	54	56	59	64	70	74	76	78	79	82	84	87	90	91	94	96	97	100	103	106
2004	52	55	57	62	68	72	74	75	77	80	82	84	87	89	91	93	94	97	100	103
2005	51	53	56	60	66	70	72	73	75	78	80	82	85	86	89	90	92	94	97	100

Note: Further information on the RPI is available from the National Statistics Website: www.statistics.gov.uk/rpi.

Source: Office for National Statistics: 020 7533 5874

1 To find the purchasing power of the pound in 2000, given that it was 100 pence in 1990, select the column headed 1990 and look at the 2000 row. The result is 74 pence. These figures are calculated by taking the inverse ratio of the respective annual averages of the Retail Prices Index (RPI).

18.8 Tax and Price Index

| | | | | | | **Tax and Price Index: January 1987 = 100** | | | | | | | | | |
| | | | | | | **DQAB** | | | | | | | | | |
	1993	1994	1995	1996	1997	1998	1999	2000	2001	2002	2003	2004	2005	2006	2007
January	128.7	132.1	137.2	141.6	143.6	147.1	150.5	152.7	156.7	156.5	161.4	166.9	172.1	175.9	..
February	129.6	132.9	138.2	142.3	144.2	147.9	150.8	153.7	157.6	157.0	162.3	167.6	172.8	176.7	..
March	130.2	133.4	138.8	143.0	144.6	148.4	151.2	154.6	157.8	157.7	163.0	168.4	173.7	177.4	..
April	131.3	135.3	140.3	141.7	143.8	149.7	151.2	155.7	156.3	158.6	164.9	168.9	174.1	178.3	..
May	131.8	135.8	141.0	142.0	144.4	150.6	151.7	156.3	157.4	159.1	165.2	169.7	174.5	179.5	..
June	131.7	135.8	141.2	142.1	145.0	150.5	151.7	156.7	157.6	159.1	165.0	170.0	174.7	180.3	..
July	131.4	135.1	140.4	141.5	145.0	150.1	151.1	156.1	156.5	158.8	165.0	170.0	174.7	180.3	..
August	132.1	135.8	141.3	142.2	146.0	150.8	151.5	156.1	157.2	159.3	165.4	170.6	175.1	181.0	..
September	132.7	136.1	142.0	143.0	146.9	151.5	152.3	157.3	157.8	160.6	166.3	171.3	175.6	181.9	..
October	132.6	136.4	141.2	143.0	147.1	151.6	152.6	157.2	157.5	160.9	166.4	171.8	175.8	182.2	..
November	132.4	136.5	141.2	143.1	147.2	151.5	152.8	157.7	156.8	161.2	166.5	172.2	176.1	182.8	..
December	132.7	137.2	142.1	143.6	147.6	151.5	153.4	157.8	156.6	161.5	167.3	173.1	176.6	184.4	..

| | | | | | | **Retail Prices Index: January 1987 = 100** | | | | | | | | | |
| | | | | | | **CHAW** | | | | | | | | | |
	1993	1994	1995	1996	1997	1998	1999	2000	2001	2002	2003	2004	2005	2006	2007
January	137.9	141.3	146.0	150.2	154.4	159.5	163.4	166.6	171.1	173.3	178.4	183.1	188.9	193.4	..
February	138.8	142.1	146.9	150.9	155.0	160.3	163.7	167.5	172.0	173.8	179.3	183.8	189.6	194.2	..
March	139.3	142.5	147.5	151.5	155.4	160.8	164.1	168.4	172.2	174.5	179.9	184.6	190.5	195.0	..
April	140.6	144.2	149.0	152.6	156.3	162.6	165.2	170.1	173.1	175.7	181.2	185.7	191.6	196.5	..
May	141.1	144.7	149.6	152.9	156.9	163.5	165.6	170.7	174.2	176.2	181.5	186.5	192.0	197.7	..
June	141.0	144.7	149.8	153.0	157.5	163.4	165.6	171.1	174.4	176.2	181.3	186.8	192.2	198.5	..
July	140.7	144.0	149.1	152.4	157.5	163.0	165.1	170.5	173.3	175.9	181.3	186.8	192.2	198.5	..
August	141.3	144.7	149.9	153.1	158.5	163.7	165.5	170.5	174.0	176.4	181.6	187.4	192.6	199.2	..
September	141.9	145.0	150.6	153.8	159.3	164.4	166.2	171.7	174.6	177.6	182.5	188.1	193.1	200.1	..
October	141.8	145.2	149.8	153.8	159.5	164.5	166.5	171.6	174.3	177.9	182.6	188.6	193.3	200.4	..
November	141.6	145.3	149.8	153.9	159.6	164.4	166.7	172.1	173.6	178.2	182.7	189.0	193.6	201.1	..
December	141.9	146.0	150.7	154.4	160.0	164.4	167.3	172.2	173.4	178.5	183.5	189.9	194.1	202.7	..

| | | | | | | **Percentage changes on one year earlier** | | | | | | | | |
	1994	1995	1996	1997	1998	1999	2000	2001	2002	2003	2004	2005	2006	2007
Tax and Price Index														
January	2.6	3.9	3.2	1.4	2.4	2.3	1.5	2.6	−0.1	3.1	3.4	3.1	2.2	..
February	2.5	4.0	3.0	1.3	2.6	2.0	1.9	2.5	−0.4	3.4	3.3	3.1	2.3	..
March	2.5	4.0	3.0	1.1	2.6	1.9	2.2	2.1	−0.1	3.4	3.3	3.1	2.1	..
April	3.0	3.7	1.0	1.5	4.1	1.0	3.0	0.4	1.5	4.0	2.4	3.1	2.4	..
May	3.0	3.8	0.7	1.7	4.3	0.7	3.0	0.7	1.1	3.8	2.7	2.8	2.9	..
June	3.1	4.0	0.6	2.0	3.8	0.8	3.3	0.6	1.0	3.7	3.0	2.8	3.2	..
July	2.8	3.9	0.8	2.5	3.5	0.7	3.3	0.3	1.5	3.9	3.0	2.8	3.2	..
August	2.8	4.1	0.6	2.7	3.3	0.5	3.0	0.7	1.3	3.8	3.1	2.6	3.4	..
September	2.6	4.3	0.7	2.7	3.1	0.5	3.3	0.3	1.8	3.5	3.0	2.5	3.6	..
October	2.9	3.5	1.3	2.9	3.1	0.7	3.0	0.2	2.2	3.4	3.2	2.3	3.6	..
November	3.1	3.4	1.3	2.9	2.9	0.9	3.2	−0.6	2.8	3.3	3.4	2.3	3.8	..
December	3.4	3.6	1.1	2.8	2.6	1.3	2.9	−0.8	3.1	3.6	3.5	2.0	4.4	..
Retail Prices Index														
January	2.5	3.3	2.9	2.8	3.3	2.4	2.0	2.7	1.3	2.9	2.6	3.2	2.4	..
February	2.4	3.4	2.7	2.7	3.4	2.1	2.3	2.7	1.0	3.2	2.5	3.2	2.4	..
March	2.3	3.5	2.7	2.6	3.5	2.1	2.6	2.3	1.3	3.1	2.6	3.2	2.4	..
April	2.6	3.3	2.4	2.4	4.0	1.6	3.0	1.8	1.5	3.1	2.5	3.2	2.6	..
May	2.6	3.4	2.2	2.6	4.2	1.3	3.1	2.1	1.1	3.0	2.8	2.9	3.0	..
June	2.6	3.5	2.1	2.9	3.7	1.3	3.3	1.9	1.0	2.9	3.0	2.9	3.3	..
July	2.3	3.5	2.2	3.3	3.5	1.3	3.3	1.6	1.5	3.1	3.0	2.9	3.3	..
August	2.4	3.6	2.1	3.5	3.3	1.1	3.0	2.1	1.4	2.9	3.2	2.8	3.4	..
September	2.2	3.9	2.1	3.6	3.2	1.1	3.3	1.7	1.7	2.8	3.1	2.7	3.6	..
October	2.4	3.2	2.7	3.7	3.1	1.2	3.1	1.6	2.1	2.6	3.3	2.5	3.7	..
November	2.6	3.1	2.7	3.7	3.0	1.4	3.2	0.9	2.6	2.5	3.4	2.4	3.9	..
December	2.9	3.2	2.5	3.6	2.8	1.8	2.9	0.7	2.9	2.8	3.5	2.2	4.4	..

Note: For further information on the TPI refer to the *Annual Supplement* in the January edition of *Monthly Digest*.

Source: Office for National Statistics: 020 7533 5874

18.9 Index numbers of producer prices

2000=100[1]

	Materials and fuels purchased (input prices) SIC 1992						
	Materials & fuel purchased by manufacturing industry[5]	Materials	Fuel[5]	Materials & fuel purchased by manufacturing industry (SA)[5]	Materials & fuel purchased by manufacturing ind. except food, beverages, tobacco & petrol (NSA)[5]	Materials & fuel purchased by manufacturing ind. except food, beverages, tobacco & petrol (SA)[5]	Materials purchased by manufacturing industry, other than food, drink and tobacco
1992 SIC							
	D				D excl DA/DF		DA
	RNNK	PLKX	RNNL	RNPE	RNNQ	RNPF	RWCJ
2000	100.0	100.0	100.0	100.0	100.0	100.0	100.0
2001	98.8	98.1	107.1	98.8	98.7	98.7	98.1
2002	94.4	93.7	103.4	94.4	94.0	94.0	93.2
2003	95.7	95.2	102.1	95.7	93.7	93.7	93.0
2004	99.5	98.7	109.9	99.5	95.4	95.4	94.2
2005	111.1	108.1	152.1	111.1	103.0	103.0	98.9
2006	121.7p	116.0p	198.3	121.7p	111.1p	111.0p	103.7
2003 Jul	94.9	95.0	94.7	95.0	93.0	93.6	92.9
Aug	96.1	96.3	94.2	95.9	93.3	94.0	93.2
Sep	95.1	95.1	95.0	95.4	93.5	94.2	93.4
Oct	96.3	95.7	103.8	96.9	93.8	94.1	93.0
Nov	96.8	95.7	112.0	96.8	94.5	94.0	93.1
Dec	97.0	95.7	114.3	96.7	94.5	93.5	92.8
2004 Jan	95.6	94.2	114.0	95.4	93.3	92.8	91.6
Feb	95.2	93.8	112.1	94.8	92.8	92.3	91.2
Mar	97.2	96.4	107.7	96.5	93.9	93.4	92.8
Apr	97.3	97.0	101.3	97.2	94.0	93.8	93.4
May	99.6	99.6	99.8	100.2	94.7	94.8	94.3
Jun	97.7	97.6	99.3	98.3	94.0	94.8	93.5
Jul	98.7	98.7	98.3	98.5	94.4	94.9	94.1
Aug	100.8	101.0	98.7	100.3	95.3	96.2	95.0
Sep	102.5	102.2	106.5	103.1	96.9	97.8	96.1
Oct	105.0	104.1	117.0	105.8	98.2	98.6	96.6
Nov	103.4	101.5	128.6	103.3	99.1	98.3	96.6
Dec	101.3	98.7	135.8	101.0	98.5	97.4	95.4
2005 Jan	104.9	102.4	139.2	104.6	100.4	99.8	97.2
Feb	105.5	103.0	140.3	105.2	100.5	99.9	97.2
Mar	107.9	105.7	138.2	107.2	101.0	100.3	97.9
Apr	107.1	105.2	132.7	106.9	100.6	100.4	97.9
May	107.0	105.2	131.7	107.8	100.9	101.0	98.3
Jun	109.3	107.5	133.2	110.1	100.8	101.8	98.1
Jul	112.7	110.9	137.3	112.2	102.5	103.1	99.6
Aug	113.9	112.0	140.1	113.1	102.3	103.4	99.2
Sep	113.2	111.3	138.4	113.9	102.2	103.4	99.3
Oct	114.4	111.1	159.5	115.2	105.1	105.5	100.5
Nov	117.5	111.1	204.8	117.4	108.8	107.7	100.7
Dec	119.6	111.5	229.3	119.1†	110.5	109.2	100.6
2006 Jan	121.4	113.7	225.7	120.9	110.8	110.0	101.2
Feb	121.3	114.0	219.9	121.0	111.2	110.4	102.1
Mar	121.7	115.1	211.7	121.1	111.1	110.3	102.7
Apr	123.4	118.3	191.9	123.2	110.5	110.4	103.7
May	121.5	117.1	181.0	122.5	109.5	109.7	103.5
Jun	121.4	117.5	174.9	122.3	109.5	110.7	103.9
Jul	124.7	120.5	181.7	124.0	111.0	111.6	105.1
Aug	123.0	118.6	181.8	122.2	110.2	111.5	104.2
Sep	119.0	114.7	176.5	120.1	109.5†	110.6†	103.9
Oct	119.6†	113.9†	196.8†	120.6	111.7†	111.9	104.5p
Nov	121.6p	114.3p	219.5p	121.3p	114.2p	112.7p	105.3p
Dec	122.0p	114.9p	217.6p	121.4p	113.5p	112.0p	104.7p

18.9 Index numbers of producer prices

continued

2000=100[1]

	Textiles	Leather	Wood and wood products	Pulp, paper and paper products	Coke, refined petroleum products and nuclear fuel	Chemicals and chemical products	Rubber products	Plastic products	Other non-metallic mineral products	Manufacture of basic metals	Machinery and equipment not elsewhere classified
1992 SIC											
	DB	DC	DD	DE	DF	DG	DH	DH	DI	DJ	DK
	RBBR	RBBS	RBBT	RABL	RAUW	RBBW	RAZZ	RBAC	RBBY	RBBZ	RBCA
2000	100.0	100.0	100.0	100.0	100.0	100.0	100.0	100.0	100.0	100.0	100.0
2001	100.5	101.9	99.2	100.7	92.0	101.1	100.4	98.8	100.5	98.5	98.9
2002	98.4	100.0	96.5	97.5	89.3	99.4	98.8	97.4	99.6	96.6	97.1
2003	99.2	101.1	96.8	96.5	95.4	103.0	101.8	99.1	100.7	99.8	97.8
2004	99.6	102.0	99.5	96.9	109.1	105.5	105.5	102.0	103.0	111.8	102.4
2005	103.3	105.7	104.1	100.7	150.9	114.4	113.2	110.4	111.8	123.0	109.3
2006	106.4p	109.5p	109.2p	106.5p	175.6	120.6p	119.9	115.5	119.9p	137.3p	116.7p
2003 Jul	98.6	101.1	97.5	95.9	94.1	102.7	101.1	98.3	100.0	98.8	97.6
Aug	98.7	101.4	97.6	96.1	98.6	102.4	101.3	98.4	100.4	99.6	97.9
Sep	98.7	101.4	96.9	96.0	89.7	102.4	101.4	98.5	100.1	100.3	98.0
Oct	98.9	101.4	97.0	96.3	95.0	103.5	102.4	99.2	101.0	100.8	98.1
Nov	99.2	101.5	97.5	96.8	92.5	104.0	103.2	99.6	101.8	102.0	98.6
Dec	99.2	101.6	97.3	96.7	93.0	104.2	103.7	99.9	101.5	102.8	98.8
2004 Jan	98.9	100.8	98.1	96.1	89.8	103.0	103.3	99.3	101.3	104.4	98.9
Feb	98.7	100.4	97.8	95.7	90.1	102.8	103.4	99.6	100.9	105.7	99.1
Mar	98.9	101.0	98.1	96.2	97.7	103.1	103.9	99.9	101.4	107.9	100.0
Apr	98.8	101.2	98.2	96.3	97.6	102.7	103.8	99.9	101.3	109.0	100.8
May	98.9	101.4	98.8	96.6	110.7	104.0	104.1	100.3	101.7	108.9	101.3
Jun	98.5	101.3	99.2	96.4	102.6	104.0	104.1	100.4	101.9	108.6	101.2
Jul	98.8	101.4	100.0	96.4	107.8	105.1	104.8	100.9	102.0	111.7	102.5
Aug	99.5	101.9	100.2	96.6	121.2	106.2	105.3	101.5	102.4	113.7	103.2
Sep	100.1	102.8	100.6	97.3	124.3	107.6	106.3	102.7	103.6	115.0	104.1
Oct	101.1	103.5	100.8	97.8	137.6	108.6	107.9	105.2	105.2	118.6	105.4
Nov	101.6	104.1	101.2	98.4	121.5	109.6	109.1	106.6	106.8	119.7	106.1
Dec	101.3	104.2	101.4	98.5	108.3	109.7	109.7	107.8	106.9	118.3	106.0
2005 Jan	102.3	104.9	102.0	99.4	120.9	112.0	111.5	108.9	108.8	120.5	107.6
Feb	102.3	104.9	102.1	99.6	124.8	112.6	111.8	109.3	109.0	121.1	107.7
Mar	102.4	105.1	102.2	99.6	138.8	113.3	112.0	109.4	109.5	121.5	107.9
Apr	102.1	104.9	103.1	99.3	137.8	112.7	111.6	109.1	109.7	122.7	108.3
May	102.1	105.1	103.1	99.1	133.1	112.9	111.7	109.1	109.7	121.5	108.3
Jun	101.9	104.7	103.2	99.2	149.6	112.4	110.8	108.0	109.2	120.8	108.6
Jul	103.0	105.4	103.8	100.0	164.5	113.4	112.0	108.3	110.8	122.0	109.4
Aug	103.0	105.3	103.7	99.7	175.9	113.7	112.5	109.8	111.0	122.5	109.2
Sep	103.1	105.7	104.2	100.0	172.9	113.8	112.9	110.3	110.6	123.2	109.2
Oct	104.3	106.3	106.2	101.6	165.9	116.1	114.9	112.9	113.0	123.9	110.4
Nov	106.4	107.8	107.4	104.6	162.0	119.1	118.0	114.9	118.7	127.1	111.9
Dec	107.2	108.6	108.1	106.2	164.1	120.5	119.2	115.1	121.8	129.0	112.8
2006 Jan	107.2	108.5	108.5	106.9	176.4	121.2	119.2	114.9	122.3	130.0	113.4
Feb	107.1	108.5	108.5	107.5	172.9	121.5	119.3	115.0	122.4	131.1	113.9
Mar	107.0	108.5	108.3	107.2	176.1	121.4	119.6	115.1	121.8	132.0	114.3
Apr	106.4	108.9	108.2	106.2	191.5	120.7	119.3	115.1	119.7	134.4	115.1
May	105.3	108.6	107.8	105.1	184.9	119.5	118.7	114.4	117.8	136.1	115.7
Jun	105.4	109.3	108.1	104.9	182.7	119.6	119.0	114.8	117.3	136.8	116.3
Jul	106.0	110.0	108.7	105.8	196.5	120.7	119.9	115.6	118.8	139.4	117.8
Aug	105.8	109.9[†]	108.8	105.6	190.2	120.6	119.5	115.6	118.2	139.0	117.8
Sep	105.6	109.7	109.3	105.4[†]	164.1p[†]	119.6[†]	119.7[†]	115.5	117.6[†]	139.7	117.9
Oct	106.4	110.2	110.5	106.8[†]	157.0p[†]	120.2	120.7p	116.4p[†]	119.7	142.2[†]	118.9
Nov	107.4p[†]	111.0p	111.6p[†]	108.1p	154.7p	121.2p	122.0p	117.2p[†]	122.1p	143.8p	119.9p[†]
Dec	107.2p	111.0p	111.6p	107.9p	160.1p	121.1p	121.8p	116.8p	121.6p	143.0p	119.5p

18.9 Index numbers of producer prices
continued

2000=100[1]

	Materials and fuel				Products of manufacturing industries except food, beverages, tobacco & petroleum manufacturing (NSA)	Products of manufacturing industries except food, beverages, tobacco & petroleum manufacturing (SA)	Products of the food, beverages and tobacco manufacturing industries	Quarterly construction output price index[2]	Monthly index of average price of new dwellings - at mortgage completion stage[3]
	Electrical and optical equipment	Transport equipment	Manufacturing not elsewhere classified	Output of manufactured products					
1992 SIC									
	DL	DM	DN	F	Part of F	2 to 4			
	RBCB	RBCC	RBCD	PLLU	PLLV	PLLW	POKH	JYYC	FCBA
2000	100.0	100.0	100.0	100.0	100.0	100.0	100.0	120	84.6
2001	97.2	99.1	99.1	99.7	99.4	99.4	101.9	124	90.3
2002	92.5	97.3	97.8	99.8	99.3	99.3	103.3	128	108.7
2003	88.8	98.1	99.9	101.3	100.6	100.6	104.6	133	126.4
2004	87.7	100.1	104.4	103.8	102.5	102.5	106.9	143	138.6
2005	90.1	105.3	109.9	106.7	104.7	104.7	108.4	150	147.6
2006	93.8p	110.4p	116.4p	109.3p	107.0p	107.0p	110.7	..	156.5
2003 Jul	88.6	98.1	99.8	101.2	100.6	100.7	104.8	..	126.6
Aug	88.5	98.2	100.1	101.4	100.7	100.7	104.9	..	129.6
Sep	88.8	98.4	100.3	101.4	100.9	100.8	104.8	134	127.6
Oct	88.0	98.4	100.4	101.6	100.8	100.8	105.1	..	132.6
Nov	88.1	98.6	100.9	101.7	100.9	101.0	105.5	..	128.8
Dec	87.6	98.7	101.1	101.9	101.2	101.2	105.7	138	132.0
2004 Jan	86.7	98.4	101.5	102.1	101.4	101.4	105.9	..	131.5
Feb	86.3	98.1	101.7	102.3	101.6	101.6	106.1	..	129.4
Mar	87.1	98.8	102.6	102.8	101.9	101.8	106.4	139	131.6
Apr	87.3	99.1	102.9	103.1	101.8	101.7	106.9	..	135.9
May	87.6	99.5	103.3	103.5	101.9	101.9	107.3	..	136.7
Jun	87.2	99.3	103.3	103.6	101.8	102.0	107.5	141	140.9
Jul	87.2	99.8	104.4	103.8	102.3	102.4	107.6	..	142.5
Aug	87.6	100.3	105.3	104.2	102.9	102.9	106.8	..	142.3
Sep	88.8	101.2	105.9	104.5	103.2	103.1	107.1	144	144.5
Oct	89.0	102.1	107.2	105.2	103.7	103.7	107.0	..	144.4
Nov	89.0	102.6	107.7	105.3	103.9	103.9	107.0	..	143.0
Dec	88.3	102.5	107.2	104.9	103.6	103.7	107.2	147	140.4
2005 Jan	89.2	104.1	108.5	104.8	104.0	103.9	107.3	..	143.9
Feb	88.9	104.2	108.6	105.1	104.1	104.1	107.7	..	144.0
Mar	89.0	104.3	108.9	105.8	104.3	104.2	108.0	148	147.4
Apr	88.9	104.4	109.3	106.5	104.5	104.3	108.3	..	144.6
May	89.2	104.7	108.9	106.3	104.4	104.4	108.3	..	146.9
Jun	89.6	104.5	108.6	106.2	104.0	104.3	108.2	149	148.0
Jul	90.5	105.5	109.5	107.0	104.6	104.7	108.5	..	149.7
Aug	90.1	105.4	109.7	107.3	104.9	104.9	108.4	..	148.8
Sep	90.1	105.4	110.2	108.0	105.4	105.3	108.7	151	148.5
Oct	91.2	106.3	111.0	107.9	105.1	105.1	108.5	..	151.1
Nov	92.1	107.3	112.2	107.7	105.3	105.3	109.2	..	146.9
Dec	92.5	107.9	112.9	107.4	105.4	105.5	109.2	151	150.9
2006 Jan	92.8	108.3	113.4	107.8	105.8	105.7	109.2	..	155.5
Feb	93.1	108.7	113.9	108.1	106.0	106.0	109.2	..	150.9
Mar	93.2	109.0	114.3	108.4	106.3	106.2	109.4	153†	156.1
Apr	93.2	109.5	115.3	109.2	106.9	106.7	109.6	..	153.7
May	92.8	109.6	115.5	109.6	106.9	107.0	110.3	..	156.3
Jun	93.4	110.0	116.0	109.8	107.1	107.3	110.7	154	156.0
Jul	94.1	111.1†	117.2	110.1	107.2	107.3	111.1	..	154.9
Aug	93.3	110.8†	117.0	110.2†	107.3	107.3	111.4	..	156.1
Sep	93.4	110.7	117.4†	110.0†	107.6	107.5	111.5	155	158.5
Oct	94.8	111.6	118.6†	109.6	107.8	107.7	111.6	..	156.0†
Nov	96.3p†	112.7p	119.3p	109.6p	107.8p	107.9p	112.0p†	..	159.1
Dec	95.7p	112.4p	119.1p	109.8p	107.8p	108.0p	112.4p	..	164.3

18.9 Index numbers of producer prices
continued

Output of selected sub-sections of industry

	Textiles and textile products[4]	Leather and leather products	Wood and wood products[4]	Pulp, paper and paper products; publish-ing and printing	Chemicals & chemical products; man-made fibres	Rubber and plastic products	Other non-metallic mineral products	Basic metals and fabri-cated metal products	Machinery and equipment not elsewhere classi-fied[4]	Elect-rical and optical equipment	Transport equipment	Furniture and other manufactur-ed goods n.e.s.
1992 SIC												
	DB	DC	DD	DE	DG(part)	DH	DI	DJ	DK	DL	DM	DN
	POKI	POKJ	POKK	POKL	POKN	POKO	POKP	POKQ	POKR	POKS	POKT	POLS
2000	100.0	100.0	100.0	100.0	100.0	100.0	100.0	100.0	100.0	100.0	100.0	100.0
2001	99.2	102.5	99.9	101.5	100.2	100.3	101.9	99.9	100.9	94.7	98.4	100.3
2002	98.8	102.7	100.0	102.1	100.5	100.4	105.0	99.5	101.8	90.0	98.8	100.9
2003	98.7	102.9	101.8	104.0	103.9	100.5	107.8	101.3	101.9	87.5	99.2	103.8
2004	98.5	102.9	105.2	106.0	106.7	101.5	109.6	108.8	103.3	86.6	100.3	104.4
2005	100.0	104.5	110.0	108.6	111.5	106.3	113.9	118.4	106.5	85.9	102.4	104.5
2006	101.2	106.5	113.1	110.7	115.8	109.7	118.2	125.5	109.2	86.8	103.9	105.7
2003 Jul	98.7	103.0	103.1	104.1	104.3	100.8	108.4	101.4	102.2	87.4	99.0	103.8
Aug	98.6	102.7	103.1	104.3	104.0	100.8	108.4	101.6	102.1	87.2	99.2	104.1
Sep	98.6	102.6	103.4	104.5	103.8	100.7	108.0	101.7	102.1	87.3	99.4	104.3
Oct	98.7	103.3	103.4	104.4	103.7	100.4	107.6	101.8	102.2	87.1	99.5	104.4
Nov	98.7	103.4	103.5	104.4	103.8	100.2	107.7	102.1	102.2	87.3	99.5	104.2
Dec	98.6	102.7	103.4	104.4	104.4	100.5	107.5	102.2	102.1	87.2	99.4	105.1
2004 Jan	98.5	102.9	103.3	104.9	104.7	100.9	107.7	103.0	102.3	86.9	99.4	105.0
Feb	98.4	102.3	103.6	105.0	104.9	101.0	108.2	103.5	102.3	86.7	99.6	104.7
Mar	98.3	102.2	103.4	105.0	105.4	100.9	109.2	104.3	102.5	86.4	99.9	104.5
Apr	98.3	101.9	103.8	105.1	105.4	101.0	109.4	105.6	102.5	86.4	100.1	104.4
May	98.3	101.9	104.7	105.3	105.7	101.0	109.6	106.5	103.1	86.4	100.1	104.4
Jun	98.3	102.5	105.0	105.4	106.2	101.1	109.7	107.9	102.9	86.4	100.0	104.7
Jul	98.4	102.8	106.1	105.7	106.5	101.2	110.2	110.3	103.0	86.5	100.2	104.6
Aug	98.5	103.2	106.4	106.9	107.0	101.4	110.2	111.0	103.5	86.4	100.4	104.1
Sep	98.6	104.2	106.7	107.1	107.5	101.6	110.2	111.8	103.8	86.8	100.8	103.8
Oct	98.7	103.6	106.2	107.3	108.4	102.1	110.1	113.5	104.9	86.6	101.1	104.1
Nov	98.7	103.8	106.5	107.5	109.1	102.5	110.5	114.0	104.3	86.7	101.0	104.2
Dec	98.8	103.7	106.7	107.4	109.8	103.7	110.4	114.7	104.4	86.7	101.1	104.2
2005 Jan	98.9	103.8	107.2	107.8	110.4	104.6	111.8	116.9	104.9	86.7	101.5	104.3
Feb	99.5	104.4	107.9	108.2	110.7	105.0	112.2	117.4	105.1	85.7	101.9	103.8
Mar	99.7	104.4	108.1	108.3	110.6	105.2	113.0	117.4	105.5	85.8	102.0	104.7
Apr	100.0	104.4	109.6	108.2	111.0	105.4	114.3	118.3	105.8	85.7	101.9	104.5
May	100.0	104.6	110.5	108.3	111.0	105.8	114.2	118.4	106.0	85.9	102.1	103.9
Jun	100.0	104.7	110.6	108.6	110.7	105.9	115.0	118.5	106.3	86.1	102.1	104.1
Jul	100.2	104.5	110.9	108.8	111.2	106.5	114.7	118.5	107.0	86.2	102.5	104.3
Aug	100.0	104.4	110.7	108.9	111.4	106.7	114.4	118.4	107.0	86.0	102.6	104.4
Sep	100.1	104.8	110.9	108.9	111.9	107.1	114.1	118.5	107.4	85.6	102.7	104.5
Oct	100.2	104.8	111.2	109.0	112.5	107.5	114.2	118.8	107.7	85.7	103.0	104.7
Nov	100.3	104.8	111.1	109.1	113.4	108.0	114.4	119.4	107.8	85.6	103.0	105.0
Dec	100.5	105.0	111.3	109.3	113.8	108.5	114.1	119.8	107.9	85.8	103.1	105.2
2006 Jan	100.7	106.0	111.5	109.6	114.9	108.8	115.8	120.6	108.1	86.3	103.3	105.0
Feb	100.6	106.1	111.4	110.1	114.5	109.2	116.3	121.1	108.3	86.3	103.4	105.3
Mar	100.8	106.2	111.6	110.3	114.7	109.4	117.6	121.6	108.5	86.4	103.7	105.4
Apr	101.0	106.8	112.0	110.4	116.0	109.4	118.0	122.9	108.8	86.3	103.6	105.6
May	101.1	106.3	112.0	110.5	115.9	109.4	118.1	124.4	109.0	86.3	103.4	106.0
Jun	101.1	106.3	112.3	110.5	115.9	109.6	118.0	124.9	109.2	87.2	103.5	105.1
Jul	101.2	106.5	112.7	110.5	115.7	109.9	118.4	126.6	109.5	87.3	103.7	105.3
Aug	101.4	106.3	113.0	110.8	116.0	110.0	118.7	127.7†	109.4	87.5	103.6	105.8
Sep	101.4	106.2	113.9	111.1	116.6	110.2	119.1	128.1	109.7	87.3	104.1	106.0
Oct	101.6p	107.0	114.8	111.6p	116.3	110.5	119.4	129.0	109.7	87.1p	104.6	106.2
Nov	101.5p†	107.0p	115.6p	111.5p	116.5p†	110.2p†	119.5p†	129.3p	109.9p†	86.9p†	104.7p†	106.4p†
Dec	101.6p	107.0p	116.1p	111.5p	116.6p	110.2p	119.3p	129.2p	109.9p	86.6p	104.7p	106.5p

1 This month's edition contains data rebased onto 2000=100. For information the rebased back data for the headline PPI series is available under related links at www.statistics.gov.uk/ppi

2 A base weighted (1995=100) combination of the separate price indices for contractors' output in the six new work sectors. For a fuller description see *Economic Trends* No 297.

3 From February 2002, data are based on a significantly enlarged return from the Survey of Mortgage Lenders, and are calculated through improved methodology. Annual and quarterly data prior to February 2002 are from the 5% Survey of Mortgage Lenders, and have been rebased to Feb 2002=100.

4 Indicates values which are considered less reliable than the remainder currently published mainly due to the lack of market coverage.

5 The Climate Change Levy was introduced in April 2001. Further information on PPI is available from the National Statistics Website: www.statistics.gov.uk/ppi.

Sources: Office for National Statistics: Tel 01633 812106; DTI (JYYC): 020 7215 1953; DCLG (FCBA): 020 7944 3325

18.10 House Price Index[1]
Analysis by Government Office Regions

Feb 2002 = 100[2]

	United Kingdom	North East	North West	Yorkshire and the Humber	East Midlands	West Midlands	East	London	South East	South West	England	Wales	Scotland	Northern Ireland
	WLPE	WLPF	WLPG	WLPH	WLPI	WLPK	WLPL	WLPM	WLPN	WLPT	WLPU	WLPV	WLPX	WLPY
2004 Nov	150.1	186.4	171.8	172.3	171.3	162.2	144.5	131.6	138.8	154.6	149.1	176.5	160.1	134.5
Dec	149.0	185.0	170.8	173.9	170.1	161.5	144.9	128.9	138.2	152.2	148.0	178.5	156.9	140.3
2005 Jan	148.9	182.7	173.5	169.8	170.2	158.3	142.4	129.6	137.7	153.7	147.5	178.4	159.9	144.1
Feb	148.1	182.2	170.3	171.9	167.5	159.5	142.8	130.5	135.5	152.5	146.9	178.2	157.1	138.1
Mar	151.3	189.9	176.7	174.7	169.4	163.7	144.4	133.0	139.2	153.6	150.0	180.7	162.1	143.5
Apr	150.1	185.6	176.2	175.2	167.7	162.8	141.9	131.2	137.4	151.8	148.4	181.4	166.8	142.4
May	150.8	189.1	177.5	179.7	168.8	162.2	142.0	131.7	136.8	151.1	148.8	185.9	169.6	144.7
Jun	152.0	192.6	180.3	180.4	170.0	162.8	143.4	133.5	137.5	150.9	150.0	183.3	173.4	146.3
Jul	153.7	194.2	181.4	180.4	170.3	166.4	145.0	134.6	139.5	153.1	151.6	186.0	174.4	154.6
Aug	153.7	194.9	182.1	185.2	171.2	165.2	144.1	134.0	138.9	153.0	151.5	188.7	175.0	152.9
Sep[3]	154.0	192.7	182.4	183.6	171.8	164.3	145.0	134.3	140.2	153.1	151.8	186.2	175.1	158.0
Oct	152.7	196.2	181.4	184.9	171.3	163.1	142.2	131.7	138.4	152.5	150.4	188.9	173.6	154.8
Nov	153.4	195.1	181.8	182.2	170.7	164.9	144.2	133.5	138.7	153.9	151.2	186.6	173.4	158.1
Dec	153.3	197.0	183.6	184.9	170.2	164.4	142.8	133.7	137.7	153.3	151.1	187.7	174.0	158.4
2006 Jan	155.1	196.1	181.1	185.2	171.0	165.8	145.6	136.1	140.9	154.6	152.8	188.7	176.3	161.2
Feb	153.2	194.6	183.4	186.1	170.9	164.6	141.3	132.5	138.0	154.2	150.8	189.0	174.9	160.1
Mar	156.2	197.5	185.0	188.0	172.0	165.7	145.1	138.4	140.1	156.3	153.8	190.2	179.5	163.7
Apr	157.6	199.5	187.5	190.2	172.2	166.0	146.3	140.0	141.2	157.2	155.1	192.6	179.7	167.3
May	159.3	203.7	188.5	190.6	173.1	168.2	147.6	141.0	143.0	159.1	156.6	194.8	186.1	170.7
Jun	160.0	203.4	189.4	192.5	173.4	170.1	147.5	140.8	143.6	158.9	157.0	198.7	188.8	174.1
Jul	162.7	205.4	192.8	195.2	177.1	171.4	150.8	144.1	145.6	160.6	159.6	199.5	190.8	186.5
Aug	165.0	208.4	194.8	198.8	178.4	174.2	153.0	144.1	148.2	163.5	161.5	202.0	197.3	193.6
Sep	166.3	207.9	196.6	199.9	180.5	175.9	153.4	146.4	148.5	165.2	162.8	205.3	197.7	195.3
Oct	165.6	208.7	195.0	199.9	180.8	173.9	152.2	145.4	147.6	165.0	162.0	201.7	197.1	205.6
Nov	166.9	207.7	197.2	200.3	179.6	176.2	153.6	146.7	148.6	165.9	163.0	205.6	196.6	216.9
Dec	168.5	213.7	197.3	203.2	180.6	177.2	154.4	149.5	149.7	165.2	164.4	204.7	200.5	221.6

Percentage change on a year earlier

	WLPZ	WLQA	WLQB	WLQG	WLQH	WLQI	WLQJ	WLQK	WLQL	WLQX	WLQY	WLRE	WLRF	WLRK
2005 Nov	2.2	4.6	5.8	5.8	−0.4	1.7	−0.2	1.5	−0.1	−0.4	1.4	5.7	8.3	17.6
Dec	2.9	6.5	7.6	6.3	–	1.8	−1.4	3.8	−0.3	0.8	2.1	5.1	10.9	12.9
2006 Jan	4.1	7.3	4.4	9.0	0.5	4.8	2.2	5.0	2.3	0.6	3.5	5.8	10.2	11.9
Feb	3.4	6.8	7.7	8.3	2.0	3.2	−1.0	1.5	1.8	1.1	2.6	6.1	11.3	15.9
Mar	3.3	4.0	4.7	7.6	1.5	1.2	0.5	4.0	0.6	1.7	2.5	5.3	10.8	14.1
Apr	5.0	7.5	6.5	8.6	2.7	2.0	3.0	6.7	2.8	3.6	4.5	6.1	7.7	17.4
May	5.7	7.7	6.2	6.0	2.5	3.7	3.9	7.1	4.5	5.3	5.2	4.8	9.7	17.9
Jun	5.3	5.6	5.1	6.7	2.0	4.4	2.9	5.5	4.5	5.3	4.6	8.5	8.9	19.1
Jul	5.9	5.8	6.3	8.2	4.0	3.0	4.0	7.0	4.4	4.9	5.3	7.3	9.4	20.6
Aug	7.4	7.0	7.0	7.3	4.2	5.5	6.2	7.6	6.7	6.9	6.6	7.0	12.7	26.6
Sep	8.0	7.9	7.8	8.9	5.1	7.1	5.8	9.0	5.9	7.9	7.2	10.3	12.9	23.6
Oct	8.5	6.4	7.5	8.1	5.4	6.7	7.1	10.6	6.8	8.2	7.9	7.0	13.6	31.6
Nov	8.8	6.5	8.4	9.9	5.2	7.0	6.4	10.1	7.2	8.0	8.0	10.5	13.4	36.6
Dec	9.9	8.4	7.4	9.9	–	–	–	–	–	–	–	–	–	–

1 Series based on prices at the mortgage completion stage collected through the Survey of Mortgage Lenders. The index takes into account the mix of properties sold.
2 The series starts at February 2002 rather than January 2002 because the required volume of completions was achieved from that date only.
3 From September 2005 the index is based on the new Regulated Mortgage Survey (CML/BankSearch).

Source: Department for Communities and Local Government

18.11 Index of purchase prices of the means of agricultural production and of producer prices of agricultural products[1,2,3]

2000=100

		Weights	2005	2005 Dec	2006 Jan	2006 Feb	2006 Mar	2006 Apr	2006 May	2006 Jun	2006 Jul	2006 Aug	2006 Sep	2006 Oct	2006 Nov
Purchase prices[4]															
Goods and services															
currently consumed	BYEA	100.0	115.9†	116.1	117.4†	117.4	117.7	119.9	120.1	120.2	121.8	123.3	121.8	119.8	119.7
Seeds	BYEB	3.3	108.1	108.5	110.9†	110.9	110.9	111.3	111.3	109.4	111.1	111.1	111.1	100.7	100.7
Energy, lubricants	BYED	8.1	137.4	135.8	142.6	144.9	150.2	161.3	165.4	159.3	156.0	157.0	143.9	138.3	133.5
Fertilizer and soil improvers	BYEE	9.1	143.3	151.0	151.5	154.6	153.8	154.5	152.5	149.7	149.7	149.1	149.6	150.0†	150.7
Plant protection products	BYEF	7.2	102.9	102.0	102.4	103.3	103.8	103.8	102.9	103.2	103.5	103.4	103.1	103.4	103.8
Animal feedingstuffs	BYEG	26.4	102.9	102.7	103.6	103.8	104.8	105.1	105.3	106.7	108.0	106.9	108.9†	110.0	112.4
Maintenance of plant	BYEI	7.9	130.3	133.3	135.0	135.3	135.9	136.2	136.7	137.4	138.2	138.4	139.0	139.7	140.8
Maintenance and repair of buildings	BYEJ	3.6	118.1	118.9	119.5	120.6	121.7	122.2	123.4	124.7	125.7	127.3†	127.9	128.9	129.6
Veterinary services	BYEK	3.2	103.9	107.9	107.9	107.9	107.9	107.9	108.4	112.0	112.0	112.0	112.0	112.3	111.6
Other goods and services[5]	BYEL	31.2	114.5†	113.8†	113.2	113.2	113.8	114.9	115.2	115.8	114.8	114.4	114.8	115.6	115.8
Goods and services contributing to investment in agriculture	BYEM	100.0	108.7	109.0	109.6	110.4	111.0	111.1	111.4	111.4	111.6	111.8†	112.0	112.8	113.8
Machinery and other equipment	BYEN	71.5	103.8	105.0	105.8	106.7	107.1	107.1	107.8	107.7	107.7	107.7	107.8	109.7	111.9
Buildings	BYEO	19.5	123.7	125.1	126.1	126.9	127.8	128.3	129.4	130.5	131.5	132.9†	133.4	134.4	134.9
Producer prices															
All products	BYEP	100.0	109.6	113.0	110.5†	110.2	113.5	114.7	112.9	114.7	112.8	116.1	116.6	114.8	117.6
All crop products	BYEQ	40.2	108.5	112.9	111.0†	110.9	114.9	116.9	115.8	118.3	117.0	123.0	121.3	118.2	123.3
Cereals	BYER	13.3	99.1	99.5	100.9	102.2	103.3	104.1	106.2	109.0	106.0	105.2	113.9	121.1	124.3
Industrial crops[6]	AE6A	4.3	113.9	115.7	114.8†	114.7	114.4	115.5	119.6	121.8	126.3	126.1	121.4	104.8	105.4
Forage plants	AE6B	1.9	110.4	101.5	102.7	101.3	106.5	109.7	111.4	111.4	99.0	93.0	95.6	102.7	105.1
Fresh vegetables	BYET	7.7	120.1	136.8	130.0	128.0	139.0	128.0	113.0	124.4†	121.7	140.7	140.0	124.9	136.2
Fresh fruit	BYEU	1.9	120.1	123.3	123.9	119.1	124.5	141.1	143.8	111.8	115.4	128.4	117.4	109.1	117.6
Potatoes	AE6C	4.5	109.2	110.2	118.3†	121.2	130.9	142.9	152.2	153.8	131.5	175.3	144.4	129.4	145.1
Flowers and plants	BYEW	5.9	105.6	107.3	109.2†	108.7	107.6	107.5	107.3	106.7	106.5	106.7	106.6	110.5	110.2
Other crop products (including seeds)	BYEX	0.7	110.8	110.6	110.1†	110.1	110.6	110.6	111.0	110.2	109.3	110.0	110.5	111.3	111.6
Animals and animal products	BYEY	59.8	110.4	113.1	110.2	109.8	112.5	113.5	110.9	112.3	110.0	110.6	112.6	112.6	113.3
Animals for slaughter	BYEZ	35.3	110.3	111.4	108.5	108.9	114.6†	118.1	117.5	119.3	113.6	113.7	113.9	112.1	112.7
Milk	BYFA	20.2	109.0	113.7	110.2	108.4	107.1	102.5	99.3	99.4	101.4†	104.0	109.1	111.2	112.3
Eggs	BYFB	3.2	121.1	125.8	128.5	128.3	128.0	124.4	123.4	123.4	125.3	126.3	125.0	126.3	126.4
Other animal products	BYFC	1.1	109.3	113.0	111.4	111.1	110.7	106.5†	103.2	101.1	101.6	103.5	106.1	107.8	108.7

1 Index numbers for the years 1983 to 2003 on 1995 = 100 base and also at a more detailed level are available from the Department for Environment Food and Rural Affairs, Room 146, Kings Pool, 1-2 Peasholme Green, YO1 7PX. Tel 01904 455253.

2 The sum of the percentages of categories included in "Goods & Services consumed" and "All Products" do not add to 100% due to the exclusion of some minor categories.

3 All data and weights have been revised to be in line with Eurostat policy that the *Agricultural Account* and the *API* should be the same.

4 A revised feedstuffs index has been calculated and incorporated in this edition. Further details are available on request.

5 Formerly *General expenses*.

6 Primerily including Oilseeds, Linseed and Protein crops.

Source: Department for Environment, Food and Rural Affairs

18.12 Average weekly and hourly earnings and hours of full-time employees on adult rates whose pay for the period was unaffected by absence: United Kingdom
April 2000 to 2006

| | Manufacturing industries[1] | | | | All industries and services | | | |
| | Gross weekly earnings(£) | Total Paid Hours | Hourly earnings(£) | | Gross Weekly earnings(£) | Total Paid Hours | Hourly earnings(£) | |
			Including overtime pay and overtime hours	Excluding overtime pay and overtime hours			Including overtime pay and overtime hours	Excluding overtime pay and overtime hours
Total								
	C7PU	C7QL	C7PV	C7PW	C7Q5	C7QX	C7Q7	C7Q9
2000	417.2	41.3	10.10	10.00	425.2	39.7	10.71	10.71
2001	439.9	41.3	10.66	10.62	449.8	39.7	11.33	11.36
2002	455.6	41.0	11.12	11.09	472.2	39.6	11.94	11.98
2003	476.5	40.9	11.65	11.62	487.1	39.5	12.32	12.34
2004[2]	485.0	41.0	11.83	11.80	498.2	39.5	12.60	12.63
	493.1		12.03	12.01	506.1		12.80	12.84
2005	508.0	40.6	12.51	12.50	516.5	39.4	13.11	13.15
2006	528.4	40.8	12.96	12.97	537.4	37.5	13.62	13.67
Men								
	C7PX	C7QT	C7PY	C7PZ	C7QA	C7QZ	C7QC	C7QE
2000	445.6	42.0	10.62	10.54	471.8	41.0	11.50	11.53
2001	469.5	41.9	11.21	11.19	498.6	41.0	12.16	12.24
2002	482.9	41.6	11.62	11.62	523.4	40.8	12.83	12.92
2003	503.2	41.5	12.13	12.12	539.3	40.8	13.21	13.28
2004[2]	511.2	41.6	12.30	12.28	548.1	40.8	13.44	13.51
	519.4		12.50	12.49	557.4		13.67	13.76
2005	533.8	41.2	12.97	12.98	568.1	40.6	13.98	14.05
2006	554.9	41.3	13.43	13.46	591.6	40.7	14.54	14.62
Women								
	C7Q2	C7QV	C7Q3	C7Q4	C7QF	C7SA	C7QH	C7QJ
2000	312.1	38.9	8.02	7.97	344.9	37.4	9.22	9.20
2001	332.2	39.0	8.52	8.50	367.1	37.5	9.79	9.79
2002	350.8	38.8	9.04	9.03	386.8	37.5	10.32	10.32
2003	372.8	38.7	9.64	9.62	400.7	37.4	10.71	10.70
2004[2]	380.8	38.7	9.83	9.84	416.8	37.5	11.11	11.12
	388.1		10.02	10.02	422.1		11.26	11.27
2005	404.3	38.4	10.52	10.51	435.7	37.4	11.64	11.65
2006	422.6	38.6	10.94	10.97	453.6	37.6	12.08	12.11

1 Results relate to Division D (SIC) 1992.

2 In 2004 a number of supplementary surveys were introduced to improve the coverage of the Annual Survey of Hours and Earnings. Data for 2004 are presented including these supplementary surveys (top). Figures are also presented excluding supplementary surveys (bottom) to give figures comparable with earlier years.

Source: Office for National Statistics: 01633 819024

18.13 Average weekly and hourly earnings of full-time employees on adult rates by industry division: United Kingdom
April 2003 to 2006
£

Full time employees on adult rates whose pay was unaffected by absence

	Agriculture, hunting, and forestry	Fishing	Mining and quarrying	Manufacturing	Electricity, gas and water supply	Construction	Wholesale and retail trade; repair of motor vehicles and personal & household goods	Hotels and restaurants	Transport, storage and communication	Financial intermediation	Real estate, renting and business activities	Public administration and defence, compulsory social security	Education	Health and social work	Other community, social and personal service activities
SIC 1992 Division	A	B	C	D	E	F	G	H	I	J	K	L	M	N	O

Average gross weekly earnings

Total

	C9EG	C9EI	C9EK	C9EM	C9EO	C9EP	C9EQ	C9ER	C9ES	C9ET	C9EU	C9EV	C9EW	C9EX	C9EY
2003	340.5	392.7	657.0	476.5	561.5	489.8	414.6	311.3	476.3	660.6	568.5	469.9	481.6	446.8	486.8
2004[1]	355.8	415.7	617.1	485.0	579.2	505.1	421.3	319.1	494.4	667.5	573.9	497.6	495.9	478.9	499.5
	362.5	419.0	633.5	493.1	591.9	509.4	433.3	323.8	504.3	696.3	590.6	496.6	493.6	474.9	515.4
2005	364.7	440.6	657.9	508.0	612.2	524.6	425.2	323.5	508.0	701.3	589.3	525.0	518.6	503.2	503.8
2006	386.6	466.8	782.0	528.4	632.7	548.0	447.4	333.4	527.6	720.2	616.6	544.3	533.7	516.7	527.1

Men

	C9F2	C9F4	C9F6	C9F8	C9FA	C9FC	C9FE	C9FO	C9FQ	C9FS	C9FU	C9FW	C9FY	C9G2	C9G4
2003	356.2	391.4	671.2	503.2	595.7	503.8	464.4	351.9	493.5	832.1	636.7	522.5	528.8	581.1	562.2
2004[1]	368.8	426.6	637.1	511.2	607.9	517.8	470.3	345.9	514.0	829.8	635.3	549.0	542.2	624.3	572.8
	375.4	433.2	653.7	519.4	626.1	521.5	483.2	352.5	522.0	869.2	652.5	547.9	539.3	614.8	593.1
2005	381.9	445.7	675.4	533.8	647.3	537.6	469.5	357.2	527.7	872.4	654.9	583.6	568.4	669.5	563.6
2006	402.2	450.4	804.2	554.9	669.1	563.3	493.6	365.3	545.0	881.9	685.5	599.4	586.7	693.0	585.4

Women

	C9G6	C9G8	C9GA	C9GC	C9GE	C9GG	C9GI	C9GK	C9GM	C9HJ	C9HL	C9HN	C9HP	C9HR	C9HT
2003	270.8	–	566.5	372.8	426.0	370.8	321.6	262.2	410.0	463.7	446.4	390.9	445.9	394.0	379.1
2004[1]	290.9	–	496.3	380.8	464.8	392.8	330.9	283.8	423.1	474.1	461.0	420.8	461.8	419.7	398.4
	297.5	–	511.0	388.1	453.0	403.8	339.3	287.2	439.0	492.5	474.5	418.9	460.2	417.9	405.7
2005	287.6	–	552.5	404.3	490.2	411.5	343.2	282.4	432.8	500.0	472.8	440.2	484.0	440.3	417.9
2006	321.7	–	668.8	422.6	505.9	416.8	363.1	293.8	464.4	526.3	492.7	463.4	498.3	451.1	441.3

Average hourly earnings (excluding overtime)

Total

	C9HV	C9HX	C9HZ	C9I3	C9I5	C9I7	C9IA	C9IC	C9IE	C9IG	C9II	C9IK	C9IM	C9IO	C9IQ
2003	7.44	9.09	14.99	11.62	13.99	11.22	10.26	7.63	11.20	18.32	14.54	11.88	13.55	11.57	12.38
2004[1]	7.87	9.41	14.60	11.80	14.29	11.68	10.44	7.74	11.73	18.44	14.67	12.62	13.96	12.39	12.60
	8.03	9.57	14.97	12.01	14.61	11.81	10.74	7.86	11.99	19.25	15.10	12.57	13.90	12.29	13.02
2005	8.27	10.14	15.56	12.50	15.33	12.15	10.57	7.93	11.89	19.54	15.14	13.28	14.64	12.99	12.65
2006	8.68	11.10	18.86	12.97	15.37	12.66	11.10	8.13	12.57	20.00	15.73	13.80	15.06	13.41	13.29

Men

	C9IS	C9IU	C9IW	C9IY	C9J2	C9J4	C9J6	C9J8	C9JA	C9JC	C9JE	C9JG	C9JI	C9JK	C9JM
2003	7.57	8.95	14.97	12.12	14.70	11.39	11.21	8.40	11.37	23.01	15.99	12.91	14.48	14.73	14.00
2004[1]	7.96	9.26	14.80	12.28	14.82	11.83	11.38	8.23	11.97	22.88	15.94	13.62	14.88	15.82	14.14
	8.11	9.47	15.16	12.49	15.29	11.95	11.71	8.38	12.20	23.99	16.39	13.56	14.79	15.57	14.68
2005	8.46	10.13	15.72	12.98	15.96	12.30	11.42	8.56	12.12	24.19	16.52	14.43	15.64	16.91	13.83
2006	8.86	10.38	19.03	13.46	15.94	12.85	11.99	8.74	12.70	24.37	17.16	14.90	16.14	17.58	14.48

Women

	C9JO	C9JQ	C9JS	C9JU	C9JW	C9JY	C9K2	C9K4	C9K6	C9K8	C9KG	C9KI	C9KK	C9KS	C9KU
2003	6.81	–	15.17	9.62	11.17	9.69	8.38	6.64	10.51	12.88	11.83	10.28	12.83	10.30	9.98
2004[1]	7.38	–	13.31	9.84	12.17	10.25	8.60	7.07	10.80	13.10	12.22	11.07	13.26	10.97	10.39
	7.55	–	13.73	10.02	11.83	10.55	8.82	7.16	11.22	13.62	12.58	11.01	13.22	10.93	10.60
2005	7.35	–	14.54	10.51	13.13	10.82	8.92	7.12	10.99	14.00	12.59	11.57	13.92	11.49	10.90
2006	7.91	–	17.95	10.97	13.37	10.83	9.40	7.33	12.06	14.69	13.06	12.13	14.32	11.82	11.48

1 In 2004 a number of supplementary surveys were introduced to improve the coverage of the Annual Survey of Hours and Earnings. Data for 2004 are presented including these supplementary surveys (top). Figures are also presented excluding supplementary surveys (bottom) to give figures comparable with earlier years.

Source: Office for National Statistics: 01633 819024

18.14 Average weekly and hourly earnings of full-time employees on adult rates by age group: United Kingdom

April 2001 to 2006
£

	Full time employees on adult rates whose pay was unaffected by absence						
	18-21	22-29	30-39	40-49	50-59	60+	All ages
Average gross weekly earnings							
Total							
	C7MV	C7MX	C7NG	C7NI	C7NK	C7OW	C7NM
2001	239.4	373.6	483.5	498.7	457.4	..	449.7
2002	247.7	390.8	507.5	523.6	477.2	..	472.1
2003	251.2	396.9	522.1	545.0	489.8	..	487.1
2004[1]	257.6	402.3	531.0	560.1	519.0	442.3	498.2
	260.7	410.2	540.7	567.3			506.1
2005	266.0	411.0	556.3	582.8	544.4	470.4	516.4
2006	273.5	421.4	575.9	610.4	568.9	509.5	537.3
Men							
	C7NO	C7NQ	C7NS	C7NU	C7O9	C7OU	C7OB
2001	256.6	401.4	520.6	560.1	510.5	..	498.6
2002	267.1	419.2	545.0	587.9	533.5	..	523.3
2003	266.3	425.0	562.5	612.0	543.4	..	539.3
2004[1]	272.4	426.4	568.8	623.9	580.6	466.6	548.2
	275.3	435.4	580.5	632.2			557.4
2005	278.9	432.8	596.4	651.2	606.8	497.3	568.0
2006	289.0	443.3	615.4	684.5	634.7	539.4	591.6
Women							
	C7OD	C7OF	C7OH	C7OJ	C7OL	C7SD	C7ON
2001	219.0	339.5	412.6	389.1	353.1	..	366.9
2002	225.7	356.4	436.1	410.4	371.6	..	386.8
2003	233.6	363.6	447.8	426.5	390.1	..	400.7
2004[1]	240.1	374.2	462.5	447.8	417.9	361.4	416.8
	243.3	380.5	468.4	452.6			422.1
2005	249.8	386.2	486.0	468.8	445.8	390.3	435.6
2006	253.7	397.1	508.7	488.8	465.6	420.0	453.6
Average hourly earnings (excluding overtime)							
Total							
	C7OP	C7OR	C7OT	C7OV	C7OX	C7PA	C7OZ
2001	5.92	9.45	12.16	12.68	11.53	..	11.35
2002	6.13	9.93	12.83	13.35	12.11	..	11.97
2003	6.25	10.11	13.20	13.84	12.39	..	12.34
2004[1]	6.42	10.22	13.47	14.20	13.24	10.92	12.63
	6.49	10.44	13.73	14.40			12.85
2005	6.62	10.48	14.16	14.85	13.96	11.72	13.14
2006	6.80	10.72	14.67	15.54	14.55	12.70	13.67
Men							
	C7P3	C7P5	C7P7	C7P9	C7PB	C7P8	C7PC
2001	6.16	9.83	12.74	13.86	12.50	..	12.24
2002	6.39	10.32	13.45	14.59	13.16	..	12.92
2003	6.40	10.48	13.85	15.12	13.35	..	13.28
2004[1]	6.57	10.48	14.05	15.40	14.38	11.29	13.52
	6.66	10.72	14.36	15.63			13.76
2005	6.75	10.68	14.77	16.13	15.10	12.10	14.05
2006	6.99	10.92	15.26	16.94	15.76	13.15	14.62
Women							
	C7PJ	C7PL	C7PN	C7PP	C7PR	C7P6	C7PT
2001	5.63	8.96	10.99	10.47	9.51	..	9.79
2002	5.82	9.43	11.61	11.04	10.01	..	10.32
2003	6.06	9.65	11.95	11.44	10.52	..	10.70
2004[1]	6.22	9.91	12.35	11.97	11.26	9.59	11.12
	6.29	10.09	12.52	12.11			11.27
2005	6.46	10.24	13.02	12.59	12.05	10.48	11.65
2006	6.56	10.50	13.62	13.13	12.55	11.24	12.11

1 In 2004 a number of supplementary surveys were introduced to improve the coverage of the Annual Survey of Hours and Earnings. Data for 2004 are presented including these supplementary surveys (top). Figures are also presented excluding supplementary surveys (bottom) to give figures comparable with earlier years.

Source: Office for National Statistics: 01633 819024

18.15 Average earnings index: by industry (not seasonally adjusted)[1,2]

Great Britain

2000 = 100

Excluding bonuses

	Agriculture, forestry and fishing	Mining and quarrying	Food products, beverages and tobacco	Textiles, leather and clothing	Chemicals and man-made fibres	Basic metals and metal products	Engineering and allied industries	Other manufacturing	Electricity, gas and water supply	Construction
SIC 1992	(A,B)	(C)	(DA)	(DB,DC)	(DG)	(DJ)	(DK, DL,DM)	(DD,DE,DF, DH,DI,DN)	(E)	(F)
	JVUZ	JVVA	JVVB	JVVC	JVVD	JVVE	JVVF	JVVG	JVVH	JVVI
2004	122.7	117.5	117.6	117.1	118.3	115.6	117.1	115.8	110.8	119.8
2005	125.3	123.1	121.9	119.3	120.0	120.9	121.6	120.2	114.1	124.0
2004 Apr	123.7	115.1	117.2	114.4	117.7	113.2	116.7	115.2	112.1	119.2
May	120.1	116.0	118.7	116.1	118.1	115.3	117.2	116.4	111.0	118.7
Jun	123.9	116.2	117.6	117.6	119.5	115.5	117.1	116.0	113.3	119.5
Jul	122.5	116.1	117.8	119.6	119.0	117.3	118.3	116.3	111.4	120.4
Aug	120.5	114.6	118.0	117.2	118.9	116.7	117.5	115.2	110.9	119.7
Sep	123.4	115.9	117.4	118.4	118.1	116.7	117.2	115.9	109.5	120.7
Oct	122.5	127.3	118.1	118.5	120.4	117.6	118.6	116.2	111.3	121.4
Nov	127.2	122.5	119.6	118.5	120.2	117.1	119.0	116.8	110.9	121.9
Dec	128.2	121.3	121.9	119.4	121.2	116.3	119.3	117.2	111.1	122.2
2005 Jan	125.1	120.4	119.4	118.1	120.9	118.5	119.0	116.2	111.2	121.8
Feb	121.5	123.6	118.3	116.1	121.0	119.1	119.5	117.3	111.6	120.4
Mar	124.8	120.4	121.8	118.3	122.0	118.4	120.0	117.5	110.9	121.7
Apr	124.3	123.1	120.7	119.0	118.8	120.9	121.2	118.8	113.4	122.3
May	120.9	123.3	121.8	118.1	118.3	120.0	121.3	119.3	113.4	123.1
Jun	125.9	122.4	120.7	121.0	119.4	121.4	121.3	120.4	115.6	124.4
Jul	122.2	122.1	121.2	119.1	118.5	122.2	122.7	120.3	115.3	125.1
Aug	122.5	122.5	122.0	117.0	119.7	122.2	121.7	121.0	115.2	123.3
Sep	131.7	123.5	122.6	118.9	119.2	123.2	122.5	122.1	113.7	125.7
Oct	130.3	125.2	123.1	121.6	119.4	122.9	123.6	122.3	115.2	126.2
Nov	126.8	125.6	125.2	121.9	121.1	122.1	123.1	122.9	116.1	128.1
Dec	127.6	125.1	126.2	122.4	121.3	120.0	123.6	124.2	117.8	126.4
2006 Jan	129.0	127.4	125.0	122.1	121.3	124.0	123.0	124.1	115.7	126.6
Feb	132.0	124.9	124.3	123.1	121.6	124.5	124.7	124.7	116.3	127.6
Mar	133.0	126.1	125.2	121.4	121.1	125.7	125.2	125.1	115.2	127.0
Apr	141.3	127.6	129.4	122.5	122.1	125.2	126.4	125.2	114.2	126.6
May	140.2	128.1	128.4	123.2	122.0	126.9	126.3	125.9	118.3	127.2
Jun	141.4	128.4	127.8	124.0	123.0	129.5	126.5	126.9	118.2	127.9
Jul	137.2	128.7	128.3	122.8	121.6	128.4	126.4	126.5	118.7	128.2
Aug	139.9	129.0	128.2	120.1	122.5	127.9	126.2	127.1	116.2	126.7
Sep	135.7	131.0	128.1	122.1	124.3	129.3	127.7	127.7	114.6	128.5
Oct	130.3†	131.3†	128.2†	122.0†	125.1†	129.2†	128.8†	127.8	113.0	129.5†
Nov	123.8	131.7	127.8	122.4	123.7	129.7	129.3	128.6	116.5	129.6

Percentage change on the year

	JVVT	JVVU	JVVV	JVVW	JVVX	JVVY	JVVZ	JVWA	JVWB	JVWC
2005 Apr	0.5	7.0	3.0	4.1	0.9	6.8	3.9	3.2	1.2	2.7
May	0.6	6.3	2.6	1.7	0.2	4.0	3.6	2.5	2.2	3.7
Jun	1.6	5.4	2.7	2.9	−0.1	5.2	3.6	3.7	2.0	4.1
Jul	−0.2	5.2	2.9	−0.4	−0.4	4.2	3.8	3.5	3.6	3.9
Aug	1.6	6.9	3.4	−0.2	0.7	4.7	3.6	5.0	3.9	3.1
Sep	6.8	6.5	4.5	0.4	0.9	5.5	4.5	5.3	3.9	4.1
Oct	6.4	−1.7	4.3	2.6	−0.8	4.5	4.2	5.3	3.5	3.9
Nov	−0.3	2.6	4.7	2.8	0.8	4.3	3.5	5.2	4.7	5.1
Dec	−0.4	3.2	3.5	2.5	−	3.2	3.6	6.0	6.0	3.4
2006 Jan	3.1	5.8	4.7	3.3	0.3	4.7	3.4	6.7	4.1	3.9
Feb	8.6	1.0	5.0	6.1	0.6	4.5	4.3	6.4	4.2	6.0
Mar	6.6	4.7	2.8	2.6	−0.7	6.2	4.3	6.5	3.9	4.4
Apr	13.7	3.6	7.2	2.9	2.8	3.6	4.2	5.4	0.7	3.4
May	16.0	4.0	5.5	4.4	3.1	5.8	4.1	5.5	4.4	3.3
Jun	12.3	4.8	5.9	2.5	3.0	6.7	4.2	5.4	2.3	2.8
Jul	12.2	5.5	5.8	3.1	2.6	5.1	3.0	5.1	2.9	2.4
Aug	14.2	5.3	5.1	2.7	2.3	4.7	3.7	5.1	0.8	2.8
Sep	3.0	6.1	4.5	2.7	4.3	5.0	4.2	4.6	0.8	2.2
Oct	−†	4.9†	4.1†	0.4	4.9	5.1†	4.3†	4.5	−1.9	2.7
Nov	−2.4	4.8	2.0	0.4	2.2	6.2	5.1	4.7	0.4	1.2

125

18.15

Average earnings index: by industry
(not seasonally adjusted)[1,2]
Great Britain

2000 = 100

	Wholesale trade	Retail trade and repairs	Hotels and restaurants	Transport, storage and communication	Financial interm-ediation	Real estate renting and business activities	Public admini-stration	Education	Health and social work	Other services
Excluding bonuses										
SIC 1992	(G:51)	(G:50,52)	(H)	(I)	(J)	(K)	(L)	(M)	(N)	(O)
	JVVJ	JVVK	JVVL	JVVM	JVVN	JVVO	JVVP	JVVQ	JVVR	JVVS
2004	112.9	114.0	122.3	118.7	115.2	117.9	118.8	119.5	126.7	112.4
2005	117.6	116.4	126.6	123.6	120.6	122.6	124.2	124.1	132.4	117.3
2004 Apr	112.7	114.6	120.6	117.4	114.9	117.4	117.6	118.8	125.6	110.3
May	113.3	114.5	121.1	117.9	115.1	118.7	118.0	119.2	126.1	110.7
Jun	112.9	114.7	121.9	119.7	115.1	117.5	118.1	119.0	130.2	111.9
Jul	112.8	114.8	123.5	119.1	114.9	118.4	118.2	119.5	128.3	114.1
Aug	113.0	115.4	124.2	119.8	115.2	118.2	119.7	123.2	128.1	114.3
Sep	113.7	115.1	122.7	120.3	115.1	118.2	121.7	123.3	128.6	113.2
Oct	113.5	114.4	124.9	121.5	116.5	118.3	120.7	121.6	128.7	112.8
Nov	114.0	113.2	123.9	120.8	116.7	118.9	122.1	120.6	129.2	115.0
Dec	115.6	114.7	128.4	120.6	117.3	120.1	121.7	121.9	129.2	113.9
2005 Jan	115.6	117.3	122.8	121.4	117.7	120.5	120.5	122.0	129.2	114.7
Feb	115.2	115.5	123.7	120.7	118.3	121.0	121.9	120.8	128.8	114.5
Mar	116.9	115.7	126.8	121.0	121.6	120.7	125.9	120.7	128.9	116.7
Apr	117.3	117.9	125.9	122.4	120.9	122.1	124.3	124.0	132.9	115.3
May	117.6	116.3	126.3	123.3	121.3	122.1	123.0	123.5	132.9	116.8
Jun	117.3	116.0	126.8	125.2	119.2	122.3	123.0	124.0	133.9	119.2
Jul	118.0	117.8	127.1	123.9	121.8	123.5	124.3	124.5	133.0	121.3
Aug	118.1	118.3	127.3	123.4	121.1	123.0	124.7	126.1	132.9	118.8
Sep	118.0	115.8	126.2	125.8	119.5	123.2	125.3	126.8	132.9	118.6
Oct	119.1	116.0	126.7	124.9	121.0	123.7	125.4	126.3	133.2	115.4
Nov	119.1	115.2	127.4	125.2	121.3	124.3	125.7	124.9	135.0	116.8
Dec	119.3	115.4	132.5	126.4	123.3	124.7	126.9	125.4	134.7	119.8
2006 Jan	119.8	117.9	127.2	124.9	123.9	126.3	126.0	124.8	135.3	120.0
Feb	119.8	115.8	127.8	124.6	123.1	125.4	129.5	125.0	135.9	118.8
Mar	119.8	116.6	130.9	125.3	123.9	126.2	127.5	125.8	136.2	120.2
Apr	120.9	117.9	131.8	127.2	126.4	127.3	127.9	127.8	136.5	122.0
May	120.9	120.0	133.1	127.5	126.5	127.3	127.9	127.1	137.2	122.3
Jun	122.1	118.5	132.1	127.9	125.7	128.0	128.4	127.6	138.7	124.6
Jul	122.0	119.2	134.0	126.8	125.8	128.0	128.5	128.8	138.7	123.0
Aug	122.1	120.1	134.1	126.8	125.6	128.1	127.2	131.6	137.7	122.7
Sep	122.4	120.5	134.7	128.3	124.9	128.3	128.4	132.2	137.7	121.4
Oct	123.6†	120.5†	136.2†	127.0†	126.3†	129.3†	128.2	131.3†	137.8†	121.2†
Nov	123.5	118.7	136.3	127.6	125.9	129.2	128.8	130.5	139.3	122.9
Percentage change on the year										
	JVWD	JVWE	JVWF	JVYJ	JVYK	JVYL	JVYM	JVYN	JVYO	JVYP
2005 Apr	4.1	2.9	4.5	4.2	5.2	3.9	5.7	4.4	5.8	4.6
May	3.9	1.6	4.3	4.6	5.4	2.9	4.2	3.7	5.4	5.5
Jun	3.9	1.2	4.0	4.5	3.5	4.1	4.1	4.2	2.9	6.5
Jul	4.6	2.6	2.9	4.0	6.0	4.3	5.1	4.2	3.7	6.4
Aug	4.5	2.5	2.5	3.0	5.1	4.1	4.2	2.4	3.8	4.0
Sep	3.8	0.7	2.9	4.6	3.9	4.2	2.9	2.9	3.4	4.8
Oct	4.9	1.4	1.5	2.8	3.9	4.6	3.9	3.9	3.5	2.3
Nov	4.5	1.7	2.9	3.6	3.9	4.5	3.0	3.5	4.5	1.6
Dec	3.2	0.6	3.2	4.9	5.1	3.8	4.3	2.9	4.3	5.2
2006 Jan	3.7	0.6	3.6	2.9	5.3	4.8	4.6	2.3	4.7	4.6
Feb	4.0	0.3	3.3	3.3	4.0	3.7	6.2	3.4	5.5	3.8
Mar	2.5	0.8	3.3	3.6	1.9	4.6	1.3	4.3	5.7	3.0
Apr	3.0	–	4.6	3.9	4.5	4.3	2.9	3.0	2.7	5.8
May	2.8	3.1	5.4	3.4	4.2	4.2	4.0	2.9	3.3	4.8
Jun	4.1	2.2	4.2	2.2	5.4	4.7	4.3	3.0	3.6	4.5
Jul	3.3	1.2	5.4	2.4	3.3	3.7	3.4	3.5	4.3	1.3
Aug	3.4	1.5	5.4	2.8	3.8	4.1	2.0	4.3	3.6	3.3
Sep	3.7	4.0†	6.8	1.9†	4.5	4.2†	2.5†	4.3	3.6	2.4
Oct	3.8†	3.9†	7.5†	1.7†	4.4	4.5†	2.3†	3.9†	3.4†	5.0†
Nov	3.7	3.1	6.9	1.9	3.8	3.9	2.4	4.5	3.2	5.2

18.15
continued

Average earnings index: by industry
(not seasonally adjusted)[1,2]
Great Britain

2000 = 100

	Agriculture, forestry and fishing	Mining and quarrying	Food products, beverages and tobacco	Textiles, leather and clothing	Chemicals and man-made fibres	Basic metals and metal products	Engineering and allied industries	Other manufacturing	Electricity, gas and water supply	Construction
Including bonuses										
SIC 1992	(A,B)	(C)	(DA)	(DB,DC)	(DG)	(DJ)	(DK, DL,DM)	(DD,DE,DF, DH,DI,DN)	(E)	(F)
	JVUF	JVUG	JVUH	JVUI	JVUJ	JVUK	JVUL	JVUM	JVUN	JVUO
2004	121.6	121.9	113.9	114.2	120.1	116.5	118.5	112.2	110.6	119.2
2005	124.5	127.2	117.3	119.5	120.4	124.2	122.2	116.8	115.5	124.3
2004 Apr	122.7	132.6	115.0	110.7	125.6	116.0	117.6	110.9	110.6	117.1
May	119.0	115.8	115.2	113.8	116.9	114.2	117.6	113.3	109.3	118.5
Jun	123.9	116.1	112.4	114.4	117.3	115.1	117.5	112.1	123.1	117.7
Jul	122.2	114.8	112.9	116.9	117.6	120.5	118.1	112.4	109.1	119.5
Aug	118.8	114.2	111.2	113.6	115.0	115.4	116.8	109.7	108.8	116.4
Sep	122.7	118.2	113.4	114.4	113.1	115.4	117.0	110.9	106.5	118.2
Oct	121.4	127.5	110.5	115.4	116.5	120.2	118.1	111.7	108.6	119.0
Nov	126.3	123.8	112.0	114.8	114.1	117.4	119.6	112.4	108.1	124.0
Dec	125.8	125.6	120.5	120.1	121.7	120.5	122.7	115.1	108.4	124.7
2005 Jan	123.4	128.8	112.3	117.0	117.9	122.6	118.7	111.8	110.0	121.3
Feb	119.5	137.2	114.2	116.7	121.6	122.3	124.4	113.5	117.3	119.8
Mar	126.0	148.9	129.2	117.2	150.3	125.0	126.2	120.3	112.0	128.8
Apr	122.0	137.9	116.9	117.1	122.5	126.3	123.4	114.2	113.6	120.5
May	118.0	119.2	114.6	116.0	115.7	119.9	119.9	115.4	114.6	122.6
Jun	122.7	120.5	113.3	120.2	116.5	121.5	121.0	115.5	124.9	123.0
Jul	119.4	117.8	117.8	120.0	115.5	126.9	121.7	116.8	115.0	124.4
Aug	120.1	120.1	116.6	117.2	115.6	122.8	119.3	115.8	112.7	120.9
Sep	143.4	125.6	118.0	118.1	115.8	125.2	120.3	116.7	110.2	124.3
Oct	127.5	121.8	115.3	126.6	115.1	128.8	121.8	118.1	112.7	124.9
Nov	125.6	123.5	116.2	121.3	116.1	124.9	122.5	119.0	111.4	127.6
Dec	125.9	124.6	122.9	126.6	122.0	124.5	126.9	124.2	130.8	132.9
2006 Jan	126.1	130.8	117.0	123.7	117.4	127.8	123.4	120.4	113.7	123.9
Feb	129.2	131.0	120.8	123.6	121.2	125.4	132.1	121.0	115.7	125.2
Mar	130.5	160.6	132.4	125.5	146.2	130.5	135.4	127.2	118.8	130.3
Apr	138.9	150.4	127.2	124.4	121.1	132.3	130.4	121.8	116.9	122.8
May	137.3	130.3	122.0	124.4	112.9	130.2	126.7	122.3	121.3	123.0
Jun	139.0	128.8	122.5	125.6	115.4	131.8	127.0	124.1	129.6	125.8
Jul	134.5	126.8	122.5	125.4	114.8	135.2	127.4	123.6	119.2	125.1
Aug	137.2	126.6	120.4	121.8	114.7	130.4	126.3	124.0	115.6	121.6
Sep	133.0	130.6	125.1	122.7	117.8	135.6	127.6	121.9	114.4	125.1
Oct	127.6[†]	130.2[†]	121.6[†]	125.1	116.5[†]	139.6[†]	129.6[†]	122.6	114.3	125.1[†]
Nov	121.2	136.9	121.1	125.5	114.4	132.4	130.8	123.4	116.4	127.4
Percentage change on the year										
	JVYQ	JVYR	JVYS	JVYT	JVYU	JVYV	JVYW	JVYX	JVYY	JVYZ
2005 Apr	−0.5	4.0	1.7	5.8	−2.4	8.9	4.9	3.0	2.7	3.0
May	−0.8	3.0	−0.5	2.0	−1.0	5.0	1.9	1.8	4.8	3.5
Jun	−1.0	3.8	0.8	5.1	−0.6	5.6	3.0	3.1	1.5	4.5
Jul	−2.3	2.6	4.4	2.6	−1.8	5.3	3.0	4.0	5.4	4.1
Aug	1.1	5.2	4.8	3.2	0.6	6.5	2.2	5.6	3.6	3.9
Sep	16.9	6.2	4.1	3.3	2.4	8.5	2.8	5.3	3.5	5.2
Oct	5.1	−4.5	4.4	9.7	−1.2	7.1	3.1	5.7	3.8	5.0
Nov	−0.5	−0.2	3.8	5.6	1.8	6.4	2.4	5.8	3.0	2.9
Dec	0.1	−0.8	2.0	5.4	0.2	3.4	3.5	7.9	20.7	6.5
2006 Jan	2.2	1.5	4.2	5.7	−0.4	4.2	4.0	7.7	3.4	2.1
Feb	8.1	−4.6	5.7	5.9	−0.3	2.5	6.3	6.6	−1.4	4.6
Mar	3.6	7.9	2.5	7.1	−2.8	4.4	7.3	5.8	6.0	1.2
Apr	13.8	9.1	8.8	6.2	−1.2	4.8	5.7	6.6	2.9	1.9
May	16.4	9.3	6.5	7.2	−2.4	8.6	5.7	6.0	5.9	0.3
Jun	13.3	6.9	8.1	4.5	−0.9	8.4	5.0	7.4	3.8	2.3
Jul	12.6	7.7	4.0	4.5	−0.6	6.5	4.7	5.8	3.6	0.6
Aug	14.2	5.4	3.3	3.9	−0.8	6.2	5.8	7.1	2.5	0.6
Sep	−7.3	4.0	6.0	3.9	1.7	8.3	6.0	4.5	3.8	0.6
Oct	0.1[†]	6.9[†]	5.4[†]	−1.2	1.2[†]	8.4[†]	6.4[†]	3.7	1.4	0.2
Nov	−3.5	10.8	4.2	3.5	−1.5	6.0	6.8	3.7	4.5	−0.2

18.15

Average earnings index: by industry (not seasonally adjusted)[1,2]

continued

Great Britain

2000 = 100

	Wholesale trade	Retail trade and repairs	Hotels and restaurants	Transport, storage and communication	Financial interm-ediation	Real estate renting and business activities	Public admini-stration	Education	Health and social work	Other services
Including bonuses										
SIC 1992	(G:51)	(G:50,52)	(H)	(I)	(J)	(K)	(L)	(M)	(N)	(O)
	JVUP	JVUQ	JVUR	JVUS	JVUT	JVUU	JVUV	JVUW	JVUX	JVUY
2004	115.4	113.9	125.7	117.7	109.8	114.3	118.4	119.3	126.6	115.6
2005	119.3	116.6	131.5	124.6	114.4	118.4	124.1	123.8	132.5	120.3
2004 Apr	113.6	114.9	122.6	115.8	99.4	113.7	116.8	118.5	125.7	111.1
May	111.1	113.2	125.1	116.5	93.9	115.1	117.4	118.9	126.0	112.4
Jun	114.7	115.1	124.0	126.1	93.3	113.4	117.3	118.7	130.1	120.9
Jul	114.1	114.0	126.2	117.0	92.1	114.8	117.5	119.3	128.3	116.4
Aug	113.2	114.1	126.6	116.8	90.9	112.7	121.2	123.0	128.0	115.3
Sep	113.9	114.6	125.6	117.3	90.5	111.5	121.1	122.9	128.5	115.6
Oct	114.1	113.8	128.5	118.3	96.3	112.5	120.1	121.3	128.7	116.2
Nov	116.5	112.4	127.8	118.8	93.2	113.4	121.4	120.5	129.2	120.0
Dec	123.7	114.8	135.6	121.0	101.7	117.7	122.3	121.6	129.3	119.1
2005 Jan	117.0	117.0	128.6	118.2	163.7	117.7	119.6	121.7	129.1	119.5
Feb	118.9	117.5	132.0	121.6	173.7	117.3	121.1	120.7	129.2	116.0
Mar	126.3	118.7	134.5	121.7	156.0	124.5	125.3	120.4	129.3	123.7
Apr	120.8	119.0	129.4	122.6	101.0	117.3	123.6	123.9	133.0	118.3
May	116.6	115.9	131.5	131.6	96.2	116.9	122.3	123.2	132.9	120.2
Jun	118.1	116.9	129.9	133.3	96.9	118.3	122.2	123.6	134.0	127.8
Jul	118.7	117.2	130.2	125.5	97.0	120.7	124.2	124.3	133.0	122.2
Aug	115.3	116.9	130.9	121.4	96.1	117.1	126.4	125.9	133.0	120.3
Sep	115.5	114.1	128.5	122.8	94.8	115.3	124.6	126.5	132.8	119.7
Oct	119.9	115.6	129.8	122.0	93.1	116.0	125.2	126.0	133.4	116.3
Nov	121.3	114.3	131.7	123.6	96.4	117.1	125.6	124.5	134.9	117.2
Dec	123.8	116.1	140.5	130.4	108.1	122.8	129.0	125.1	134.8	122.8
2006 Jan	121.1	118.0	129.9	123.6	168.7	120.9	125.5	124.4	135.2	121.1
Feb	121.4	115.6	134.7	124.1	209.8	121.1	129.1	124.8	135.9	121.1
Mar	129.6	122.2	136.5	125.7	175.6	129.8	127.5	125.5	137.1	123.3
Apr	121.0	119.3	134.5	124.3	105.4	122.9	127.9	127.4	136.4	123.2
May	120.2	119.7	138.4	139.0	103.4	122.3	127.7	126.8	137.0	125.4
Jun	123.0	120.8	134.7	138.2	113.2	124.7	129.1	127.3	138.5	124.9
Jul	123.9	121.3	136.5	127.5	103.4	124.9	131.2	128.7	138.5	123.9
Aug	121.3	119.0	136.9	124.6	99.3	122.2	130.1	131.3	137.4	123.2
Sep	121.9	119.6	137.6	124.6	96.7	122.3	128.6	131.9	137.4	121.6
Oct	124.6†	120.2†	139.4†	122.9†	97.7†	122.6†	128.6	130.9†	137.6†	120.6
Nov	125.8	118.0	141.2	124.5	100.3	122.8	129.1	130.4	139.1	124.2
Percentage change on the year										
	JVZA	JVZB	JVZC	JVZD	JVZE	JVZF	JVZG	JVZH	JVZI	JVZJ
2005 Apr	6.3	3.5	5.5	5.9	1.6	3.2	5.8	4.5	5.7	6.5
May	5.0	2.4	5.1	13.0	2.4	1.6	4.2	3.6	5.5	6.9
Jun	2.9	1.6	4.8	5.7	3.9	4.3	4.1	4.1	2.9	5.7
Jul	4.0	2.8	3.2	7.3	5.3	5.1	5.6	4.2	3.7	5.0
Aug	1.8	2.4	3.4	4.0	5.8	3.9	4.3	2.3	3.9	4.3
Sep	1.5	−0.4	2.3	4.6	4.8	3.4	2.9	3.0	3.3	3.5
Oct	5.1	1.5	1.0	3.1	−3.3	3.1	4.3	3.9	3.7	0.1
Nov	4.1	1.7	3.1	4.0	3.4	3.3	3.4	3.4	4.4	−2.3
Dec	0.1	1.2	3.6	7.7	6.3	4.3	5.5	2.9	4.3	3.1
2006 Jan	3.5	0.9	1.0	4.6	3.1	2.8	4.9	2.3	4.7	1.3
Feb	2.1	−1.6	2.0	2.0	20.8	3.2	6.6	3.4	5.2	4.4
Mar	2.6	2.9	1.5	3.3	12.6	4.3	1.7	4.2	6.1	−0.3
Apr	0.2	0.2	4.0	1.4	4.3	4.8	3.5	2.9	2.6	4.2
May	3.1	3.3	5.2	5.6	7.6	4.6	4.4	3.0	3.1	4.3
Jun	4.2	3.3	3.7	3.7	16.7	5.4	5.6	3.0	3.4	−2.2
Jul	4.3	3.5	4.9	1.6	6.6	3.5	5.7	3.5	4.1	1.4
Aug	5.2	1.8	4.6	2.6	3.3	4.3	2.9	4.3	3.3	2.4
Sep	5.5	4.8	7.0	1.5	2.0	6.1	3.1	4.2	3.5	1.6
Oct	3.9†	4.0†	7.4†	0.7†	4.9†	5.8†	2.7†	3.9†	3.2†	3.8†
Nov	3.8	3.2	7.2	0.7	4.1	4.8	2.8	4.7	3.1	5.9

1 The above table of 20 industries was first published in the Monthly Digest in May 2002 (as table 18.11). The new set of 20 industry sectors was introduced as it better reflects the current state of the economy. Data are available in two formats: excluding bonus and including bonus, with each available as an index value and as an annual percentage change. An article covering the reasons for change can be found on our website: www.statistics.gov.uk/labour.

2 Users should note that the data contained in the previous set of 26 industry sectors are not comparable with the new set of 20 industry sectors.

Source: Office for National Statistics: 01633 819024

18.16 Average earnings index[1]: main industrial sectors
Great Britain

2000 = 100

	Whole economy				Public sector				Private sector			
	Actual	Seasonally adjusted	Single month[2]	3 month average[2]	Actual	Seasonally adjusted	Single month[2]	3 month average[2]	Actual	Seasonally adjusted	Single month[2]	3 month average[2]
SIC 1992												
	LNMM	LNMQ	LNMU	LNNC	LNNI	LNNJ	LNKW	LNNE	LNKX	LNKY	LNKZ	LNND
1996	83.3	83.3	..	..	87.8	87.6	..	..	82.2	82.3	..	..
1997	86.8	86.8	..	..	89.7	89.6	..	..	86.2	86.2	..	..
1998	91.3	91.3	..	..	92.6	92.5	..	..	90.9	91.0	..	..
1999	95.7	95.7	..	..	96.4	96.4	..	..	95.5	95.5	..	..
2000	100.0	100.0	..	..	100.0	100.0	..	..	100.0	100.0	..	..
2001	104.4	104.5	..	..	105.1	105.0	..	..	104.2	104.3	..	..
2002	108.1	108.2	..	..	109.6	109.3	..	..	107.8	107.9	..	..
2003	111.7	111.9	..	..	115.0	114.8	..	..	111.0	111.3	..	..
2004	116.7	116.8	..	..	120.4	119.8[†]	..	..	115.9	116.0	..	..
2005	121.4	121.5	..	..	125.9	125.5[†]	..	..	120.4	120.6	..	..
2002 Dec	111.3	109.5	3.4	3.9	113.2	112.1	4.9	4.7	110.9	108.6	2.9	3.6
2003 Jan	109.9	109.5	3.1	3.6	111.6	112.5	5.1	5.0	109.5	108.7	2.7	3.2
Feb	113.8	109.7	2.7	3.1	111.6	112.8	5.2	5.1	114.3	108.9	2.2	2.6
Mar	116.8	110.7	4.2	3.3	112.2	113.3	5.1	5.2	117.9	110.0	3.9	2.9
Apr	110.0	110.7	2.5	3.1	114.6	113.9	5.1	5.2	109.0	110.0	1.9	2.7
May	110.0	111.3	3.1	3.3	114.5	113.9	4.7	5.0	109.0	110.9	2.9	2.9
Jun	111.2	111.6	3.2	3.0	115.7	114.7	5.4	5.1	110.2	111.0	2.7	2.5
Jul	111.8	112.5	3.7	3.4	116.7	115.6	5.3	5.1	110.7	111.9	3.3	3.0
Aug	110.2	112.5	3.5	3.5	117.2	115.6	6.0	5.5	108.5	111.8	3.0	3.0
Sep	110.4	113.2	3.9	3.7	116.0	116.1	5.5	5.6	109.0	112.6	3.5	3.3
Oct	110.9	113.5	3.9	3.8	115.8	116.1	4.7	5.4	109.7	113.0	3.7	3.4
Nov	111.2	113.8	3.3	3.7	116.6	116.3	4.2	4.8	110.0	113.2	3.1	3.4
Dec	114.7	114.2	4.4	3.9	117.8	116.9	4.2	4.4	114.0	113.9	4.9	3.9
2004 Jan	118.2	116.3	6.2	4.6	116.1	117.1	4.1	4.2	118.7	115.3	6.0	4.7
Feb	118.1	113.6	3.6	4.7	116.5	117.8	4.4	4.2	118.5	112.7	3.5	4.8
Mar	122.2	115.4	4.3	4.7	117.0	118.2	4.4	4.3	123.5	114.7	4.3	4.6
Apr	115.0	115.7	4.5	4.1	119.4	118.6	4.1	4.3	114.1	115.1	4.6	4.1
May	114.8	115.9	4.1	4.3	119.9	119.1	4.6	4.4	113.6	115.5	4.2	4.4
Jun	116.1	116.2	4.1	4.2	122.3	119.8	4.5	4.4	114.6	115.5	4.1	4.3
Jul	115.4	116.4	3.4	3.9	121.0	119.8	3.7	4.2	114.2	115.6	3.3	3.8
Aug	114.8	117.3	4.2	3.9	123.0	120.8	4.5	4.2	112.9	116.5	4.1	3.8
Sep	114.9	117.8	4.0	3.9	122.5	121.2	4.4	4.2	113.1	117.0	3.9	3.8
Oct	115.7	118.6	4.5	4.2	121.7	121.7	4.8	4.6	114.4	117.9	4.4	4.1
Nov	116.2	119.0	4.6	4.4	121.9	121.8	4.7	4.7	114.9	118.3	4.5	4.3
Dec	119.5	119.0	4.2	4.4	123.3	122.0	4.4	4.6	118.6	118.4	3.9	4.3
2005 Jan	123.3	120.9	4.0	4.2	122.1	122.7	4.8	4.6	123.7	119.8	3.9	4.1
Feb	124.9	119.8	5.4	4.5	122.2	123.2	4.6	4.6	125.6	119.1	5.7	4.5
Mar	127.5	120.0	4.0	4.5	123.0	123.1	4.1	4.5	128.6	119.2	3.9	4.5
Apr	119.9	120.6	4.2	4.5	125.6	124.5	5.0	4.6	118.6	119.7	4.0	4.5
May	119.2	120.6	4.1	4.1	128.9	128.4	7.7	5.6	117.0	119.4	3.4	3.7
Jun	120.4	120.6	3.8	4.0	126.9	124.9	4.3	5.7	119.0	119.9	3.8	3.7
Jul	120.5	121.7	4.6	4.2	125.9	124.9	4.3	5.4	119.3	120.9	4.6	3.9
Aug	119.0	122.1	4.1	4.2	126.8	125.9	4.3	4.3	117.2	121.2[†]	4.1[†]	4.2
Sep	118.8	122.3	3.8	4.2	126.2	126.1	4.0	4.2	117.1	121.3[†]	3.7[†]	4.1
Oct	119.1	122.4[†]	3.2	3.7	126.5	126.6	4.1	4.1	117.4	121.4	3.0	3.6
Nov	119.9	123.2	3.5	3.5	127.0	127.2	4.4	4.2	118.3	122.2	3.3	3.3
Dec	124.6	124.0	4.2	3.7[†]	129.2	127.8	4.7	4.4	123.5	123.1	3.9	3.4
2006 Jan	127.2	124.5	3.0	3.6	126.8	127.8	4.1	4.4	127.4	123.4	2.9	3.4
Feb	131.6	125.8	5.0	4.1	128.5	128.2	4.1	4.3	132.5	125.5	5.3	4.1
Mar	133.2	125.3	4.4	4.1	128.0	128.5	4.3	4.2	134.6	124.4	4.4	4.2
Apr	124.1	124.8	3.5	4.3	129.3	128.1	2.9	3.8	122.8	124.1	3.7	4.5
May	124.5	125.9	4.4	4.1	133.8	133.1	3.7	3.6	122.3	125.0	4.7	4.2
Jun	126.4	126.6	4.9	4.3	131.3	129.5	3.7	3.4	125.3	126.2	5.3	4.5
Jul	125.2	126.5	3.9	4.4	131.7	130.0	4.1	3.8	123.6	125.5	3.8	4.6
Aug	123.5	126.8	3.8	4.2	131.1	129.9[†]	3.2	3.6	121.7	126.0	3.9	4.3
Sep	123.7[†]	127.3	4.1[†]	3.9	130.7	130.4[†]	3.4[†]	3.5	122.1[†]	126.5	4.3	4.0
Oct	123.9[†]	127.6	4.2[†]	4.1	130.7	130.7	3.3[†]	3.3	122.3[†]	126.8	4.4	4.2
Nov	124.5	128.0	3.9	4.1	131.5	131.2	3.1	3.2	122.9	127.1	4.0	4.2

18.16 Average earnings index[1]: main industrial sectors
Great Britain
continued

2000 = 100

	Production industries				Manufacturing industries				Service industries				of which Private sector services			
	Actual	Seasonally adjusted	Single month[2]	3 month average[2]	Actual	Seasonally adjusted	Single month[2]	3 month average[2]	Actual	Seasonally adjusted	Single month[2]	3 month average[2]	Actual	Seasonally adjusted	Single month[2]	3 month average[2]
SIC 1992	LNMO	LNMS	LNMW	LNNF	LNMN	LNMR	LNMV	LNNG	LNMP	LNMT	LNMX	LNNH	JJGF	JJGH	JJGI	JJGJ
1996	84.9	84.9	..	..	84.3	84.3	..	..	83.0	83.0	..	..	81.4	81.4	..	..
1997	88.3	88.4	..	..	87.8	87.9	..	..	86.6	86.6	..	..	85.5	85.5	..	..
1998	92.2	92.3	..	..	91.8	91.9	..	..	91.1	91.1	..	..	90.6	90.6	..	..
1999	95.8	95.9	..	..	95.6	95.6	..	..	95.7	95.7	..	..	95.4	95.4	..	..
2000	100.0	100.0	..	..	100.0	100.0	..	..	100.0	100.0	..	..	100.0	100.0	..	..
2001	104.2	104.2	..	..	104.2	104.3	..	..	104.4	104.4	..	..	104.1	104.2	..	..
2002	107.8	107.9	..	..	107.9	108.0	..	..	108.1	108.1	..	..	107.6	107.8	..	..
2003	111.6	111.7	..	..	111.7	111.9	..	..	111.7	112.0	..	..	110.6	111.0	..	..
2004	115.8	115.8	..	..	115.9	116.0	..	..	116.8	116.8	..	..	115.5	115.7	..	..
2005	120.0	120.0	..	..	120.1	120.2	..	..	121.6	121.7[†]	..	..	120.1	120.4	..	..
2002 Dec	111.7	109.8	4.3	4.1	112.0	109.9	4.2	4.1	111.0	108.9	3.0	3.8	110.2	108.1	2.5	3.5
2003 Jan	108.9	109.8	3.7	4.1	109.1	110.0	3.8	4.0	110.1	109.2	2.9	3.5	109.6	108.0	2.1	3.0
Feb	110.7	110.3	4.2	4.1	111.0	110.6	4.3	4.1	114.9	109.4	2.4	2.8	115.9	108.2	1.4	2.0
Mar	118.2	110.7	3.4	3.8	117.9	110.7	3.6	3.9	116.3	110.3	3.8	3.0	117.5	109.1	3.3	2.3
Apr	110.7	110.2	3.1	3.6	110.5	110.4	3.1	3.7	109.9	110.7	2.6	2.9	108.2	109.6	1.7	2.1
May	110.4	111.3	3.5	3.3	110.5	111.4	3.5	3.4	110.0	111.5	3.3	3.2	108.5	110.8	2.9	2.6
Jun	110.9	111.5	3.2	3.3	110.4	111.5	3.2	3.3	111.3	111.7	3.3	3.1	109.8	110.8	2.6	2.4
Jul	111.6	111.9	3.4	3.4	111.8	111.9	3.4	3.4	111.9	112.9	4.0	3.5	110.3	111.9	3.5	3.0
Aug	109.7	112.4	3.4	3.4	109.8	112.4	3.4	3.3	110.4	112.7	3.8	3.7	108.1	111.7	3.1	3.1
Sep	110.4	112.7	3.6	3.5	110.6	112.8	3.6	3.5	110.1	113.3	4.0	4.0	108.1	112.3	3.5	3.4
Oct	111.2	113.0	3.5	3.5	111.5	113.1	3.4	3.5	110.6	113.6	4.0	4.0	108.8	112.7	3.7	3.4
Nov	112.0	113.6	3.9	3.6	112.3	113.9	4.0	3.7	110.7	113.9	3.3	3.8	108.7	112.9	2.9	3.4
Dec	114.9	113.4	3.3	3.6	115.4	113.7	3.4	3.6	114.3	114.6	5.1	4.1	113.0	113.4	4.9	3.8
2004 Jan	112.6	114.1	3.9	3.7	112.8	114.3	3.9	3.8	119.8	116.1	6.4	4.9	121.0	116.4	7.7	5.2
Feb	115.1	114.2	3.5	3.6	114.9	114.3	3.4	3.6	119.0	113.2	3.4	5.0	119.7	111.5	3.1	5.2
Mar	122.1	114.5	3.4	3.6	122.1	114.7	3.6	3.7	122.0	115.5	4.7	4.8	123.7	114.4	4.8	5.2
Apr	115.9	115.2	4.5	3.8	115.6	115.3	4.5	3.8	114.7	115.5	4.4	4.2	113.1	114.5	4.4	4.1
May	115.2	116.1	4.3	4.1	115.5	116.4	4.5	4.2	114.4	115.7	3.7	4.3	112.6	114.8	3.6	4.3
Jun	115.3	115.9	4.0	4.3	114.9	116.1	4.1	4.3	116.1	116.1	3.9	4.0	114.0	114.9	3.7	3.9
Jul	115.7	116.0	3.7	4.0	116.1	116.2	3.8	4.1	115.1	116.3	3.0	3.5	113.1	114.9	2.7	3.3
Aug	113.4	115.9	3.2	3.6	113.6	116.1	3.3	3.7	115.0	117.4	4.1	3.7	112.3	116.1	4.0	3.5
Sep	113.9	116.2	3.1	3.3	114.2	116.4	3.2	3.4	114.8	118.0	4.1	3.7	112.2	116.9	4.0	3.6
Oct	115.4	116.8	3.3	3.2	115.4	116.9	3.4	3.3	115.6	118.9	4.7	4.3	113.5	117.9	4.6	4.2
Nov	115.6	117.0	3.0	3.1	115.7	117.3	2.9	3.2	115.7	119.2	4.7	4.5	113.6	118.1	4.7	4.5
Dec	119.5	117.5	3.6	3.3	119.8	117.9	3.7	3.3	119.1	119.3	4.2	4.5	117.6	118.2	4.3	4.5
2005 Jan	116.3	117.7	3.2	3.2	116.3	117.8	3.0	3.2	125.0	120.9	4.1	4.3	125.9	120.7	3.7	4.2
Feb	119.6	118.4	3.7	3.5	119.2	118.4	3.6	3.4	126.4	120.1	6.1	4.8	127.8	118.8	6.5	4.8
Mar	126.6	118.6	3.6	3.5	126.6	119.2	3.9	3.5	127.6	120.2	4.1	4.8	129.1	119.0	4.0	4.7
Apr	120.2	118.8	3.1	3.4	120.0	119.0	3.2	3.5	119.8	120.8	4.6	4.9	117.9	119.5	4.4	5.0
May	117.4	118.5	2.0	2.9	117.5	118.7	2.0	3.0	119.4	121.0	4.6	4.4	116.3	119.2	3.8	4.1
Jun	118.5	119.1	2.7	2.6	118.2	119.4	2.8	2.7	120.7	120.9	4.1	4.4	118.7	119.5	4.0	4.1
Jul	119.6	120.0	3.5	2.7	119.9	120.3	3.5	2.8	120.5	122.0	4.9	4.5	118.8	120.8	5.1	4.3
Aug	117.9	120.7	4.1	3.4	118.1	121.0	4.3	3.5	119.2	122.2	4.1	4.4	116.7	121.0	4.2	4.4
Sep	118.9	121.2[†]	4.3[†]	4.0	119.2	121.5[†]	4.4[†]	4.1	118.3	122.2	3.5	4.2	115.7	120.9	3.4	4.2
Oct	120.1	121.8	4.3	4.2[†]	120.4	122.0	4.4	4.4	118.5	122.4	2.9	3.5	115.9	120.9	2.5	3.4
Nov	120.1	122.1	4.4	4.3	120.5	122.5	4.5	4.5[†]	119.4	123.2	3.4	3.3	116.9	121.8	3.1	3.0
Dec	125.3	123.2	4.8	4.5	125.1	123.1	4.4	4.4	123.8	124.0	3.9	3.4	122.1	122.7	3.8	3.2[†]
2006 Jan	121.7	123.4	4.8	4.7	121.9	123.7	5.0	4.6	128.6	124.3	2.8	3.4	129.2	123.5	2.3	3.1
Feb	125.2	123.9	4.7	4.8	125.5	124.5	5.2	4.9	133.4	126.2	5.1	3.9	135.1	125.2	5.4	3.8
Mar	133.0	124.4	4.9	4.8	133.0	124.9	4.8	5.0	133.5	125.5	4.4	4.1	135.3	124.4	4.5	4.1
Apr	126.9	125.7	5.8	5.1	126.8	126.0	5.9	5.3	123.5	124.7	3.2	4.2	121.5	123.4	3.3	4.4
May	124.1	125.4	5.8	5.5	124.1	125.5	5.7	5.5	124.6	126.1	4.2	3.9	121.6	124.8	4.7	4.2
Jun	125.6	126.1	5.9	5.9	125.2	126.4	5.9	5.9	126.6	126.7	4.8	4.1	125.1	125.8	5.3	4.4
Jul	125.3	125.9	4.8	5.5	125.5	126.1	4.8	5.5	125.1	126.6	3.8	4.3	123.0	125.3	3.7	4.6
Aug	124.0	127.0	5.2	5.3	124.4	127.4	5.3	5.3	123.5	126.7[†]	3.7	4.1	121.0	125.6	3.8	4.3
Sep	125.2	127.6	5.3	5.1	125.6[†]	128.0	5.4	5.2	123.3[†]	127.3[†]	4.2	3.9	120.8[†]	126.3	4.5	4.0
Oct	126.0[†]	127.9	5.0	5.2	126.5[†]	128.3	5.1	5.3	123.4[†]	127.7	4.3[†]	4.1	121.0[†]	126.6[†]	4.7	4.3
Nov	125.9	128.1	4.9	5.1	126.0	128.3	4.7	5.1	124.1	128.1	4.0	4.2	121.7	127.1	4.3	4.5

1 The most recent month's data is subject to revision.
2 Single month and 3-month averages show the percentage change year on year.

Source: Office for National Statistics: 01633 819024

19 Leisure

19.1 Television licences

Thousands

	Television licences current			Television licences current	
	End of period			End of period	
	Monochrome	Colour		Monochrome	Colour
	BTAA	BTAB	Jun	71	23 934
1999	232	22 205			
2000	169	22 373	Jul	70	23 951
2001	124	22 896	Aug	69	23 994
2002	98	23 191	Sep	68	23 984
2003	79	23 523	Oct	67	24 051
			Nov	63	23 926
2004	62	23 948	Dec	62	23 948
2005	51	24 213			
2006	42	24 364	2005 Jan	61	23 977
			Feb	60	24 026
2002 Aug	110	23 188	Mar	58	24 103
Sep	108	23 174	Apr	58	24 130
Oct	107	23 377	May	57	24 144
Nov	100	23 146	Jun	56	24 170
Dec	98	23 191			
			Jul	56	24 173
2003 Jan	96	23 244	Aug	55	24 202
Feb	95	23 308	Sep	55	24 249
Mar	94	23 392	Oct	54	24 300
Apr	91	23 400	Nov	52	24 168
May	90	23 430	Dec	51	24 213
Jun	89	23 443			
			2006 Jan	50	24 247
Jul	88	23 483	Feb	50	24 297
Aug	86	23 490	Mar	49	24 370
Sep	85	23 550	Apr	49	24 377
Oct	84	23 626	May	48	24 410
Nov	80	23 465	Jun	47	24 442
Dec	79	23 523			
			Jul	47	24 460
2004 Jan	78	23 601	Aug	46	24 462
Feb	77	23 685	Sep	44	24 506
Mar	75	23 824	Oct	44	24 506
Apr	74	23 875	Nov	42	24 373
May	73	23 897	Dec	42	24 364

Source: Capita Business Services Ltd.: 0117 3021003

19.2 UK cinema statistics[1,2]

	Sites (number)	Screens (number)	Total number of admissions (millions)	Gross box office takings (£ million)	Revenue per admission (£)	Revenue per screen (£ thousand)
	JMHX	JMHY	JMHZ	JMIA	JMIB	JMIC
1997	747	2 383	138.9	486.2	3.50	204.0
1998	761	2 638	135.2	504.9	3.73	191.4
1999	751	2 825	139.1	549.7	3.95	194.6
2000	754	3 017	142.5	572.8	4.02	189.9
2001	766	3 248	155.9	645.0	4.14	198.6
2002	775	3 402	175.9	755.3	4.29	222.0
2003	776	3 433	167.3	742.0	4.44	216.1
2004	773	3 475	171.3	769.6	4.49	221.4
2005	771	3 486	164.7	770.3	4.68	221.0

1 Includes Isle of Man and the Channel Islands.
2 Admissions are based on all cinemas taking advertising.

Source: CAA/Gallup/Nielsen EDI

Leisure

19.3 Average issue readership of national daily newspapers
rolling 12 months' periods ending

Thousands

		2003 Dec	2004 Mar	2004 Jun	2004 Sep	2004 Dec	2005 Mar	2005 Jun	2005 Sep	2005 Dec	2006 Mar	2006 Jun	2006 Sep
The Sun	WSDV	8 824	8 872	8 185	8 059	8 825	8 584	8 185	8 157	8 138	8 059	8 071	7 874
Daily Mail	WSEI	5 784	5 647	5 681	5 666	5 740	5 818	5 686	5 682	5 635	5 456	5 455	5 364
Daily Mirror/Daily Record	WSEH	6 153	6 008	6 190	6 026	5 913	5 813	5 455	5 435	5 322	5 138	4 980	4 945
Daily Mirror	WSEM	4 785	4 737	4 274	3 956	4 657	4 587	4 274	4 214	4 148	3 956	3 857	3 825
The Daily Telegraph	WSEN	2 208	2 217	2 170	2 081	2 181	2 227	2 170	2 156	2 159	2 081	2 033	2 140
Daily Express	WSEP	2 045	2 088	2 063	1 876	2 132	2 114	2 063	2 064	1 977	1 876	1 838	1 751
The Times	WSES	1 729	1 628	1 738	1 853	1 655	1 681	1 738	1 781	1 811	1 853	1 791	1 772
Daily Star	WSEQ	1 777	1 936	1 848	1 682	1 965	1 941	1 848	1 825	1 778	1 682	1 617	1 533
The Guardian	WSET	1 272	1 072	1 175	1 175	1 068	1 132	1 175	1 217	1 222	1 175	1 189	1 190
The Independent	WSEU	560	627	617	731	643	606	617	672	705	731	766	741
Financial Times	WSEY	465	494	444	348	453	485	444	394	391	348	346	384
Any national morning	WSEZ	23 723	23 789	22 917	22 686	23 680	23 200	22 917	23 085	23 068	22 686	22 411	22 007

Source: National Readership Surveys Ltd.

19.4 Overseas travel and tourism

Not seasonally adjusted

	Visits by overseas visitors to the UK (thousands)	Expenditure by overseas visitors to the UK (£ million)	Visits by UK residents abroad (thousands)	Expenditure by UK residents abroad (£ million)	Net earnings in UK (£ million)
	GMAA	GMAK	GMAF	GMAM	GMAO
2001	22 835	11 306	58 281	25 332	−14 026
2002	24 180	11 737	59 377	26 962	−15 225
2003	24 715	11 855	61 424	28 550	−16 695
2004	27 755	13 047	64 194	30 285	−17 238
2005	29 970	14 248	66 441	32 154	−17 906
2006	32 170	15 400	68 410	33 385	−17 985
2001 Q1	4 863	2 406	10 842	4 888	−2 481
Q2	6 279	2 815	15 662	6 574	−3 760
Q3	7 100	3 819	19 652	8 921	−5 102
Q4	4 593	2 266	12 125	4 949	−2 683
2002 Q1	4 525	2 025	10 943	5 047	−3 022
Q2	6 375	2 885	15 611	6 945	−4 060
Q3	7 555	4 002	19 729	9 254	−5 251
Q4	5 724	2 825	13 094	5 717	−2 892
2003 Q1	4 944	2 150	11 506	5 446	−3 296
Q2	6 073	2 744	16 297	7 086	−4 342
Q3	7 534	4 041	20 330	10 018	−5 977
Q4	6 165	2 919	13 291	5 999	−3 080
2004 Q1	5 449	2 229	11 817	5 729	−3 500
Q2	7 022	3 231	16 911	7 602	−4 370
Q3	8 501	4 390	21 273	10 437	−6 047
Q4	6 783	3 196	14 194	6 517	−3 321
2005 Q1	6 172	2 644	12 821	6 411	−3 767
Q2	7 868	3 562	17 417	8 110	−4 548
Q3	8 858	4 474	21 767	10 912	−6 438
Q4	7 072	3 568	14 436	6 722	−3 154
2006 Q1	6 369	2 753	13 121	6 657	−3 904
Q2	8 371	3 928	18 589	8 663	−4 735
Q3	9 970	5 050	21 952	11 215	−6 165
Q4	7 460	3 670	14 740	6 850	−3 180

Source: International Passenger Survey, Office for National Statistics

20 Weather

20.1 District summary[1] for June 2006

| | Air temperature (degrees celsius) | | | | | | Difference from average | | Percent of average | | |
| | | | Difference from average | | | | | | | | |
	Highest maximum[2]	Lowest minimum[2]	Maximum	Minimum	Mean	Mean 30cm soil temperature (degrees celsius)	Raindays[3]	Rainfall	Sunshine	Sunshine (hours)
June 2006										
District:										
0 Scotland N	27.8	-0.1	1.8	1.3	1.6	1.0	-2	90	110	165.6
1 Scotland E	26.4	1.6	2.1	1.3	1.7	1.1	-3	70	112	181.5
2 England E & NE	29.9	2.9	2.6	1.4	2.0	1.5	-5	32	115	204.2
3 East Anglia	30.5	1.1	2.3	1.2	1.8	1.7	-4	39	123	241.0
4 Midlands	29.6	2.7	3.0	1.7	2.3	2.0	-5	28	120	216.0
5 England SE	31.2	1.8	2.6	1.2	1.9	2.0	-5	38	136	274.2
6 Scotland W	27.3	2.1	1.7	1.2	1.4	1.3	-4	97	110	186.3
7 England NW & N Wales	28.9	2.4	2.3	1.5	1.9	1.5	-5	49	113	198.6
8 England SW & S Wales	28.0	2.0	2.5	1.2	1.8	1.5	-6	52	134	252.6
N Ireland	26.8	1.6	2.2	1.2	1.7	1.5	-6	50	120	190.8
Scotland	27.8	-0.1	1.9	1.3	1.6	1.1	-3	87	110	176.5
England	31.2	1.1	2.6	1.4	2.0	1.7	-5	40	123	230.7
Wales	28.9	2.1	2.5	1.4	1.9	1.6	-5	42	126	223.3
England & Wales	31.2	1.1	2.6	1.4	2.0	1.7	-5	40	124	229.7

Anomalies are with respect to the 1961-90 averaging period.

June 2006 - A very warm, dry and sunny month across most areas. Mean temperatures generally 1-2 degrees C above average. Rainfall exceptionally below average across the Midlands. Sunshine levels well above average across southern England.

1 District values for each element are computed using all available climate stations, excluding rooftop sites for minimum air temperature. The values in the table may not be compatible with other time series (eg. Central England Temperature, England and Wales Rainfall).
2 Highest maximum and lowest minimum air temperatures for each district are determined by calculating 95 percentiles.
3 Raindays are the number of days during which the total precipitation is at least 0.2mm.

Source: Met Office

20.2 UK Annual Summary

| | Max Temp | | Min Temp | | Mean Temp | | Sunshine | | Rainfall | |
	Actual (degrees celsius)	Anomaly (degrees celsius)	Actual (degrees celsius)	Anomaly (degrees celsius)	Actual (degrees celsius)	Anomaly (degrees celsius)	Actual (hours/ day)	Anomaly (%)	Actual (mm)	Anomaly (%)
	WLRL	WLRM	WLRO	WLRP	WLRR	WLRS	WLRX	WLRY	WLSH	WLSI
1985	11.3	−0.6	4.4	−0.4	7.8	−0.5	1 276.1	95.4	1 072.7	97.6
1986	11.2	−0.6	4.2	−0.6	7.7	−0.6	1 361.5	101.8	1 182.9	107.6
1987	11.4	−0.4	4.7	−0.2	8.1	−0.3	1 249.5	93.4	1 034.6	94.1
1988	12.2	0.3	5.4	0.6	8.8	0.5	1 324.3	99.0	1 131.2	102.9
1989	13.1	1.2	5.5	0.7	9.3	1.0	1 563.8	116.9	1 018.5	92.6
1990	13.1	1.2	5.8	0.9	9.4	1.1	1 490.7	111.4	1 172.8	106.7
1991	12.1	0.3	5.1	0.2	8.6	0.3	1 302.0	97.3	998.2	90.8
1992	12.3	0.4	5.2	0.4	8.7	0.4	1 290.8	96.5	1 186.8	107.9
1993	11.8	−0.1	5.0	0.1	8.4	..	1 218.6	91.1	1 121.1	102.0
1994	12.4	0.5	5.5	0.6	8.9	0.6	1 366.9	102.2	1 184.7	107.7
1995	13.0	1.1	5.4	0.6	9.2	0.9	1 588.5	118.7	1 023.7	93.1
1996	11.7	−0.1	4.7	−0.1	8.2	−0.2	1 403.5	104.9	916.6	83.4
1997	13.1	1.3	5.8	1.0	9.4	1.1	1 430.3	106.9	1 024.0	93.1
1998	12.6	0.8	5.8	1.0	9.1	0.8	1 268.4	94.8	1 265.1	115.1
1999	13.0	1.1	5.9	1.0	9.4	1.1	1 419.4	106.1	1 237.2	112.5
2000	12.7	0.8	5.6	0.8	9.1	0.8	1 367.5	102.2	1 335.6	121.5
2001	12.4	0.6	5.3	0.5	8.8	0.5	1 411.9	105.5	1 049.9	95.5
2002	13.0	1.1	6.0	1.2	9.5	1.2	1 304.0	97.5	1 280.5	116.5
2003	13.5	1.6	5.6	0.7	9.5	1.2	1 587.4	118.7	901.5	82.0
2004	13.0	1.2	6.0	1.2	9.5	1.2	1 361.4	101.8	1 210.1	110.1
2005	13.1	1.2	5.9	1.1	9.5	1.1	1 399.2	104.6	1 083.0	98.4

Anomalies are with respect to the 1961-90 averaging period.

Source: Met Office

Index — figures indicate tables numbers

Index